Voice of the Customer: Capture and Analysis

Other Books in the Six Sigma Operational Methods Series

Parveen S. Goel, Rajeev Jain, and Praveen Gupta • *Six Sigma for Transactions and Service*

Praveen Gupta • *The Six Sigma Performance Handbook*

Thomas McCarty, Lorraine Daniels, Michael Bremer, and Praveen Gupta • *The Six Sigma Black Belt Handbook*

Alastair Muir • *Lean Six Sigma Statistics*

Andrew Sleeper • *Design for Six Sigma*

Kai Yang • *Design for Six Sigma for Service*

Voice of the Customer: Capture and Analysis

Dr. Kai Yang

New York   Chicago   San Francisco   Lisbon   London   Madrid
Mexico City   Milan   New Delhi   San Juan   Seoul
Singapore   Sydney   Toronto

The McGraw·Hill Companies

Cataloging-in-Publication Data is on file with the Library of Congress

McGraw-Hill books are available at special quantity discounts to use as premiums and sales promotions, or for use in corporate training programs. For more information, please write to the Director of Special Sales, Professional Publishing, McGraw-Hill, Two Penn Plaza, New York, NY 10121-2298. Or contact your local bookstore.

Voice of the Customer: Capture and Analysis

1 2 3 4 5 6 7 8 9 0   DOC DOC   0 1 9 8 7

ISBN: 978-0-07-146544-1
MHID: 0-07-146544-8

Sponsoring Editor
  Ken McCombs

Editorial Supervisor
  Patty Mon

Project Manager
  Vasundhara Sawhney,
  International Typesetting
  and Composition

**Acquisitions
Coordinator**
  Laura Hahn

Developmental Editors
  Andy Carroll
  Brian MacDonald

Copy Editor
  Margaret Berson

Proofreader
  Bev Wiler

Indexer
  Broccoli Information
  Management

Production Supervisor
  Jim Kussow

Composition
  International Typesetting
  and Composition

Illustration
  International Typesetting
  and Composition

Art Director, Cover
  Jeff Weeks

ABOUT THE AUTHOR

Kai Yang, Ph.D., has wide experience in quality and reliability engineering. The Executive Director of Enterprise Excellence Institute, a renowned quality engineering organization based in West Bloomfield, Michigan, he is co-author of the influential *Design for Six Sigma: A Roadmap for Product Development*. He is also Professor of Industrial and Manufacturing Engineering at Wayne State University, Detroit.

Contents

1

Value, Innovation, and the Voice of the Customer

In today's global economy, several business functions have become global commodities and subject to stiff price competition: first manufacturing activities, then IT, and most recently traditional R&D, such as routine engineering design work. To find success in this competitive reality, your business needs to take the high ground in value creation. Value is a measure of how much the customer really appreciates a product or service, and how much customers are willing to purchase this product or service. In Figure 1.1, you can see that there are some legendary products or services, such as Hollywood movies, Intel CPUs, and Microsoft operating systems, that solidly command the market place—these products creates enormous profits for their companies.

Figure 1.1 reveals the secrets of how these products and companies create value. There are two success models for companies to create high values. One type of company has a commanding lead either in technology or brand recognition. Their products or brands dominate the marketplace and become industry standards. Examples of this kind of company include Microsoft, Intel, Cisco, Google, and so on. The driver for value creation for this type of company is technical or brand dominance, or technology-driven innovation. The other type of company develops products or services that capture the heart of customers; examples of this kind of companies or products include Starbucks and Apple's iPod. The driver for value creation for this type of company is customer-centric innovation. However, these two types of companies or products are few; most companies are mediocre in both technical or brand dominance and customer value position. However, in order to survive in the global economy of the 21st century, it is wise for a company to excel in at least one of the above two aspects; that is, either the technical/ brand dominance, or the

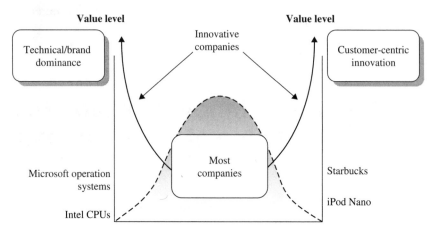

Figure 1.1 Value and innovation

customer-centric innovation. In both aspects, thorough understanding of voice of the customer is very important. The main topic of this book is about the voice of the customer (VOC). I will discuss thoroughly the role of voice of the customer in these two success models.

1.1 Defining Customer Value

In the two aforementioned successful models, the real key for success is that the product or brand brings exceptional value to customers, so that many customers crave the product and are willing to pay good prices. In this subsection, the following issues are discussed:

> What is value and how do you create it? What is the role of value for long term commercial success?

Profitability is one of the most important factors for long term commercial success. High profitability is determined by strong sales and overall low cost in the whole enterprise operation. It is common sense that:

$$\text{Business profit} = \text{Revenue} - \text{Cost}$$

In addition:

$$\text{Revenue} = \text{Sales volume} \times \text{Price}$$

Here, *price* means "sustainable price," that is, the price level that customers are willing to pay. Many researchers (Sheridan 1994, Gale 1994) have found that both sales volume and sustainable price are mostly determined by *customer value*—customers' opinions determine a product's fate. Customers' opinions will decide the price level, the size of the market, and the future trend of this product family. A product that has high customer value is often featured by increasing market share, increasing customer enthusiasm towards the product, word of mouth praise, reasonable price, and a healthy profit margin for the company that produces it, as well as increasing name recognition.

Sherden (Sherden 1994) and Gale (Gale 1994) provided a good definition for customer value. In their assessment, customer value is defined as perceived benefit (benefits) minus perceived cost (liabilities):

Customer Value = Benefits − Liabilities

Here the benefits are the factors that increase the customer value, and the liabilities are the factors that decrease the customer value.

Benefits in this equation include the following categories:

Functional benefits   This category includes the actual benefits of a product or service delivered to customers. In other words, functional benefits are "what the product or service does for customers". For example, the functional benefits of a food item include "providing energy, providing nutrition, providing taste" and so on. For investment services, the functional benefits include economic benefits and revenue to customers. The reliability, quality and durability of the product are also part of functional benefits. For example, a kitchen knife delivers the function of "cutting food". In comparing a high quality knife with a low quality knife, both of them can cut food, but the high quality knife can cut better, last longer, so it will deliver more functional benefits.

Psychological benefits   In addition to functional benefits, psychological benefits are a very important component for a product or service. A very obvious example is that a plain color T-shirt will sell at much lower price than that of a T-shirt with a famous logo. A famous logo will give customers emotional and self-expression benefits on top of the regular T-shirt's functional benefit, "covering body". Brand image is also a part of psychological benefit. A known good brand brings confidence to customers about products quality, reliability and durability. A famous brand may even raise customers' social status, such as a Mercedes Benz car. If a product is the first of its kind, it usually brings psychological awe to customers, such as the first copy machine. A "first-of-its-kind" product will usually command a high price and often create a brand name. Competitions in the market place also brings psychological effects into customers' value judgments, if there are many competitors producing the same or similar a product, that usually will create a perception that this product is a commodity and it doesn't carry much value.

Service and convenience benefits   This category includes availability, which is the ease of accessing the product or service; it also includes service, which is the ease of getting help in case of product problems or failure.

Liabilities in this equation include the following categories:

Economic liabilities   This category includes all monetary expenses incurred for owning the product or service. The price that a customer pays in order to buy the product or service is certainly a major part of economic liabilities. But, in addition to the price of the product or service, there are many other indirect ownership costs associated with owning this product or service, which include:

- Acquisition costs: This includes transportation cost, shipping cost, time and efforts spent to obtain the service

- Usage costs: This includes additional cost to use the product or service in addition to the purchase price, such as installation, training cost, and so on.

- Maintenance costs: This includes repair cost, routine maintenance costs, regular upgrades and so on.

- Ownership costs: This includes financing cost, licensing cost and so on.

- Disposal costs: This includes disposal cost of hardware, environmental regulation compliance cost and so on.

Psychological liabilities   This category includes the negative psychological effects of the product or service. An unknown brand name may make a customer feel uncertain about the dependability of the product or service. A "cheap brand" may cause self-esteem liability for some customers. A poorly performed product or service not only delivers low functional benefits, but will also make the customer feels bad.

Service and convenience liability   This category includes all the negative effects related to service and convenience. Even if the original product is very good, lack of service, or poor service will use customer issues when the product encounters problems. For example, lack of repair facilities, high cost of spare parts and high repair cost are among the major concerns for some potential buyers of foreign made cars. Poor availability, such as long delivery time, is also an example of service and convenience liability.

This customer value definition is comprehensive, and explains what types of product or services that customers are willing to pay a premium price to buy.

There are numerous cases where customers are willing to pay a higher price to buy a product with better brand name image, because brand name image is a psychological benefit for the customer. Toyota and

General Motors have a joint venture in California that produces an identical car model. Some of the units have a Toyota brand, and some have a GM brand; however, the units with the Toyota brand can sell for a few hundreds dollars more than the identical cars with the GM brand, because the public perception is that Toyota has the preferable brand image. Brand name image is an important portion of customer value.

As another example, a neighborhood store will sell an item at higher prices than the identical item in discount chain stores. This is because of the perceived convenience in obtaining these items in neighborhood stores. This is a part of the service and convenience benefit defined in customer value.

Therefore, to gain business profit, creating products with high customer value is a must. Many business enterprises often fail to see the multiple aspects of customer value. They may create a product with tremendous functionality, but one that is very poor in customer service, accessibility, and psychological aspects, and therefore, the product will fail.

In a competitive marketplace, success will become more difficult to achieve. Your competitors can learn from and copy your product, learn your customer value proposition, and ultimately offer a similar product at a lower price. There are also disruptive innovations, which change the whole landscape of the competitive situation. Thus, achieving a high value position is becoming more and more a moving target. For example, video renting stores, such as Blockbuster, were everywhere in the United States and their business was very good. With the emergence of Internet downloading and Internet-based video rental businesses, such as Netflix, video rental stores lost a lot of their attractions to customers. Blockbuster has suffered huge losses in recent years. Another example is the sport utility vehicles (SUVs) made by American automobile manufacturers. SUVs helped GM, Ford, and Chrysler a lot in their revenue and profitability in 1990s and early 2000s. However, with the drastic improvement of fuel efficiency of Toyota vehicles, such as Prius, and a surge of gasoline prices in the years 2005 to 2006, Ford and GM lost a big chunk of the automobile market share due to the consumers' abandonment of SUVs and Trucks.

In any case, the key to value creation is the trend-setting innovation, either technology-driven innovation or a customer-centric innovation, because innovation will change the rules of game, and if you do it right, it will put you on the top of the competition. If you are a business leader, then you have to learn how to lead innovation. You will need to lead this kind of innovation many times in your business tenure.

1.2 Innovation Roadmap

What is innovation? What are the key factors for successful innovation? Based on Amabile et al (1996), "All innovation begins with creative ideas ...

We define innovation as the successful implementation of creative ideas within an organization. In this view, creativity by individuals and teams is a starting point for innovation; the first is necessary but not sufficient condition for the second."

Specifically, innovation has two aspects. One is creativity; better creativity generates out-of-the-ordinary ideas to create "first-of-its-kind" products. Another aspect of creativity is to make new ideas into commercial successes.

In developing innovative products or services, there are two kinds of driving forces. The first is the technology push; the phenomenal development of the IT industry from the 1990s up to now started with new technological breakthroughs—take Microsoft and Google for example. The other driving force is customer or market pull; this innovation is usually started by discovering a hidden market need. Starbucks and the iPod are such examples. Starbucks is not simply a regular coffee shop; its vision is to become someplace other than home and the workplace. It not only provides coffee, but also provides a casual meeting places, free electrical outlets, and wireless Internet access.

The driver for technological push innovation is creativity and the ability to generate new ideas, and connecting this innovation with customer and market needs. The driver for customer-centric innovation is the ability to discover a hidden market. You need great vision to discover some hidden unmet customer need and accurately identify a customer value proposition, that is, what things customers really crave. In both technology push and market pull innovative product development scenarios, capturing the voices of customers, especially the unarticulated voice of the customers, is really a key factor. A creative mind may bring new ideas, but a new idea alone is not enough; the new idea needs to catch customers' hearts. Without knowing the real voice of the customer, you cannot catch the customers' heart, so you cannot succeed.

Another key factor for successful innovations is an effective product development process. You need an effective product development process to make creative ideas and customer-centric innovation into products at low cost and with high quality.

Finally, innovation is not only a product and technology matter. Business model, brand strategy, services, and so on are all parts of a grand road map for innovation.

1.3 Voice of the Customer: Mining for the Gold

In the previous section, I discussed the key factors for value-based innovation. Clearly, capturing the voice of the customer is a very important factor, whether you are dealing with a technology push-or-market pull oriented product or service development.

If you had some magic power and were able to discover exactly what customers are craving, and if you also knew how to produce their dream product at a low price, then you would be guaranteed to get rich! Therefore, capturing the exact voice of the customer is like striking gold.

Of course, this kind of lucky chance is very rare. In the natural world, you need to mine for gold. You need to explore and search. Sometimes you find bits and pieces of gold mixed into other minerals, which you then need to distill and purify to get pure gold. The same is true for the voice of customer. You need to search for a good source of customer information by finding bits and pieces. The real voice of customer may hide deep in the customer's mind, and many customers are humble people. If you only use traditional customer data collection methods, you may only get inaccurate and incomplete information. It takes a lot of effort and sophisticated methods to mine the voice of the customer accurately and distill this VOC information into valuable inputs for the product development process.

What strategy and methods for voice of customer capturing and analysis that we use depends very much on what this VOC information is used for. The purposes for the voice of customer information capturing and analysis are the following:

- You need the right kinds and sufficient amount of accurate voice of customer information to provide necessary inputs for all stages and levels of the product design work. The stages include product design and manufacturing process design, and the levels refer to the system level, subsystem level, and component level.

- You need a good set of VOC information to learn what are the key customer value factors for this kind of product; what factors, such as price and functions or really excite customers; and what the product's customer value position is.

- You may need to explore the possibility of shifting your current customer value proposition to another one, so you can develop a new innovative product that is different from competitors. In this case, you need to capture enough VOC information for decision making.

- We may want to capture the right kinds and a sufficient amount of voice of customer information to provide necessary inputs for product improvements in limited scope, instead of a complete new product design.

This book will show you how to capture and analyze the voice of the customer in all these four scenarios.

1.4 Overview of This Book

Since VOC capturing and analysis is a very important part of the product development process, we will not cover the process of VOC capturing and analysis in isolation. Chapter 2 discusses all aspects of product development processes. I'll cover some state-of-the-art product development theories and advanced product development processes. During this discussion, the roles of VOC capturing and analysis in the product development process are thoroughly discussed. After reading this chapter, you should have a good idea of how voice of customer information can help your product development process, and what kinds and what amount of VOC information are really needed for this.

Chapter 3 discusses issues related to the customer value. Because the customer value position determines the market position of the product, you need to know about the relationship between VOC and customer value, and how to obtain key customer value information from VOC. I will also discuss how a customer value position can be modified to create a new product market position; this strategy is called the "blue ocean strategy" (Kim and Mauborgne 2005). Chapter 3 also discusses how to link VOC information to product design specifics in a very clear and exact manner.

Chapter 4 discusses conventional VOC capturing methods, including customer surveys, interviews, and Internet surveys, in great detail. Chapter 5 discusses an anthropology-based VOC capturing method, the ethnographic method. Because this VOC capturing method can capture hidden, unarticulated VOC information much better than conventional VOC capturing methods, it is becoming very popular and many customer-centric innovation practices are heavily relying on the ethnographic method. In this chapter, I give very detailed descriptions of the method and provide several examples and case studies.

Chapter 6 explains how to process raw voice of customer data and transform them into clearly defined customer data. Chapter 7 discusses the method of quality function deployment (QFD). The QFD method is a systematic method to transform VOC data into product functional requirements, and then design specifications. This method serves as the interface between VOC data and the product design process.

Customer value creation and improvement is related to other methodologies, such as brand development (Chapter 8), value engineering (Chapter 9), and the theory of inventive problem solving (Chapter 10). These methodologies are thoroughly discussed in the relevant chapters. The last chapter, Chapter 11, provides some necessary background in statistics.

2

The Product Development Process

This book is about how to capture and analyze the voice of the customer. However, the major purpose of capturing and analyzing the voice of the customer is to provide vital input for the product development process. In order to derive the best approach to capturing and analyzing the voice of the customer, you need to know how the product development process really works. This chapter will provide you with that information. I'll introduce the key stages in production development and then discuss the first four stages in detail. I'll then discuss the nature of the product development process from an information perspective. You'll learn about the development and evolution of the theories and best practices for the product development process. You'll discover the leading framework for the product development process, called the lean product development process. Finally, I'll discuss the relationship between the voice of the customer and the product development process.

2.1 Defining Product Cost and Development

A *product* is anything that can be offered to a market that might satisfy a want or need. A product is one of two types: *tangible* (physical) or *intangible* (nonphysical). Tangible products are what most people think products are; examples of tangible products include bicycles, laptop computers, printer paper, cars, and airplanes. Intangible products are related to service industries, for example, vacation packages, insurance policies, and medical treatment.

Every product is sold in the market for a price. For good products, the customers are willing to pay higher prices. The prices that customers are willing to pay depend on a supply-and-demand relationship. How much the customers will demand a certain product depends on the value of the product. Based on the research of Sherden (1994) and Gale (1994), the value of the product can be expressed as the following equation:

$$\text{Value} = \text{Benefits} - \text{Liabilities}$$

Because the customers are really buying the benefits offered by a product, not just the physical entity of the product, in a real marketplace, the product itself, or *generic product*, could be bundled with everything that is needed to deliver the benefits to customers. This bundle is called the *whole product*. The whole product typically augments the generic product with training and support, manuals, cables, additional software, online help, warranty, installation instructions, professional services, and so on. The concept of the whole product is illustrated in Figure 2.1.

As this analysis shows, customers demand products purely for their benefits. The liabilities, such as the prices they have to pay to buy the products, are necessary evils that customers have to live with. For a product, if the gap between benefits and liabilities is large, then customers will perceive that the benefits they get from the product are much larger than the liabilities they have to accept, and the product will sell quickly and easily. Besides the benefits offered by the product, market competition can also greatly affect the selling of the product. The supply-and-demand equation is very different in a competitive environment, with one supplier versus multiple suppliers. Competition will increase the supply of the same or similar products, which will give customers more choices. Competitors can also offer products with the same or better values for their products, by offering more benefits, having lower liabilities (mostly lower prices), or both. The product with a better value will take more market share. When competition in the marketplace becomes fierce, one common tactic is to reduce the selling price of the product in order to improve the product's value position in the marketplace and

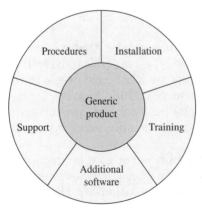

Figure 2.1   The Whole Product

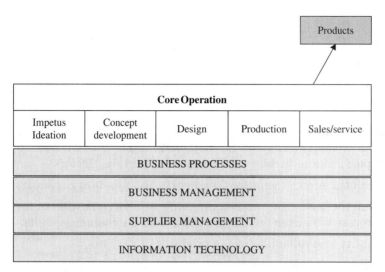

Figure 2.2 Business Operation of Product Producing Companies

keep the market share. However, there is a limit to how much you can lower the price—the limit is the cost of providing the product, because the selling price has to be higher than cost in order to make a profit.

Figure 2.2 illustrates how a typical product is produced. There is always a core operation, which consists of the product development and the actual production of the product. There are also other business operations, however, such as marketing, finance, personnel, and so on.

From Figure 2.2, you can see that the cost of providing a product has the following components:

- Cost of product development
- Cost of production
- Cost of running supporting operations

Besides these three cost components, the success of the product development and production system also depends on other factors, such as the quality and value of the product, the time to market, and so on. The economic model for a product development and production system is illustrated in Figure 2.3.

The ultimate goal for any product development and production system is to make a profit, and the profit is the output of the system. There are three important inputs in the system:

- **Cost** The total cost includes the product development cost, production cost, and the cost of running supporting operations. Lower costs lead to higher profit.

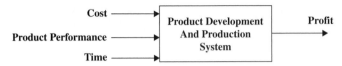

Figure 2.3 The Economic Model of Product Development and Production System

- **Product Performance**   Product performance depends on how much benefit the product delivers to the customers, and how well it does so. Product performance further depends on the following factors:

  - How well the product captures the hearts and minds of customers

  - How well the technology and engineering work is conducted in the design process to deliver the performance that the customer wants

  - The quality and reliability of the product

  - The delivery and service of the whole product

- **Time**   The product should be introduced to the market with the right timing. In many cases this time factor depends on how quickly and flexibly the company can introduce a new product into the marketplace. If the market demands a product, and you are the first one to deliver this product, you will have a dominant position in the marketplace.

Based on this analysis, you can see that the total cost, product performance, and time are three dominant factors in the product development process. We can derive the following specific performance metrics for product development processes:

- **Product development lead time**   Most companies consider their product development lead time to be exceptionally important for determining the performance of their product development activities. Product development lead time is particularly important because this metric determines the speed with which new products can be introduced into the marketplace. Companies that have high speed in product development can introduce new products more often and adapt more quickly to changes in customer tastes. This ultimately translates into a larger market share for the company. Lead time is usually measured in months, and can range from fractions of a month to tens of months, depending on the complexity of the product and the skill of a company's development staff.

- **Efficiency**   In attempting to reduce product development lead times, however, few companies can afford to ignore the efficiency of their product development. In product development, efficiency is the cost of manpower and other resources required for the product development.

- **Robustness**   In addition, the robustness (including quality, reliability, and flexibility—how well the product does what it was meant to do) of the design is particularly important for evaluating any product development process.

- **Life-cycle cost**   Life-cycle costs, including development costs; production costs; sales and distribution costs; service, support, and warranty costs; and disposal costs may be included in computing the life-cycle cost for a product. Some companies even include the costs due to pollution during the production and use of the product as part of the holistic analysis of the life-cycle cost. Product development has a particular vested interest in keeping the life-cycle cost for any product as low as possible.

On a longer time scale, product development lead time, efficiency, robustness, and life-cycle costs will contribute a great deal to the level of customer satisfaction, market share, and revenues that the company will have. These will in turn translate into profitability and will influence the organization's long-term business viability.

2.1.1   Product Development Process Flowchart

Product development is a complicated process. There are many subprocesses, such as capturing what customers really want, creating product concept designs and detailed product designs, and designing the manufacturing process. These subprocesses focus on issues such as the functionality of the product, as well as aesthetics and ease of manufacture. Other tasks in product development may include building and testing of prototypes using various pieces of test equipment, and analysis of candidate designs. Common goals in product design and development processes include shortening the time required to design new products, introducing new designs more frequently, producing designs that are more innovative while meeting the customer's needs, and reducing the cost required to design new products.

In general, product development involves the activities that are used to

1. Determine that a new product is required to serve some need
2. Conceive of a concept for the product based on the wants and needs of customers
3. Develop all the technical specifications for the product
4. Devise a production process
5. Validate both the design and the production process

Figure 2.4 illustrates a simple process flowchart for the product development process. In this process, there are eight stages:

- **Stage 0** Impetus/ideation
- **Stage 1** Customer and business requirements study
- **Stage 2** Concept development
- **Stage 3** Product/service design/prototyping
- **Stage 4** Manufacturing process preparation/product launch
- **Stage 5** Production
- **Stage 6** Product/service consumption
- **Stage 7** Disposal

The earlier stages of the cycle are often called *upstream*, and the later stages are often called *downstream*. Since this book deals with voice of the customer, and the voice of the customer is mostly related to the first four stages of the product life cycle. This chapter will cover the first four stages. There is quite a bit of other literature that covers the last four stages.

2.2 The Product Development Process–End to End

Many companies manage their product development process by using flowcharts similar to that in Figure 2.4. Managing product development by stages is also called the *stage-gate* process. For different companies, and different kind of products, the number of stages, and the name of each stage, might be somewhat different. Some product development is simply a modification of existing product. However, in even the simplest product development process, there are several distinct stages. For different stages, the objectives and the type of work performed in each stage will be distinctively different. In this section, I will discuss the first four stages in a typical product development process in detail. It is crucial for all the people involved in the product development process, be they engineers, marketing people, or managers, to understand this process in order to do their job well.

2.2.1 Opportunity Identification and Idea Generation: Stage 0

Opportunity identification and idea generation is the very first stage of the product development process. It is Stage 0 in the flowchart illustrated in Figure 2.4. This stage may also be called the "fuzzy front end," because this stage is at the front end of the product development process, and it is not clear what exactly will be accomplished in this stage. Typically, an opportunity for a new or improved product is

---

Stage 0: Impetus/Ideation

- New technology, new ideas, competition lead to new product/service possibilities
- Several product/service options are developed for those possibilities

↓

Stage 1: Customer and business requirements study

- Identification of customer needs and wants
- Translation of voice of the customer into functional and measurable product/service requirements
- Business feasibility study

↓

Stage 2: Concept Development

- High-level concept: general purpose, market position, value proposition
- Product definition: Base-level functional requirement
- Design concept generation, evaluation, and selection
- System/architect/organization design
- Modeling, simulation, initial design on computer or paper

↓

Stage 3: Product/Service Design/Prototyping

- Generate exact detailed functional requirements
- Develop actual implementation to satisfy functional requirements, i.e. design parameters
- Build prototypes
- Conduct manufacturing system design
- Conduct design validation

↓

Stage 4: Manufacturing Process Preparation/Product Launch

- Finalize manufacturing process design
- Conduct process testing, adjustment, and validation
- Conduct manufacturing process installation

↓

Stage 5: Production

- Process operation, control, and adjustment
- Supplier/parts management

↓

Stage 6: Product/Service Consumption

- After-sale service

↓

Stage 7: Disposal

Figure 2.4 A typical product/service life cycle

identified by marketing people or upper management, and then the product development team will explore this opportunity. However, the outcomes of this stage can be very diverse; sometimes, the team may find this opportunity is worth pursuing, so it develops a follow-up product development plan. Sometimes the team may find it is not feasible to develop an appropriate product at this time. This stage is also called the "discovery" stage. Whatever name people may call this stage, this is where you design the DNA for your product and plant the seeds. This stage might be the highest leverage point in product development because it will define what the product will be.

The fuzzy front end is the messy "getting started" period of the new product development process. The front end is where the organization formulates a concept of the product to be developed and decides whether to invest resources in the further development of an idea. It includes all activities, from the search for new opportunities, to the formation of a germ of an idea, to the development of a precise concept. The fuzzy front end ends when an organization begins development.

Although the fuzzy front end may not be an expensive part of product development, it can consume 50 percent of development time (Smith and Reinertsen 1998), and it is where major commitments are typically made involving time, money, and the product's nature. Consequently, this phase should be considered as an essential part of development rather than something that happens "before development," and its cycle time should be included in the total development cycle time.

There are three different elements in this fuzzy front end:

1. Opportunity identification. In this element, large or incremental business and technological changes are identified in a more or less structured way. As discussed in Chapter 1, there are two major driving forces for new product development: the market pull and technology push. The *market pull* simply means that there are new market trends, new voices of customers that lead to new product opportunities, and so on. *Technology push* means that new discoveries and inventions in science and technology may lead to new product opportunities. Overall, in opportunity identification, there are two voices that need to be heard: the voice of the customer and the voice of technology. There are many methods in capturing these two voices, such as voice-of-the-customer (VOC) capturing, which is the main theme of this book. The methods of capturing the voice of technology include technology roadmaps, technology trees, and the theory of inventive problem solving (TRIZ).

2. Opportunity analysis. The second element is the opportunity analysis. It is done to translate the identified opportunities into the business- and technology-specific context of the company. In this element, there are also two kind of analyses: one is VOC analysis, and the other

is technology analysis. Chapter 6 of this book will discuss voice-of-the-customer analysis in detail. The technology analysis can be supported by technology road maps, technology trees, and the theory of inventive problem solving.

3. Idea generation and handling. The third element is the generation and handling of ideas. The most important objective of Stage 0 is to generate viable product ideas that can be further developed into great products to fit the market opportunities. Until the 1990s, most companies were satisfied with regular product extensions and hunches as new ideas. In today's global economy, with much greater global competition, the ability to generate breakthrough innovations becomes so important that this "light bulb glow" or hit-and-miss type of idea generation approach is no longer adequate. A well-organized idea generation and handling system becomes necessary. A good idea generation and handling system should have the following features:

- You need to generate many ideas, because the majority of initial ideas will be discarded due to some kind of infeasibilities. Generating many ideas will ensure that some good ideas will survive at the end.

- You need to develop ideas from many different sources, from different people, some ideas from your own research and development people, some ideas from your customers or suppliers, some ideas from university researchers, and so on. Comparing, combining, improving, and cross-pollinating ideas from these sources will often generate some very strong, unique, mature new ideas.

- All incoming ideas need to be screened, analyzed, reviewed, grouped, improved, and finally sorted by a responsible person (usually a product development manager) or team. Some obviously good ideas will be sorted out and can be sent to the next product development stage. The ideas that can't pass the screening and sorting are not necessarily bad ideas; they might be good ideas, but the time is not ripe to fully develop them. These ideas should be put in an "idea vault" and periodically retrieved and reviewed for their viabilities. After each idea gets reviewed, the feedback should be sent to the person who generated the idea, so he or she can further improve his or her ideas. Other people in the company should be able to get access to this idea vault.

2.2.2 Customer and Business Requirement Study: Stage 1

In Stage 1, the information captured in the VOC study by using methods such as surveys, focus groups, and ethnographic research is further analyzed to gain deeper understanding of its meaning.

The raw voice of customers is often vague, confusing, and not well defined. With the help of some methods such as the affinity diagram method (KJ method), this raw voice-of-the-customer data will be translated into a new set of relatively well-defined, quantitative quality metrics. The affinity diagram method will be thoroughly discussed in Chapter 6. Usually, people in the industry call these metrics *top-level specifications*. In Six Sigma practice; these top-level specifications are called Critical-to-Quality characteristics (CTQ). The translation of the raw voice of the customer data into CTQs will help the product development group to establish a clear picture of exactly what kind of product customers really want. This CTQ information can be further used to estimate the cost of developing and producing such a product; thus it will help to build the business feasibility study. The methods for translating the raw voice of the customer to CTQs will be discussed in detail in Chapter 6.

Another key analysis of the raw voice-of-the-customer data is to conduct customer value analysis. A customer value analysis identifies key customer value factors, such as key performance characteristics and price, and it evaluates the competitive comparison of your product with your competitors' product in all these customer value factors. Customer value analysis is very useful in determining how well your product can compete in the marketplace, and identifying the key areas to improve to increase your competitiveness. Customer value analysis is discussed thoroughly in Chapter 3.

During this stage, the business case will be developed based on estimates of the total available market, customer needs, investment requirements, competition analysis, and project uncertainty. Based on your business case, you will make the decision of launching the product development project.

2.2.3   Concept Development: Stage 2

Concept development (Stage 2 in Figure 2.4) is an important stage in product development. It is a process in which the concept of a new product, or product platform, is developed gradually, from fuzzy to specific; from a higher level, such as the system level, to the lower level of the product architecture, such as subsystems and components. The concept development stage comes before prototype building and physical testing (which is Stage 3).

Concept development is an extremely important stage; the cost incurred in this stage is lower than later stages, because you haven't started expensive prototype building, validation testing, manufacturing tooling tests, and so on. However, the impact of this stage is huge, much like conceiving a new baby. If the new baby has perfect DNA, the future for the new baby will look great; if the baby has DNA flaws, it is nearly impossible to make it up at a later stage. A study by Fredrikson

(1994) concluded that the design decisions made during the early stages of the design life cycle have the largest impact on total cost and quality of the system. It is often claimed that up to 80 percent of the total cost is committed in the concept development phase (Fredrikson 1994). In my experience, at least 80 percent of the design quality is also committed in the early phases, as depicted in Figure 2.5. In the early design stage, the cost committed is low, but the impact on product design is high; in the later design stage, the cost committed is very high (after buying production machineries, facilities, and so on), but at this later phase, design changes for corrective actions can only be achieved at high costs, including customer dissatisfaction, warranty, marketing promotions, and in many cases under the scrutiny of the government (such as recall costs).

In the concept design stage, where few final design decisions have been made, there is a lot of freedom to choose and try out different design concepts. Different concepts could lead to vastly different product performances, quality, and cost. So it is important to front-load this concept design stage, that is, put a lot more effort into this stage.

The concept design stage can be further divided into the following steps:

- High-level concept development
- Product definition and base-level functional requirement
- Design concept generation, evaluation, and selection
- System/architect/organization design
- Modeling, simulation, initial design on computer or paper

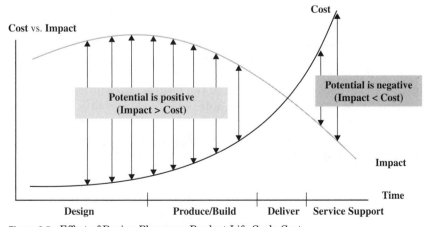

Figure 2.5   Effect of Design Phases on Product Life Cycle Cost

There are several methods that can be used to support this stage, which I'll discuss here.

2.2.3.1 User Innovation Sometimes the best product concept ideas come from customers. *User innovation* refers to innovations developed by consumers and end users, rather than manufacturers. Eric von Hippel of MIT discovered that most products and services are actually developed by users, who then give ideas to manufacturers. This is because products are developed to meet the widest possible need; when individual users face problems that the majority of consumers do not, they have no choice but to develop their own modifications to existing products, or entirely new products, to solve their issues. In using this method, you should focus on *lead users*. Lead users are users who are at the leading edge of the applications of the product, or they are "ahead of their time." For example, some high-end product research and development engineers use computer-aided design software to analyze some extremely challenging problems; they really use the software to its limits, so they often supplement the existing software with their own subroutines, driver programs, and so on. On the other hand, most regular users, such as draft engineers, just use this software as drafting software; their application of the software is far from reaching the software's functional limits. Lead users usually have pretty good ideas about the weaknesses of the current product and what additional functions should be added to make the product more perfect.

2.2.3.2 Theory of Inventive Problem Solving TRIZ (Teoriya Resheniya Izobreatatelskikh Zadatch) is the Russian acronym for "Theory of Inventive Problem Solving" (TIPS). TRIZ has developed over 1,500 person-years of research and study over a significant portion of the world, including the most successful solutions of problems from science, mathematics, and engineering, and systematic analysis of successful patents from around the world, as well as the study of the internal, psychological aspects of human creativity.

Dr. Genrich S. Altshuller, the creator of TRIZ, started an investigation of invention and creativity in 1946. Based on the fact that almost all human inventions are documented in the form of patents, Dr. Altshuller theorized that studying patents could lead to uncovering the secrets of inventions and inventors. Dr. Altshuller initially reviewed 200,000 patent abstracts; in subsequent years, many millions of patents were further studied by many TRIZ researchers. Based on the extensive studies of those patents, these are the major initial findings of TRIZ:

- Though there are millions of patents and inventions, all innovations are based on a very small number of inventive principles and strategies, and those inventive principles and strategies can be taught and learned. By learning and using these principles and strategies, people can "shortcut" the invention process.

- The bottlenecks for breakthroughs in technology or product are often caused by contradiction—a *contradiction* is usually caused by an improvement to one technical attribute of a system that leads to deterioration of other technical attributes. For example, as a container becomes stronger, it becomes heavier, and faster automobile acceleration reduces fuel efficiency. Inventions are often related to overcoming contradictions; many contradictions can be resolved by using inventive principles.

- By studying the history of inventions, TRIZ developed sophisticated prediction methods for technology evolution trends, that is, predicting what will be next in a given branch of technology. This TRIZ technical prediction tool can be used to guide a company to steps ahead of their competitors.

Over the next 40 years (1950–1990), TRIZ developed into a system of philosophy, a problem-solving process, and methods. TRIZ has five key philosophical elements:

- **Ideality**   The ultimate criterion for system excellence; this criterion is the maximization of the benefits provided by the system, and minimization of the harmful effects and costs associated with the system.

- **Functionality**   The fundamental building block of system analysis; it builds models about how a system works, and how it creates benefit, harm, and costs.

- **Resources**   Maximum utilization of resources is one of the keys to achieve maximum ideality.

- **Contradictions**   Contradiction is a common inhibitor for increasing functionality; removing contradiction usually greatly increases the functionality and raises the system to a totally new performance level.

- **Evolution**   The evolutionary trend of the development of technological systems is highly predictable, and can be used to guide further development.

The TRIZ problem-solving process is a four-step process, consisting of problem definition, problem classification and tool selection, solution generation, and evaluation.

The TRIZ methods include inventive principles, separation principles, system simplification, and system functional analysis and improvement methods, as well as a technical prediction method.

After the breakup of the Soviet Union in the 1990s, TRIZ became known to the rest of the world. The bulk of TRIZ research, development, and implementation also moved to the West. Every year, TRIZ researchers will study new patents to update the TRIZ knowledge base. However, TRIZ researchers found that while many millions of patents are being filed, the number of new inventive principles is growing at a much, much slower speed. This fact actually proves TRIZ is a valuable, stable knowledge base that is not going to be obsolete quickly. The application of TRIZ is also on the rise—based on the TRIZ Journal, 105 companies are currently using TRIZ, including Dupont, Mobil, 3M, Honeywell, Ford, Toyota, BMW, Samsung, Pfizer, and Lockheed Martin. TRIZ has helped these companies to develop many breakthrough inventions and greatly improved their competitive positions. Noticeably, Samsung is an exemplar company for using TRIZ. According to *Fortune* magazine, TRIZ has helped Samsung to greatly improve its research and development capabilities and aided Samsung in surpassing Sony as a leading company in consumer electronics.

Overall, TRIZ is a systematic innovation method that is based on the distilled fundamental knowledge of millions of successful patents and business case studies. Learning TRIZ will make product development team members significantly more creative, and shorten the process of innovation. Successful TRIZ applications can greatly improve their competitive advantage.

In the concept development stage, TRIZ is a powerful tool that can be used to generate creative product ideas. TRIZ will be discussed in detail in Chapter 10.

2.2.3.3 Value Engineering

From the customer's point of view, a product's value is the functions and benefits it delivers. From the manufacturer's point of view, the product consists of a bundle of parts and interfaces, and these parts and interfaces cost money. Value engineering is a systematic method that can trace relevant parts and interfaces to each product's functions and benefits, and calculate what is the lowest cost to deliver each function and benefit. From this information, the product development team can find ways to improve the product design by delivering the same functions and benefits with the lowest possible cost. Value engineering is very useful to improve product concepts. Value engineering is thoroughly discussed in Chapter 9.

2.2.3.4 Quality Function Deployment

Quality function deployment (QFD) is a flexible and comprehensive group decision-making technique used in product or service development. QFD can strongly help an organization focus on the critical characteristics of a new or existing product or service from the separate viewpoints of the customer market segments, company, or technology development needs. The results of the technique yield transparent and visible graphs and matrices that can be reused for future product developments.

QFD transforms customer needs (the voice of the customer) into engineering characteristics of a product or service, prioritizing each product characteristic while simultaneously setting development targets for product or service development.

The most important tool in the quality function deployment is the House of Quality (HOQ). The House of Quality is a graphic tool for defining the relationship between customer desires and the product's capabilities. It utilizes a planning matrix to relate customer wants to how a firm (that produces the products) is going to meet those wants. The HOQ provides and organizes the information that the product team needs to refine the product concept. It looks like a house with a correlation matrix as its roof, customer wants versus product features as the main part, competitor evaluation as the porch, and so on. Figure 2.6 shows what a house of quality looks like. The QFD method will be discussed in detail in Chapter 7.

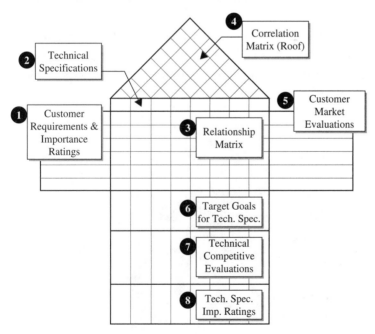

Figure 2.6 House of Quality

2.2.3.5 Pugh Concept Selection Pugh concept selection is a design evaluation for several design alternatives. Pugh concept selection uses a scoring matrix called the Pugh matrix. The Pugh matrix is related to the QFD method and is a form of prioritization matrix. By using the Pugh concept selection method, every design alternative will get a score, so the best design alternative can be selected.

2.2.3.6 Set-Based Design and Modularity *Modular design* is a design practice in which a product is broken into smaller subsystems. The subsystems are connected via standard interfaces. In this case, the subsystems become *decoupled*, that is, the designing of one subsystem is not dependent upon other subsystems. Therefore, the design work for each subsystem can be conducted in parallel. Under modular design, set-based design practice can be used. *Set-based design* means that instead of generating one design concept for each subsystem, multiple design concepts are generated for each subsystem by different teams.

For example, a bicycle can be divided into the following subsystems, as illustrated by Figure 2.7: frame, drive, wheels, brakes, and suspension. A possible scenario for set-based design is that three different design concepts could be developed for each subsystem. Because these subsystems are designed to fit standard interfaces, the possibilities add up as follows:

$$3 \text{ frames} \times 3 \text{ drives} \times 3 \text{ wheel sets} \times 3 \text{ brakes} \times 3 \text{ suspensions}$$
$$= 243 \text{ combinations}$$

This huge amount of design information will give you sufficient choices to end up with an extremely solid final design.

Figure 2.7 Subsystems of a Modular Design

Set-based design should only be used in the concept development stage, where the design cost is low and before you build prototypes and conduct tests. Set-based design is an important tool in lean product development, which will be discussed in detail later in this chapter.

2.2.4 Product Design and Prototype: Stage 3

After the concept development stage, the product design needs to be refined. The specific product design should be developed for each subsystem and components. Design parameters of every level should be developed, analyzed, verified, and tested, if necessary. In this stage (Stage 3), physical prototypes will be built and tests will be conducted.

The product design and prototype stage can be further divided into the following steps:

1. Generate exact detailed functional requirements

2. Develop actual implementation to satisfy functional requirements, that is, design parameters

3. Build prototypes

4. Conduct manufacturing system design

5. Conduct design validation

The product design and prototype stage is also a very important stage. However, the freedom to change the design will be rather limited, because the design concept, product architecture, and technologies selection for the product design would have been finalized at the concept development stage. For example, in an automotive design project, the choice of using a traditional internal combustion engine or a hybrid engine would have been firmly finalized in or before the concept development stage. If you make a decision to use a traditional internal combustion engine, then in the product design stage, you have to stick with this choice, even if gas prices are suddenly raised to $5 per gallon. If you change the design at this point, everything else will be affected and you would essentially have to redesign the whole thing. In the product design and prototype stage, you still can change many design parameters, without changing the product concept, technology, and architecture.

There are several methods that can be used to support this stage, specifically Design of Experiments (DOE), Taguchi method and robust design, and computer simulation models.

2.2.4.1 Design of Experiments   Design of Experiments (DOE) is also called *statistically designed experiments*. The purpose of the experiment and data analysis is to find the cause-and-effect relationship between the output and experimental factors in a process. The process model of DOE is illustrated in Figure 2.8.

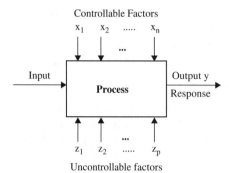

Figure 2.8 A Process Model of DOE

In any DOE project, you deliberately change the experimental factors and observe their effect on the output. The data obtained in the experiment will be used to fit empirical models relating output (y) with experimental factors. In product design practice, y is often used to represent a product performance metric. x_1, x_2, ..., x_n are often relevant product design parameters or process factors. For example, in a chemical process, the performance metric, y, could be the yield of the process, and the key process factors, such as temperature and pressure, could be x_1, and x_2. Mathematically, you are trying to find the following functional relationship:

$$y = f(x_1, x_2, ..., x_n) + \varepsilon \tag{2-1}$$

where ε is experimental error, or experimental variation. The existence of ε means that there may not be an exact functional relationship between y and $(x_1, x_2, ..., x_n)$. This is because

- Uncontrollable factors $(z_1, z_2, ..., z_p)$ will influence the response y, but are not accounted for in equation (2-1).

- There are experimental and measurement errors on both y and $(x_1, x_2, ..., x_n)$ in the experiment.

A DOE project requires many steps, as described in the following sections.

Step 1: DOE Project Definition   This is the first step, but certainly not a trivial one. You need to identify the objective of the project and find the scope of the problem. For example, in a product design, you need to identify what you want to accomplish. Do you want to reduce defects? Do you want to improve the current product's performance? What is the performance? What is the project scope? Do you work on a subsystem, or a component?

Step 2: Selection of Response Variable (Output)   After defining the project, you need to select the response variable, y. In selecting the response variable, the experimenter should determine if this response variable could provide useful information about the process under study. Usually, the response variable is a key performance measure of the process. You also want y to be

- A continuous variable, which would make data analysis much easier and more meaningful
- A variable that can be easily and accurately measured

Step 3: Choice of Factors, Levels, and Ranges   Actually, steps 2 and 3 can be done simultaneously. It is desirable to identify all the important factors that have significant influence on the response variable. Sometimes, the choice of factors is quite obvious, but there are cases where a few very important factors are hidden.

There are two kinds of factors: the continuous factor and the discrete factor. A *continuous* factor can be expressed by continuous real numbers. For example, weight, speed, and price are continuous factors. A *discrete* factor is also called a category variable, or attributes. For example, types of machines, types of seed, and types of operating system are discrete factors.

In a DOE project, each experimental factor will be changed at least once. That is, each factor will have at least two settings. Otherwise, that factor will not be a variable, but rather a fixed factor in the experiment. The numbers of settings of a factor in the experiment are called *levels*. The choice of number of levels in the experiment also depends on time and cost considerations. The more levels you have in experimental factors, the more information you will get from the experiment, but there will be more experimental runs, leading to higher cost and longer time to finish the experiment.

Step 4: Select an Experimental Design   The type of experimental design you select will depend on the number of factors, the number of levels in each factor, and the total number of experimental runs that you can afford to run.

The most frequently used experimental designs are full factorial designs and fractional factorial designs. If the number of factors and levels are given, then a full factorial experiment will need more experimental runs, thus becoming more costly, but it will also provide more information about the process under study. The fractional factorial will need a smaller number of runs, thus costing less, but will also provide less information about the process.

I will discuss how to choose a good experimental design in subsequent sections.

Step 5: Perform the Experiment   When running the experiment, you must pay attention to the following:

- Check performance of gauges/measurement devices first.
- Check that all planned runs are feasible.
- Watch out for process drifts and shifts during the run.
- Avoid unplanned changes (such as swapping operators at the halfway point).
- Allow some time (and backup material) for unexpected events.
- Obtain buy-in from all parties involved.
- Preserve all the raw data.
- Record everything that happens.
- Reset equipment to its original state after the experiment.

Step 6: Analysis of DOE Data   Statistical methods will be used in data analysis. From the analysis of experimental data, you can obtain the following results:

- Identification of significant and insignificant effects and interactions. (However, it is also possible that none of the experimental factors will be found to be significant, in which case the experiment is inconclusive. This situation may indicate that you may have missed important factors in the experiment.)
- DOE data analysis, which can identify significant and insignificant factors by using Analysis of Variance (ANOVA)
- Ranking of relative importance of factor effects and interactions
- Empirical mathematical model of response versus experimental factors
- Identification of best factor-level settings and optimal output performance level

Step 7: Conclusions and Recommendations   Once the data analysis is completed, the experimenter can draw practical conclusions about the project. If the data analysis provides enough information, you might be able to recommend some changes to the process to improve its performance. Sometimes, the data analysis cannot provide enough information, in which case you may have to do more experiments.

When the analysis of the experiment is complete, you must verify whether the conclusions are valid or not. These are called *confirmation runs*.

2.2.4.2 Taguchi Method and Robust Design The Taguchi method is a comprehensive quality strategy that builds robustness into a product or process during its design stage. The Taguchi method is a combination of sound engineering design principles and Taguchi's version of DOE, which is called an *orthogonal array experiment*.

In regular DOE, usually only controllable factors are studied. In the Taguchi method, both controllable factors and noise factors are studied. Noise factors are also called *uncontrollable factors*; they are factors that are not under the manufacturer's control, such as usage environment (temperature, humidity, vibration, and so on), customer's misuse, incoming material, and component variation. Clearly, both controllable factors (design parameters or process factors) and noise factors will affect the product performance (y). In the Taguchi method, a special type of experimental design, called *inner-outer arrays*, is used to study both control factors and noise factors.

In regular DOE, the common objective is to optimize y; in the Taguchi method, the objective is to achieve robustness. For example, in a traditional DOE study on a chemical process, you may try to find optimal control-factor levels, that is, the best temperature and best pressure to maximize the yield. Under these best temperature and pressure levels, the average yield may be high, but it might be very sensitive to other noise factors, such as incoming materials. When incoming material properties are changing because of batch-to-batch material variations, your chemical quality and effective yield may fluctuate a lot. In this case, the "optimal" settings are not robust. The Taguchi method looks for controllable factor-level settings that can deliver consistent performance even under the influence of significant noise factor variation. In the chemical process example, the Taguchi method tries to find the best temperature and pressure levels that not only deliver reasonably high yield, but also deliver stable quality and yield under noise factor variation, such as variation of incoming material.

Robust design is a very important approach in the product design stage to ensure that the product has consistent quality and reliability.

2.2.4.3 Computer Simulation Models Increasingly, choices of computer simulation models are available for many applications. There are two categories of computer simulation models. They are mechanism-based simulation models and Monte Carlo simulation models.

Mechanism-Based Simulation Models   Mechanism-based simulation models are usually commercial computer packages that are designed for specific fields of applications, such as mechanical, electrical, and electronics. The computer packages follow established scientific principles or laws to model the relationships for specific classes of applications.

Mechanical components and system design can be created and analyzed using computer-aided design (CAD) software. Analysis capability includes geometrical dimensions and tolerances (GD&T), three-dimensional views and animation, and so on. CAD software is usually the starting point for many computer-aided engineering (CAE) analyses, such as stress and vibration analysis.

Finite element analysis (FEA) is a computer-based technique that can evaluate the many key mechanical relationships, such as the relationships among force, deformation, material property, dimension, and structure strength, in a design. In FEA, the original continuous 3-D geometrical shapes are subdivided into many small units, called *finite element meshes*. The precise differential equation forms of basic relationships in mechanics are approximated by a large-scale linear equation system. FEA methods can study mechanical stress patterns in static or dynamic loaded structures, strain, response to vibration, heat flow, and fluid flow.

The use of FEA to study the fluid flow is called computational fluid dynamics (CFD). CFD can be used to analyze air flow, fluid flow, and so on; thus, it can be used to analyze aerospace design, automobile engine design, ventilation systems, and others.

Electrical and electronic circuits can be analyzed by using electrical design automation (EDA) software that is based on the mechanism models in electrical engineering, such as Ohm's law, logical circuit models, and so on. In EDA software, component parameters, input electrical signals, such as power, waveform, and voltage, and circuit diagrams, are inputted into the program. The program can perform many different analyses, such as evaluation of circuit performances, sensitivity analysis, and tolerance analysis.

Electromagnetic effects can be analyzed by using a special kind of FEA software that is based on Maxwell's law.

There are also numerous mechanism-based computer simulation models for other applications, such as chemical engineering, financial operations, and economics.

Monte Carlo Simulation Model   Monte Carlo simulation is a technique that simulates large numbers of random events. For example, transactions in a bank branch office are random events. A random number of customers enter the office at random arrival times, and each one will make a random type of transaction, such as deposit, withdrawal, or loan, with a random amount of money. A Monte Carlo simulation model can simulate such events. A Monte Carlo simulation starts with generating a large number of random numbers that follow prespecified probability distributions. For example, the customer arrival time to the bank branch can be assumed to be a *Poisson arrival process*, in which the key parameter, that is the mean inter-arrival time, can be set by the analyst or estimated with old data. After you set all the parameters for the probability models, such as mean and variance of the transaction amount, average transaction time, and system interaction relationships, the Monte Carlo simulation can generate virtual bank transaction processes for a large number of virtual customer arrivals. The simulation model can then provide key system performance statistics in the form of a chart and or histogram. In the bank example, these key performance measures could include customer waiting time, percentage of time that a clerk is idle, daily transaction amount, and so on.

A Monte Carlo simulation is very helpful in analyzing service processes, factory flows, banks, hospitals, and so on.

2.2.4.4 Prototypes   Prototypes are trial models for the product. Ulrich and Eppinger (2000) defined a prototype as "an approximation of the product along one or more dimensions of interest." In this context, a prototype can be a drawing, a computer model, a plastic model, or a fully functional prototype fabricated at the pilot plant.

Ulrich and Eppinger also further defined two types of prototype: analytical or physical. An *analytical prototype* represents the product in a mathematical or computational form. Many aspects of the product can be analyzed in this form. For example, a finite element model (FEM) can be used to analyze a mechanical part with respect to force stress, deformation, and so on. A Monte Carlo simulation model can be used to simulate service flow, waiting time, and patient process rate in a clinic. A *physical prototype* is a real look-alike prototype made of either substitute materials or actual materials designed for the product. For example, a prototype of an automobile manifold may be made of easy-to-form plastic instead of the aluminum that the finished product would be made of. In this case, researchers are only interested in studying the geometrical shape aspect of the manifold, not the heat resistance and strength aspects of the manifold.

A prototype can be also *focused* or *comprehensive* (Ulrich and Eppinger 2000). A focused prototype only represents a part of, or a subset of, the product's functions or attributes. A comprehensive prototype represents most of the product functions and attributes, and some prototypes represent all product functions and attributes.

All analytical prototypes are focused prototypes, at least at the current technological level, because it is impossible to build a virtual model that can represent all of a product's functions and attributes. For example, an FEM model can only represent the mechanical aspect of the product's characteristics; it cannot represent chemical properties such as corrosion. Physical prototypes can be very comprehensive; a preproduction prototype has all of the product's required functions and attributes. Physical prototypes can also be very focused; a plastic prototype may only be able to represent the geometrical aspect of a part, not material properties or functions.

There are four commonly used physical prototypes: experimental prototype, alpha prototype, beta prototype, and preproduction prototype.

Experimental prototypes are very focused physical prototypes; they are designed and made to test or analyze a very well-defined subset of functions and attributes. For example, a plastic prototype of an automobile manifold is built so that the engineering team can study the geometrical aspects of manifold functions.

Alpha prototypes are used in product functional performance validation. An experimental prototype made in the lab, or a concept car model, is one such example. Usually an alpha prototype can deliver all the intended functions of a product. The materials and components used in alpha prototypes are similar to those that will be used in real production. But they are made in a prototype process, not by the manufacturing process to be used in mass production. The rapid prototyping technique is often used to make alpha prototypes because of its speed. Rapid prototyping is the automatic construction of physical objects using solid free-form fabrication. The first techniques for rapid prototyping became available in the 1980s and were used to produce models and prototype parts. Today, they are used for a much wider range of applications and are even used to manufacture production-quality parts in relatively small numbers.

Beta prototypes are used to analyze reliability requirements validation, usage requirements validation, product specification validation, and so on. They may also be used to test and debug the manufacturing process. The parts in beta prototypes are usually made with the actual production process or supplied by the intended part suppliers, but usually they are not produced at an intended mass production facility. For example, an automobile door panel beta prototype might be made by the same machinery as that of the assembly plant, but it is made by a selected group of validation engineers and technicians, not by hourly workers.

Preproduction prototypes are the first batch of products made by the mass production process. At this point in time the mass production process is not operating at full capacity. These prototypes are usually

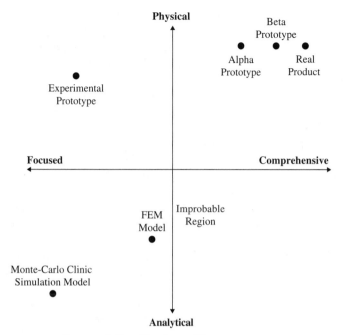

Figure 2.9 Types and Classifications of Prototypes

used to verify production process capability, as well as test and debug the mass production process.

Figure 2.9 illustrates the types and classifications of prototypes.

So far, I have discussed Stage 0 to Stage 3 of the product life-cycle process. These stages cover the product development process, which is also called the *upstream* of the product life cycle. The remaining four stages cover the production, usage, and disposal of the product, which are called the *downstream* of the product life cycle. Since the main objective of this book is capturing and analyzing the voice of the customer, which is primarily related to the product development process, I will not discuss the last four stages of the product life cycle in detail in this chapter.

2.3   The Nature of Product Development: Information and Knowledge Creation

In the last two sections, we discussed the common features of product development processes. There are many companies that develop and manufacture products, but the results of their product development processes are vastly different. Even within the same industry, the product development process of the best company can be vastly better than that of its peers. Recently, the product development process used by

Toyota (Morgan and Liker 2005) received a lot of attention. Compared with other automotive manufacturers, Toyota not only has a highly efficient lean production process, but also Toyota's key product development performance metrics, such as product development lead time (how long it takes from the beginning of design of a new vehicle to launch into production), the quality of design (how competitive the designed vehicle is in the marketplace), and total development cost, are much better than its competitors.

These facts lead to several questions. What makes a product development process a better process? Can people learn, understand, and develop a better product development process? What is the nature of product development? Is there a scientific basis of product design and development? If yes, can we use this scientific basis to guide product development processes? I will try to answer these questions in this part of the chapter.

Research on the product design and product development process gradually became popular in recent years. Research in the product design arena started in Europe. Primarily, the Germans developed some design guidelines that kept improving at a consistent pace. A huge body of research has been published in German on design practice. Most of these efforts are listed in Hubka (1980) and Phal and Beitz (1988). The German design schools share common observations. For example, a good design entity can be judged by its obedience to some design principles; the design practice should be decomposed to consecutive phases; and methods for concept selection are required. Besides the Germans, the Russians also developed an empirical inventive theory which promises to solve difficult and seemingly impossible engineering problems, the so-called TRIZ or "Theory of Inventive Problem Solving" (TIPS) (Altshuller 1988, 1990; Rantanen 1988; Tsourikov 1993).

In the United States, over the last three decades, there have been progressive research efforts in the product design area, particularly in the field of mechanical design. Dixon (1966) and Penny (1970) are among the pioneers in this research. Ullman (1992) stated that the activities of design research were accelerating along the following trends: artificial intelligence computer-based models, design synthesis (configuration), cognitive modeling, and design methodologies. Suh (1990) developed the *axiomatic design approach*; in this approach; the design practice is modeled as mappings from one domain to another domain. Several design principles are derived to characterize and benchmark the design practice. This approach enables evaluation of design solutions based on different principles. McCord and Eppinger (1993) and Pimmler and Eppinger (1994) developed a theoretical framework to optimize the sequence of product development projects. The stage-gate system was developed by Cooper

(1990) to manage the product development process. In the stage-gate system, the stages are the product development stages that I illustrated in the preceding section; the *gate* means a rigorous evaluation process at the end of each stage, where the product development team tries to make sure that all necessary pieces of work in this stage are performed satisfactorily, and unnecessary projects are discontinued. In Reinertsen's excellent book, *Managing Design Factories* (1997), the product development process is treated as an information generation process; how to generate information effectively becomes an important goal for a good product development process. To improve the product development lead time, queue theory is used to guide the rules for design project sequencing. Huthwaite (2004) proposed his "lean design" approach, which emphasizes customer values and design complexity reduction. In recent years, Toyota's product development system has been getting a lot of attention (Morgan and Liker 2006, Kennedy 2003).

In the literature I just cited, the generic product development process is studied from different perspectives. In this section, I am proposing a comprehensive theoretical framework to model and analyze the product development process. I call this framework the *information and knowledge-based product development theory*. Before discussing this theory fully, I will first discuss axiomatic design theory and a part of Reinertsen's work as necessary foundations to my theory.

2.3.1 Axiomatic Design

Motivated by the absence of scientific design principles (Suh 1984–2001) tried to find a scientific basis for good design practice. He did this by looking into many successful and unsuccessful design case studies, and attempted to determine whether there are common features for good design practices. At first, Suh found that there were 12 such common features; later he found that these 12 common features of good design can be further reduced to just two. Suh called these two common features for good design practices axioms, so his scientific design principles are based on these two axioms. An *axiom* is any sentence, proposition, statement, or rule that forms the basis of a formal system. As an equally important result, he modeled the design processes as "mappings from one domain to another domain." This is a very important observation, and it helped to give deep insight into the design process. In this subsection, I will first discuss the mappings and domains, and then I will discuss the two design axioms.

2.3.1.1 Design as a Mapping Process   If we look into "what designers do all the time," you would find that they always treat design as a mapping process, as illustrated by Figure 2.10.

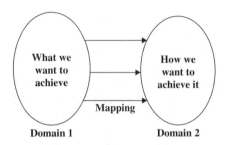

Figure 2.10   Design as a Mapping Process

For example, if what you want to achieve is "move people from one place to another on the ground without consuming fossil fuel power," then the design engineers will have to figure out "how we want to achieve it." The design engineers may give you a design solution of "bicycle." A product design project can be divided into many tasks from "given whats, find hows." Each of these tasks can be seen as a mapping from one domain to another domain, as illustrated in Figure 2.10.

In Suh's point of view (2001), the design world consists of four domains: the customer domain, the functional domain, the physical domain, and the process domain. The *customer domain* is characterized by the needs that customers want the product to fulfill. In the *functional domain*, the customer domain elements are translated into functional requirements. All products are developed to provide functions, that is, the job the product must do to satisfy customers. *Functional requirements* are specific requirements for functions. For example, in automotive design, in the customer domain, a customer may say that he wants a "faster car." This is one customer attribute; the engineer needs to figure out what the customer means by "faster car." Finally, by comparing the performance of the competitors' product, the engineer may figure out that faster car may mean "ability to accelerate from a speed of 0 km/hour to a speed of 100 km/hour within 10 seconds"—this is a functional requirement. Clearly, a product may have many functional requirements. Specifically, functional requirements (FRs) are a minimum set of independent requirements that completely characterize the functional needs of the design solution in the functional domain.

The *physical domain* is also called the *design parameter domain*. Design parameters (DPs) are the elements of the design solution in the physical domain that are chosen to satisfy the specified FRs. Design parameters (DPs) mean "what specific physical design can be used to deliver all required functions." For example, if a functional requirement for a car is to accelerate from 0 km/hour to 100 km/hour in 10 seconds, the design parameter for that would include necessary powertrain design specifications, including engine design, transmission design, and so on. The process domain deals with process variables. The *process domain*

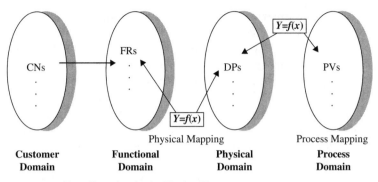

Figure 2.11   Four Domains in the Design Process

refers to the relevant manufacturing process specifications that can build the production facility to produce the designed product. *Process variables (PVs)* are the elements in the process domain that characterize the process that satisfies the specified DPs. Figure 2.11 illustrates the four domains in the design process.

Table 2.1 shows the specific contents of the four domains for different industries.

2.3.1.2   The Independence Axiom   The design process involves three mappings between four domains (Figure 2.11). The first mapping involves the mapping between customer attributes and the functional requirements (FRs). This mapping can be performed by the means of Quality Function Deployment (discussed in Chapter 7). The physical

TABLE 2.1   Four Design Domains for Different Industries

	Customer Domain (CN)	Functional Domain (FR)	Physical Domain (DP)	Process Domain (PV)
Manufacturing	Attributes that customers desire	Functional requirements specified for the product	Design parameters that satisfy FRs	Process variables that fulfill DPs
Materials	Desired performances	Required properties	Microstructures	Processes
Software	Attributes desired in software	Output specifications of the program codes	Inputs, algorithms, methods	Machine codes, compilers, modules
Organization	Customer satisfaction	Function of the organization	Programs, office, activities	People and other resources
Business	ROI	Business goals	Business structures	Human/ financial resources

mapping involves the functional requirements domain and the design parameter domain (DPs). The process mapping involves the design parameter domain and the process variable domain. Out of the three mappings, the first mapping involves translating the raw voice of the customer to CTQs, and translating CTQs to functional requirements. The last two mappings, that is, the physical mapping and process mapping, are very challenging and important. The equation $y = f(x)$ is used to reflect the relationship between the domain, array y, and array x, in the related mapping, where the array $\{y\}_{mx1}$ is the vector of requirements with m components, $\{x\}_{px1}$ is the vector of design parameters with p components, and A is the sensitivity matrix representing the physical mapping with $A_{ji} = \partial y_j / \partial x_i$. In the process mapping, matrix B represents the process mapping between the DPs and the PVs. The overall mapping is matrix $C = AB$, the product of both matrices. When matrix A is a square diagonal matrix, that is, $m = p$ and $A_{ji} \neq 0$ when $i = j$ and 0 elsewhere, the design is called *uncoupled*; i.e., each y can be adjusted or changed independent of the other y. An uncoupled design is a one-to-one mapping and is represented as follows:

$$
\begin{Bmatrix} y_1 \\ \cdot \\ \cdot \\ y_m \end{Bmatrix} = \begin{bmatrix} A_{11} & 0 & \cdot & 0 \\ 0 & A_{22} & & \cdot \\ \cdot & & \cdot & 0 \\ 0 & \cdot & 0 & A_{mm} \end{bmatrix} \begin{Bmatrix} x_1 \\ \cdot \\ \cdot \\ x_m \end{Bmatrix} \qquad (2\text{-}2)
$$

In the *decoupled* design case, matrix A is a *lower / upper triangle* matrix; that is, the maximum number of nonzero sensitivity coefficients equals $p(p-1)/2$ and $A_{ij} \neq 0$ for $i = 1, j$ and $i = 1,p$. A decoupled design is represented as follows:

$$
\begin{Bmatrix} y_1 \\ \cdot \\ \cdot \\ y_m \end{Bmatrix} = \begin{bmatrix} A_{11} & 0 & \cdot & 0 \\ A_{21} & A_{22} & 0 & \cdot \\ \cdot & \cdot & \cdot & 0 \\ A_{m1} & A_{m2} & \cdot & A_{mm} \end{bmatrix} \begin{Bmatrix} x_1 \\ \cdot \\ \cdot \\ x_m \end{Bmatrix} \qquad (2\text{-}3)
$$

The decoupled design may be treated as an uncoupled design when the x's are adjusted in some sequence conveyed by the matrix. Uncoupled and decoupled design entities possess conceptual robustness; that is, the x's can be changed to affect other requirements in order to fit customer attributes. A coupled design results when the matrix has the number of requirements, m, greater than the number of x's, p, or when the physics is binding to a great extent, so off-diagonal sensitivity elements are nonzeros. The coupled design may be uncoupled or decoupled by *smartly* adding $m\text{-}p$ extra x's to the problem formulation. A *coupled* design is represented as follows:

$$\begin{Bmatrix} y_1 \\ \cdot \\ \cdot \\ y_m \end{Bmatrix} = \begin{bmatrix} A_{11} & A_{12} & \cdot & A_{1p} \\ A_{21} & A_{22} & & \cdot \\ \cdot & & \cdot & A_{(m-1)p} \\ A_{m1} & \cdot & A_{m(p-1)} & A_{mp} \end{bmatrix} \begin{Bmatrix} x_1 \\ \cdot \\ \cdot \\ x_p \end{Bmatrix} \qquad (2\text{-}4)$$

After these mathematical preparations, I can now present the first axiom of the axiomatic design:

Maintain the independence of FRs: In an acceptable design, the DPs and the FRs are related in such a way that a specific DP can be adjusted to satisfy its corresponding FR without affecting other FRs.

Mathematically, the first axiom states that if a design matrix linking functional requirements and design parameters, A, is a diagonal matrix, then this physical mapping is uncoupled, or completely independent. In this case, you can adjust one design parameter value to match the corresponding functional requirement, independently. If the design matrix A is a lower or upper triangular matrix, such as the one in equation (2.3), then this physical mapping is decoupled. In this case, you can adjust the design parameters in a fixed sequence to match corresponding functional requirements one by one. If the design matrix is a full matrix, then the physical mapping is coupled. In this case, you cannot easily adjust design parameters to match the corresponding functional requirements.

Axiom 1 of axiomatic design basically states that, out of three kinds of design mappings, uncoupled, decoupled, and coupled design, uncoupled design is an ideal design. The decoupled design is not as good as uncoupled design, but it is better than coupled design, so it is acceptable. Coupled design is undesirable. Figure 2.12 illustrates these three kinds of design matrices. The same things can also be said for the process mapping, that is, the mapping from physical domain to process domain.

We will use Example 2.1 to further explain the concept of axiomatic design.

Example 2.1 Water Faucet Design   Water faucets are used in kitchens or bathrooms to control the water temperature and flow rates. From an axiomatic design point of view, the faucet has two important functional requirements, FR_1 = control the flow of water, FR_2 = control

$$\begin{bmatrix} X & 0 & 0 \\ 0 & X & 0 \\ 0 & 0 & X \end{bmatrix} \quad \begin{bmatrix} X & X & X \\ 0 & X & X \\ 0 & 0 & X \end{bmatrix} \quad \begin{bmatrix} X & X & X \\ X & X & X \\ X & X & X \end{bmatrix}$$

uncoupled   decoupled   coupled
design   design   design

Figure 2.12   Three Kinds of Design Matrices

the water temperature. There are two popular designs for faucets used in kitchens or bathrooms, as illustrated in Figure 2.13.

One design is called the *two-handed design*. In this design, the two design parameters are: DP_1 = valve angle for the hot water valve, and DP_2 = valve angle for the cold water valve. Clearly, when you adjust DP_1, that is, the hot water valve, it will affect both the flow rate (FR_1) and water temperature (FR_2). The same will be true for cold water valve (DP_2). So whenever you adjust these two valves, you will simultaneously change both the flow rate (FR_1) and water temperature (FR_2). This design definitely violates Axiom 1. From the design matrix shown in Figure 2.13, this design is a coupled design. You need to do a lot of adjustments to get both flow rate and temperature right.

The other design is called the *one-handed design*; in this design the two design parameters are: DP_1 = valve angle for mixed water (done by adjusting the vertical angle of the valve), DP_2 = mix ratio of cold/hot water (done by adjusting the horizontal angle of the valve). So flow rate (FR_1) can be

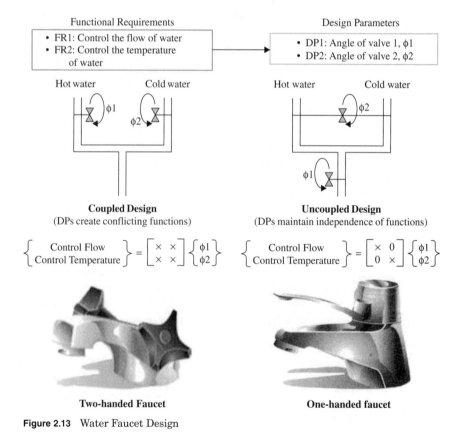

Figure 2.13   Water Faucet Design

adjusted by DP_1 alone, and water temperature (FR_2) can be adjusted by DP_2 alone. This design is an uncoupled design; its design matrix is shown in Figure 2.13. This design satisfies Axiom 1.

Implications and Benefits of the Independence Axiom in Product Development Practice   The independence axiom will definitely favor modular design practice, because the design parameters in different modules are most likely to be independent of each other. In addition, the independence axiom will also encourage maintaining parametric independence of design parameters within each module. In applying axiomatic design, there is a common misconception stated as follows: "We cannot use the independence axiom, because it means I have to make many physical parts, so each part can behave independently of the others." The truth is, a design parameter is not same as a physical part. In a single physical part, you could have many design parameters. Figure 2.14 illustrates a two-headed wrench that has different wrench units on the same physical part. Both wrench units are different, independently adjustable design parameters.

There are many benefits of applying the independent axiom:

1. Parallel developments and testing. As discussed earlier, the independence axiom favors modular design practice. If a system is developed based on modular design practice, it will enable parallel development and testing of different modules, or subsystems. Modular design and testing is illustrated in Figure 2.15. Suppose you are trying to design and build a personal computer. A personal computer can be subdivided into many subsystems, such as CPU, memory, and so on. If you have a coupled design, that is, there are a lot of dependencies among subsystems, then changing the design of a subsystem will affect the design of other systems. In this case, you have to design one subsystem at a time. Only after you freeze the design of one subsystem can you start the next. This will give you a lengthy and sequential design process. If you apply the independence axiom and adopt a modular design practice, you can develop several subsystems in parallel, which can speed up the product development process greatly.

2. Easier engineering change. Engineering change is a situation where the goals, objectives, and functional requirements of a product will be forced to change due to a variety of causes, such as market changes, government regulation changes, and so on. Clearly, an independent

Figure 2.14   A Two-Headed Wrench

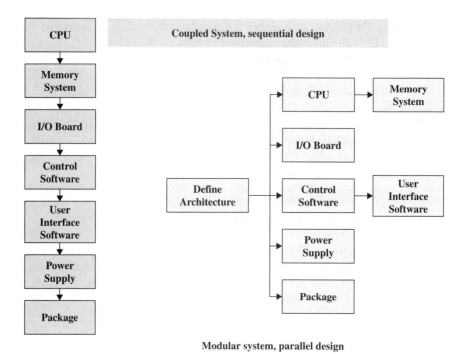

Figure 2.15 Coupled System vs. Modular System

design will have a simpler relationship between functional requirements and design parameters. When the functional requirements
have to change, it is much easier to figure out how you have to change
design parameters of uncoupled designs to match the new functional
requirements than that of coupled designs.

3. Simpler operation. An independent design will also have a simpler relationship between adjustment factors and system function, because they are a part of functional requirements and design
parameters. Example 2.1 illustrates this fact quite well.

2.3.1.3 The Information Axiom   The second axiom of axiomatic design
deals with the complexity in design. The formal statement of Axiom 2
is as follows:

*Minimize the information content: Among alternative designs that satisfy
Axiom 1, the best design has the minimum information content.*

In Suh's *Principles of Design* (1990), information content is defined
as "the total number of bits of information that is needed to store the
complete information for the design." In recent years (Suh 2001), the
definition of information content by Suh's school is mostly focused on

the probability of success for delivering all the FRs. Suh used the following entropy model to define the information content:

The information content, that is, entropy I_i, for a given FR_i, is defined as follows:

$$I_i = \log_2 \frac{1}{P_i} = -\log_2 P_i \qquad (2\text{-}5)$$

Where P_i stands for the probability for FR_i to be delivered successfully.

For a group of m independent FR_is, the success probability for the whole system, P, is

$$P = \prod_{i=1}^{m} P_i \qquad (2\text{-}6)$$

Then the total information content, that is, total entropy I, is

$$I = \log_2 \frac{1}{P} = -\log_2 P = -\log_2 \left(\prod_{i=1}^{m} P_i \right) = -\sum_{i=1}^{m} \log_2 P_i \qquad (2\text{-}7)$$

Intuitively, Axiom 2 basically states: if you have several alternative designs, and every design can deliver the functional requirements and satisfy the independence axiom, then the simplest design is the best design.

In recent axiomatic design practices, however, most real-world case studies are related using Axiom 1. The information contents defined by equations (2.5) through (2.7) are difficult to quantify in practical circumstances.

I believe that Axiom 2 is a very fundamental and powerful design principle, and more comprehensive and practical complexity metrics should be developed to foster the application of this axiom. Later in this chapter, we will use Axiom 2 to guide the lean product development process.

2.3.2 Design as an Information Production Factory

In his excellent book *Managing the Design Factory* (1997), Reinertsen stated that the only purpose of the design process is to produce useful information economically. In this book, information is what design engineers are producing in their work, such as drawings, test results, specifications, and so on. Reinertsen also used entropy to describe the information as follows:

$$I = \log_2 \frac{1}{P_e} \qquad (2\text{-}8)$$

where P_e is the probability for a event to occur. Clearly, I will be large if P_e is small. Reinertsen stated that if an event has a very small probability of occurring, but it does occur, then this event will contain a lot of information. Specifically, Reinertsen applied this entropy formula to engineering test applications as in equation (2.9):

$$I_{\text{Test}} = P_{\text{Failure}} \bullet I_{\text{Failure}} + P_{\text{Success}} \bullet I_{\text{Success}} \qquad (2\text{-}9)$$

where I_{Test} stands for the information contents of a test, P_{Failure} and P_{Success} refer to the probability that a test is a failure or a success, respectively, and I_{Failure} and I_{Success} are the corresponding information contents for a failed test, or a successful test, respectively.

By analyzing equation (2.9), Reinertsen stated that if a test has a high probability of failure, and the real test result is a failure, then the information content in this test is low, because you don't learn much from this test. On the other hand, if a test has a high probability of success and the actual test does yield a success, the information content is also low for the same reason. On the contrary, if a test has a high probability of failure, but the actual test yields a success, based on equation (2.8), it can be proved that the information content of this test is high. This makes sense because a success in a high-risk test contains some rare information. Similarly, if a test has a high probability of success, but the actual test yields a failure, then this test must have high information content because it will shed some light on rare failure causes. However, for most tests, if the probability of failure is high, most likely, the test will yield a failure, and if the probability of success is high, then most likely the test will yield a success. So high success-rate tests and high failure-rate tests are not likely to provide a lot of information. Mathematically, it can be proved that if the failure probability of a test is 50 percent, (so the probability of success is also 50 percent), then the expected information content in this kind of test will be the highest. Figure 2.16 illustrates this fact.

Based on this analysis, Reinertsen further proposed that in engineering design practice, you should control the risk level for each project at around 50 percent. That is, a project, or a test, will have an equal probability of success or failure, assuming you can learn useful lessons from both successes and failures. A low-risk level strategy will make learning things too slow, because you don't learn much from a successful project if the project is very easy to do. A high-risk-level strategy will also slow down the learning process, because each success is so hard to get that you will be buried under a pile of failures. Obviously, this medium-risk, medium-difficulty level learning strategy makes a lot of sense and it will enable us to gain information at the highest speed.

Reinertsen also stated that information has time value; the earlier you get the information, the better. If you get the information too late in the product development process, the necessary design change will be

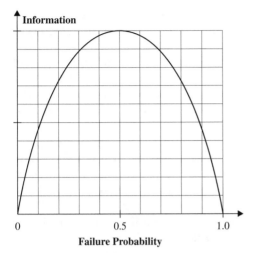

Figure 2.16 Failure Probability and Information Content

very difficult to implement. In Reinertsen's point of view, early information can be obtained by conducting small-scale tests early in the product development process. He also stated that not every piece of information has the same value—some information is critical, some information is not as critical. In the product development process, you would like to get critical information earlier than noncritical information. This fact forms the basis for design job sequencing.

2.3.3   Information and Knowledge Mining

What is the nature of the product development process? What are the key factors for a superior product development process? How can you improve your product development process? We are trying to dig out answers for these questions in this subsection.

The answer to the question "what is the nature of the product development process?" can be found by looking into "what is the final result of the product development process?" What are design engineers doing every day? In general, what is the extended product development team (including marketing, management, research and development scientists, and technicians) doing every day? They are creating documents, compiling testing reports, doing design analysis, drafting graphs, calculating survey data statistics, creating specifications, building prototypes, designing and making tools for producing the product, and developing assembly operations. In general, they are generating all kinds of information and knowledge. When are their jobs done? Their jobs are done after the production people have obtained enough useful information to produce the product effectively, reliably, economically, and with good quality, and the products shipped to customers are free of after-sale problems. Clearly, the nature of the product development process is an information generation factory.

2.3.3.1 Defining Information and Knowledge

Data, information, and knowledge are closely related. *Data* consists of numbers, characters, images, and signals. Data is usually unorganized. *Information* is the result of processing, manipulating, and organizing data in a way that creates meaning to the person receiving it.

However, data and information alone cannot fully explain the product development process. Without human involvement, the information is static and not dynamic. At least for now, you cannot build a 100 percent computerized product development system that just takes care of itself with zero human involvement. *Knowledge* plays an important role. There are many different definitions of knowledge. According to Ballard (1987), knowledge is information plus theory; other researchers define knowledge as "information combined with experience, context, interpretation, and reflection. It is a high-value form of information that is ready to apply to decisions and actions" (Davenport et al. 1998). In my point of view, knowledge is a personal or organization-based information-processing mechanism; knowledge is a combination of an information bank, interpreter, organizer, enabler, improver, and creator; knowledge should represent truth, and it should be actionable. All of this is illustrated in Figure 2.17.

A product development organization can be modeled as shown in Figure 2.18. Inside the product development organization, there are many knowledge cells, which can be individual professionals, and many knowledge centers, which can be teams. All kinds of information flow around, into, and out of knowledge centers and cells. Each knowledge center and cell will process information and create modified or new information after processing.

In the product development process, there are three dominant types of information processing: information mining, information transformation, and information/knowledge creation.

2.3.3.2 Information Mining

Information mining is the extraction of valuable information from raw information sources. For example, in research and publication work, the literature survey is information mining work. The raw information sources include professional journals, the Internet, professional conferences, and leading professionals. The process of a literature survey is to extract relevant and high-quality information, and

Figure 2.17 Knowledge and Information

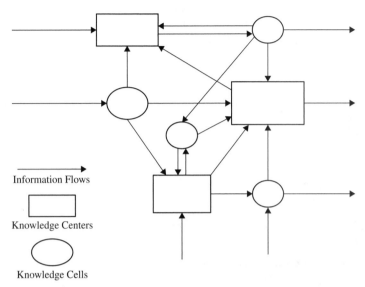

Figure 2.18 Product Development Process as a Connected Network of Knowledge Centers and Cells

then process and compile it. The literature survey report is the extracted, condensed information. Similar to metal that has been mined as compared to ore, the literature survey report will serve as a stepping-stone for further research.

In the product development process, there are two major types of information mining work: mining of the voice of the customer, and mining of technological information. Figure 2.19 shows how these two types of information mining fit into the product development process. In addition, as the figure shows, the voice of the customer can come from both external and internal customers. In general, the steps in information mining are as follows:

1. Prospecting to locate information sources

2. Exploration to find and then define the extent and value of information sources

3. Exploitation to extract information from information sources

4. Processing and refining the extracted information for the information users

Information Mining of the Voice of the Customer   According to the theory of axiomatic design, product development is a sequence of mapping processes. The very first step is mapping from the customer domain to the function domain; this step is really a rigorous product definition step. The whole value of the product is largely determined by whether the product will be welcomed by potential external customers, or buyers.

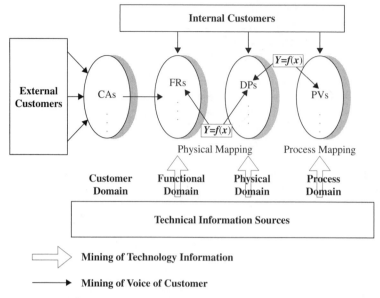

Figure 2.19   Information Mining in the Product Development Process

If customers don't want your product, it will be a failure. So the ability to accurately mine potential customers' minds is critically important in the product development process. There is a saying, "garbage in, garbage out"—if you totally ignore what customers want, or you did not do a superb job in discovering what customers really want, your very first input to the product development process might be totally flawed, and whatever you do in the subsequent steps, you will produce a garbage product. It is not an exaggeration that accurately capturing the voice of the customer is like striking gold.

There are two kinds of customers: external customers and internal customers. The *external customers* are the customers who buy or use your products. The *internal customers* are people inside your company who are the users of your work. For example, production people are the customers of the product development team, because they are the downstream users of product design and manufacturing process design.

The specific steps in mining the voice of the external customer are described as follows:

1. Prospecting to locate information sources. The most important information source is the prospective real customers. You have to get most of this information from the people, not from the literature, electronic media, and so on.

2. Exploration to find and then define the extent and value of information sources. At first, you need to determine which customer group

and what kind of customers you need to locate, and then how to select sample customers to work with. The value of the information source is similar to the value of ore in metallurgy—the value of the information depends on how much valuable information you can get from this source, and how easy it is to extract the real valuable information. For example, lead users and competitors' customers might be really valuable information sources. The lead users can provide a lot more in-depth opinions than regular users, and provide information on the future market, because what lead users are doing with the product might be the mainstream application of the product in the near future. The competitors' customers can provide information about why some people don't like your products. This information is very difficult to get from your existing customers.

3. Exploitation to extract information from information sources. This step deals with what kind of information you really want to get, and how to extract high-quality information from customers.

For the first question, the purpose of extracting the voice of the customer is to support the whole product development process. For example, you need to get enough VOC information to derive full sets of product functional requirements, based on Figure 2.19. Common problems for this step include:

- You did not get the right type of information.

- You did not get enough information or clear enough information to support your job.

For example, suppose that I eat in a restaurant and I am given a customer survey form to fill out. However, all the questions are something like: "Do you like our food?" or "Do you like our drink?" and ask me to fill in numerical ratings. How much can this information help to improve the service? If I fill in "the food is bad," then the question is, which menu item is bad? And how can it be improved? This response doesn't provide the kitchen with the right type of information, or enough information, to improve the food. Chapter 3 of this book will provide detailed information about "what kind of information do I really want to get?"

For the second question, "how do you extract high-quality information from customers?" you first need to know what defines high-quality information. In my opinion, high-quality VOC information should have some of the following characteristics:

- It accurately reflects what customers really want deep in their hearts, not just what they say.

- It uncovers some unmet customer needs, or unarticulated customer's needs in the current marketplace.

- It provides a basis for predicting future or near-future market trends and needs.

- It provides accurate and sufficient information to feed the product development process.

There are many effective methods for extracting the voice of the customer, such as customer surveys, focus groups, and ethnographic approaches. These methods will be discussed thoroughly throughout this book.

The information users of the voice of the customer are the product development team, mostly design engineers. So they are the customers of the people who collect the VOC information. At first, you need to know what kind of information the product development team really needs, so that the people who collect VOC information will collect relevant and sufficient VOC information. Secondly, the VOC information should be translated into language that information users will understand. Chapter 6 of this book will discuss how to process VOC data. Quality function deployment (QFD) is also an effective tool to relate the voice of the customer to design requirements. QFD will be discussed in Chapter 7.

Mining of Technology Information and Knowledge   Whether you develop a technology push-type product or a market pull-type product, you need to have technological know-how and information to make the voice of the customer into reality. In this information mining work, the quality and speed of the information mining are very important.

The quality of this information mining means that you are getting the best technology in terms of performance, the cost of the technology, and the robustness of the technology. The performance of the technology means that this technology can create the right functions, and it performs better than other competing technologies. The cost of technology means that its value-to-cost ratio is high; you should not adopt technology that has top-notch performance but is prohibitively expensive. The robustness of the technology is also very important; *robustness* means that this technology can deliver its functions consistently under various usage conditions and does not require excessive operator or user requirements. Many very new technologies have excellent performance, but their technical bugs have not been fully cleaned up yet; introducing too much immature technology into the product could create potential quality and reliability problems.

The speed of this information mining is also critically important; the ideal technology information mining process has the ability to provide the right technology information to the right people at high speed. In many companies, this information mining work is a slow and ineffective process, with a lot of unnecessary information and a lot of reinvention. Later in this chapter, we are going to discuss how to improve this process.

The specific steps in mining the technology information are described as follows:

1. Prospecting to locate technology information and knowledge sources. As discussed earlier, information is organized data that delivers meaning. Knowledge is the distilled information that is understood, believed to represent the truth, and actionable.

   The sources of this kind of information could be internal and external. Internal sources of information include:

   - **Company databases, company libraries, and document storage**   This type of information is relatively easy to retrieve.

   - **Everywhere in employees' computers, binders, and document folders**   Usually, this kind of information is difficult to retrieve; people don't even know where it is. This information is constantly missing, falling into cracks, misplaced, and distorted. In the product development process, a lot of time and manpower is lost in retrieving and re-creating this kind of information. I will discuss how to improve the management of this information source in a lean product development process.

   The external sources of information include:

   - **Suppliers**   Many companies contract out a lot of product development tasks, so the suppliers have a lot of this technology information.

   - **Universities and research labs**

   - **Individual researchers and small-sized providers**   This includes a lot of new startup research-oriented companies, consultant companies, and even individuals.

   - **Public domain, such as publications, patents, and the Internet**

   The sources of knowledge are more complicated, because you cannot isolate the knowledge from the people who have the knowledge. Based on Nonaka and Takeuchi (1995), there are two kinds of knowledge: explicit knowledge and tacit knowledge. *Explicit knowledge* is knowledge that can be explicitly expressed in a stand-alone form, such as oral language, written language, computer codes, mathematical formulas, and/or visual media. Written reports, presentations, research papers, books, instructional manuals, procedures, sound and visual records, and software are examples of explicit knowledge. On the other hand, *tacit knowledge* is knowledge that cannot be expressed explicitly in a stand-alone form. Tacit knowledge is buried deeply inside people and it is difficult to express and transmit it explicitly. However, it usually can be transmitted only through a lot of people-to-people interaction. The simplest example of the nature and value of tacit knowledge is that one does not learn how to ride a bike or swim by reading a textbook, but only through personal experimentation, by observing others, and/or being guided by an instructor. Similarly, nobody can learn NBA basketball skills by reading or by watching audio/video media. In product development and in manufacturing in general, a significant portion of knowledge is of this type—chief engineers, high-level technical experts, experienced senior managers, and high-level technicians all possess a great deal of tacit knowledge.

The sources of explicit knowledge are very similar to the sources of information that we just discussed. The sources of tacit knowledge are more complicated, including:

- The company's own valuable employees

- High-caliber new hires from outside the company

- Training and coaching practice through internal and external sources

2. Exploration to find and then define the extent and value of information sources. A world-class company needs an effective information and knowledge management system. This management system includes the personnel department, training, the senior management of product development, IT support, and so on. An ideal system would be the system that hires the right people, conducts the right training at the right time, and acquires the right external technical information and explicit knowledge. In many companies, however, there is enormous waste in this information and knowledge acquisition process—for example, the massive layoff of experienced personnel by American automobile manufacturers in recent years had an enormous cost in product development capabilities. The attitude that engineers and technical specialists are disposable commodities is very harmful to information and knowledge management practice, because the right kind of tacit knowledge takes a lot of time, experience, and management to create.

3. Exploitation to extract information from information sources. This step deals with the question of "what kind of information do I really want to get?" and how to extract high-quality information from the information sources. A technology roadmap (Phaal et al. 2001) can be used to support this step.

4. Processing and refining the extracted information for the information users. The users of this information are the product development

team, mostly design engineers. The people who acquire and process this information are also often the product development team. The quality of this step highly depends on the quality of product development engineers.

2.3.4   Information Transformation

As you can see, information mining is a key part of the product development process; however, the mined information is only the input to the design process. In the design process, it is only the first step; you need to transform this input into all necessary design specifications. I call this process the *information transformation process*. Information transformation is similar to the mappings in the axiomatic design. In the fuzzy front end of the product development process, you are mostly doing information mining, mining of VOC information, and mining of technology information. You do some information transformation in the fuzzy front end; this information transformation is basically processing, refining, and clarifying the raw voice-of-the-customer information. Based on axiomatic design, there are three important mappings in the product development process. The first mapping is the transformation of customer attributes into product functional requirements. The second mapping is the physical mapping, which is the transformation from functional requirements to design parameters. The third mapping is the process mapping, which is the transformation from design parameters to process variables.

Another kind of information transformation is working from the high-level system to subsystems, to components, and finally to all detailed specifications. In system engineering, this transformation process is also called *flow down*.

Unless there is some information and knowledge creation work, that is, creating new ideas, new technology and so on, most of the information transformation work deals with existing knowledge. For example, in automobile product development, body design and assembly is a big chunk of design work. Body styles have to change to make cars catch current fashion trends. But there is very little new knowledge needed in this design work—style designers design the body style, then the stamping dies have to be made, then stamped metal panels are welded to form subassemblies, and then subassemblies will be welded to form the whole automobile body-in-white (body without paint). Though there is no new technology or knowledge involved, simply the new combination of the parts and the new shape of the parts, there will still be a lot of unknowns to be worked out. In this automobile body design example, you still need to work out all the issues about fit and finish, that is, can these subassemblies fit together seamlessly?

The types of work in this information transformation usually include:

- Given a need, pull design solutions. For example, if you need to bind parts together, then we can pull many kinds of fastener solutions. This category is very broad; selection of material, selection of modules, and so on all belong to this category.
- Design of interfaces
- Shape and form design
- System flow down and integration
- Design analysis and simulation
- Testing
- Prototype building

For complicated products with thousands of parts or more, the scope of this information transformation work can be quite extensive. You need a whole organization with many people, with different tasks, different knowledge backgrounds, and different experience to work together, as illustrated in Figure 2.18. The information and knowledge need to flow from one team, or one person, to another team or teams, another person, and so on. There are several potential dangers in this "knowledge and information flow network":

- With the flow of information from one place to another, the information can be misunderstood and distorted. Because people have different backgrounds, a lot of meaning can be lost in the translation.
- The information may not flow as it should. Some teams may have done some test, or some analysis, and made some design changes, but the people on other teams didn't get this information. This will create a lot of technical bugs and inconsistencies in the design.
- The flow of the information could be very slow, or very ineffective.

Based on the preceding discussion about this information transformation process, we can summarize the following key factors for a good process:

- A good bank of information and knowledge, from which people can pull all the necessary information and explicit knowledge to create design solutions as needed. This information bank should be a live one that keeps improving and enriching itself. This information and knowledge bank can exist in many different formats. It can be a well-managed computer information system with a nice structure, containing high-quality information that is constantly updated with any new information by product team members. It can also be a "war room"—a big room, in which all important design engineers and managers are present, with a good "best-practice database." This best-practice database records the best design practices for all kinds of parts, subsystems, and so on, as well as best testing practices. These practices are mature, verified practices that have very low technical risks. A design engineer can pull solutions either by asking other engineers in the war room, or pulling them from the best-practice database. This combination of war room and best-practice database is one of the key features of Toyota's product development process.

- Well-managed training programs for product development team members. The training should enable the engineers to know all necessary methods and tools, and learn best design practices, best design principles, teamwork, and so on.

- Good design support tools and methods. One way to preserve knowledge is to build the knowledge into computer packages and similar systems. For example, the statistical software MINITAB contains so much knowledge in statistical methods that you can pull and use any of those methods very readily. There are many counter-examples. There are many fancy research consortiums with many participants from top-notch universities and multiple companies, which may create a lot of new knowledge during their operations. However, after the projects are done, this hard-earned knowledge usually gets lost very fast. Many methods can help this information transformation work quite well, such as design of experiments and robust design. These methods can accelerate the design improvement process and work out the technical bugs very effectively.

- Application of good design principles, such as axiomatic design, system engineering, and modular design. You could design every part perfectly, but if the system integration is bad, you still create very complex vulnerable designs. In a later part of this chapter, I will discuss some details of these good design principles.

- A well-designed and well-managed product development process. Just like any product, the product development process is also a man-made process that needs to be well designed. This process needs to be well managed as well. This product development process should have the following features:

  - **A customer-value–based design process**   You need to mine the voice-of-the-customer information adequately and accurately, and translate that VOC into correct engineering requirements.

  - **A front-end–loaded process**   The earlier you get the critical information, the better, so you should work more in the concept development stage. I will discuss detailed approaches for this later in this chapter.

- **A simple and easy-to-manage process** This product development process should enable high-volume, accurate, and fast information flow across the product development organizations. This
product development process should enable the jobs and projects
to be performed smoothly, quickly, and with high quality.

2.3.5 Information and Knowledge Creation

In the product development process, after you've done all the required
information mining and transformation, you may find that there are
still some design tasks for which no ready solutions can be pulled from
anywhere. These tasks involve creating new information and new
knowledge. Here are some of the scenarios:

- **Resolution of some technical bottlenecks that nobody has
accomplished before** For example, the fuel efficiency of the internal combustion engine is low, and with increasing petroleum prices,
this technical difficulty needs to be resolved.

- **Development of the new generation of product** You want to
drastically improve your product's performance, cost, and so on in
order to move ahead of competition. This improvement is not merely
a fine-tuning of existing product.

- **Develop a product with a new marketing concept** For example,
the development of sport utility vehicles in the 1980s and 1990s.

- **Technology pushes product development** Many research
results are coming out from universities, research institutions, and
so on, and many patents are created every year. Manufacturers are
bringing these new technologies into their products.

The following list summarizes some common difficulties in this information and knowledge creation work:

- Difficulties in finding new technical solutions
- Excessive cost and lead time in developing new technical solutions
- Customers' rejection of the new product
- Excessive time in ironing out technical bugs and wrinkles
- Quality and reliability problems due to immature technologies

There are many methods and approaches to overcome these difficulties; I'll discuss some of these methods and approaches in the following
sections.

2.3.5.1 Theory of Inventive Problem Solving (TRIZ) I discussed TRIZ
earlier in this chapter. Basically, the TRIZ method is based on the

studies of patents and inventions. The basic idea of TRIZ is to reuse the ideas from similar patents and inventions to solve your own problems. Because the similar patents and inventions are verified ideas and proven to work, the solutions derived from TRIZ may have a higher probability of success than ideas drawn out of the blue. TRIZ is used to resolve the difficulties in finding new technical solutions. TRIZ will be discussed in detail in Chapter 10.

2.3.5.2 Knowledge Creation Spirals   Knowledge creation not only involves idea generation, but also involves gaining deeper understanding, perfecting the concepts, ironing out technical bugs, and extending the successful new design concepts to other products. In their book *The Knowledge-Creating Company*, Nonaka and Takeuchi (1995) stated that a successful knowledge management and knowledge creation system is crucial for success for many Japanese companies. As I mentioned previously, there are two kinds of knowledge: explicit knowledge and tacit knowledge. A good knowledge management program needs to convert internalized tacit knowledge into explicit codified knowledge in order to share it, but also for individuals and groups to internalize codified knowledge and make it personally meaningful once it is retrieved from the knowledge management system.

In a traditional information system, tacit knowledge does not get enough recognition because it is difficult to codify and share it. However, tacit knowledge is very valuable and it shouldn't be wasted. Polanyi (1967) is one of the pioneers of tacit knowledge. One of Polanyi's famous aphorisms is: "We know more than we can tell."

Even for engineers and researchers, tacit knowledge is everywhere. If a group of people takes an identical set of training courses, their actual work performances will differ. You cannot reproduce a top scientist by reading exactly the same material, taking exactly the same training, and working on exactly the same projects that the scientist did. The top scientist will always have some more hidden and difficult-to-share knowledge.

The creation of human knowledge usually goes through a spiral as illustrated in Figure 2.20.

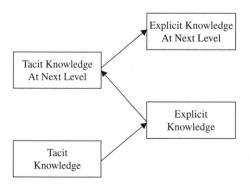

Figure 2.20 Knowledge Creation Spiral

For example, a person starts a job as a quality technician. He works in a production line, making measurements, keeping data, and observing what people are doing in the plant, so he starts to find out what it feels like to be a quality technician. This is the start of a tacit knowledge buildup. After he works for a year, he has gained a lot of insight, such as how the situation in the production line keeps changing; sometimes the quality is good, sometimes it is bad, especially when incoming material has changed, or operators change in shift. The data may fluctuate a lot, even when everything works smoothly. After one year of work, he starts his Master's program in a local university, and takes a lot of courses, such as statistical process control, quality management, and so on. After his training, he fully appreciates the explicit knowledge he got from the courses; now, not only he can do measurement and plot the data, but also he can perform analysis, do some troubleshooting, and lead some quality improvement projects. Also, after he fully appreciates and understands some advanced statistical concepts, he starts to perceive the data differently; he can sense something such as a mean shift, excessive variation, or lot-to-lot differences, which never crossed his mind before. This step is the jump from explicit knowledge to tacit knowledge of the next level.

In Nonaka and Takeuchi's book (1995), this spiral-of-knowledge-creation mechanism can be used to extend personal knowledge into team knowledge, and gradually into organizational knowledge. There are four kinds of knowledge creation spirals as shown in Table 2.2 (Nonaka and Takeuchi 1995).

In Nonaka and Takeuchi's point of view, tacit and explicit knowledge are two sides of a coin, and they are complementary to each other. In addition, both tacit and explicit knowledge can be at a personal level, as well as an organizational level. Since product development is teamwork, it is essential to spread knowledge from the personal level to the organizational level. Nonaka and Takeuchi believe that there are four modes of knowledge conversion, that is, socialization (from individual tacit knowledge to group tacit knowledge), externalization (from tacit knowledge to explicit knowledge), combination (from separate explicit knowledge to systemic explicit knowledge), and internalization (from explicit knowledge to tacit knowledge).

Socialization is a process of spreading personal tacit knowledge to organizations or other individuals. This process is often characterized

TABLE 2.2   Knowledge Creation Spirals

	To tacit knowledge	To explicit knowledge
From tacit knowledge	Socialization	Externalization
From explicit knowledge	Internalization	Combination

by well-planned and organized people-to-people interaction. For example, a top chef does "show and tell" to his coworkers to share his tricks in cooking. *Externalization* is a process of transforming tacit knowledge into explicit knowledge by observing, analyzing, and coding the tacit knowledge. This transformation can be in the form of person to him/herself, person to person, or person to a team. Observing and digitally recording NBA players' play, extracting their playing patterns, and coding them into a computer game is such an example. *Combination* is a process of combining many individual explicit knowledge pieces into a system of explicit knowledge. One example of combination is to compile a "quality control handbook" by combining many new individual method modules into an existing handbook. *Internalization* is the process of transforming explicit knowledge into tacit, operational knowledge. Using "standard operating procedures" to train new employees in real work environments is an example of such internalization.

2.3.5.3 Customer-Centric Innovation

Creativity is about the generation of new ideas. However, new ideas may not always be good ones in the marketplace. Innovation is about making new ideas into commercial success. Therefore, creativity and knowledge creation have to work hand-in-hand with the voice of the customer. This approach is called *customer-centric innovation*. Ethnographic methods and testing markets are powerful tools to support customer-centric innovation. Ethnographic methods are thoroughly discussed in Chapter 5 of this book.

2.3.5.4 Robust Technology Development

Robust technology development means building robustness into newly developed generic technology, or new technology at its infant stage. The examples of such new generic technology include new memory chips, new electronic bonding technology, new materials, and so on. New technologies are usually developed at research laboratories with ideal conditions, small batches, and on a small scale. After a generic new technology is developed, product developers will try to integrate it into new products. But usually there are a lot of hiccups in this integration process; the new technology that works well in the lab may not work well after integration, and its performance may not be stable and up to people's expectations. It usually takes a great deal of trial and error to make it work right.

Robust technology development is a strategy that tries to reduce the hiccups and make new technology integrate faster with products and production. Robust technology development proposes to conduct robust parameter design on the new technology when the new technology is still in the research lab. Robust technology development is discussed by Yang and El-Haik (2003).

2.3.5.5 Kaizen for Product Development   In the product development process, if your design involves new technology, new parts, new interfaces, and so on, there will be some technical bugs and wrinkles. These bugs and wrinkles will bring some quality and reliability problems down the road. Therefore, fixing all the bugs and ironing out all technical wrinkles quickly is a critical factor to achieve quick product development lead time and low product development cost. In addition to some systematic methods such as robust design and design of experiments, a steady step-by-step improvement effort, such as Kaizen, can also play a big role in bringing out a quality product in a short time. Product development Kaizen has been used effectively in such Japanese companies as Toyota and Honda, and has achieved great success. One simple approach is visual management. In this approach, a designated area inside the company is devoted to visually displaying the status of progress for a product development project. This area is often called a "project control room" (*Oobeya* in Japanese). In this project control room, a big product development project is subdivided into a hierarchy of subtasks of different levels, and wall spaces are color-coded and are allocated to subtasks so that each project subarea can post its own "dashboards" and project reports. In each dashboard, key tracking metrics for product development tasks, such as task progress versus schedules, project tracking data, and testing results are posted. This project control room approach is a center of project information sharing and learning. People can learn from each other, and duplicated tasks or overlapping work can be easily discovered and eliminated. Both the management and team members can see the big picture and discover the "bottlenecks" in the product development project. This approach helped Japanese auto manufacturers to cut the lion's share of product development lead time and cost.

2.3.6 The Ideal Product Development Process

In the theory of inventive problem solving (TRIZ), there is an approach called the "ideal final solution." The ideal final solution is an extreme description of what an ideal product is. For example, what is an ideal pen? The ideal final solution for an ideal pen would be: "A mark-making device that can make any kind of mark you wish, make any change you wish, without trace of corrections. This device uses no resources, produces no harm and no waste, and costs no money." This kind of ideal final solution may sound unrealistic, but it provides an ultimate goal. By comparing the ideal final solution with your current solution, you can find the gaps in performance and visualize where you need to improve.

2.3.6.1 The Ideal Final Solution   For the product development process, what would be the ideal final solution? We can easily list the following features:

- The customer's reception of the product is overwhelming! It is absolutely a hit! Everything is designed exactly the way the customer wants!

- The product performance is absolutely amazing; it beats all the competition right there!

- The cost is extremely low, much lower than all competitors' cost!

- The product is so unique that nobody else delivers a similar product. It is the first and only one in the market; it will take competitors tremendous effort to catch up.

- The quality and reliability is the best in the class; it is much better than what customers are expecting.

2.3.6.2 Reality Gaps   Obviously, this kind of product does exist, but there are very few. Based on the previous description of the nature of the product development process, and based on the history of many successful and unsuccessful product development stories, you can figure out what can go wrong in each of the five categories in the preceding list:

1. Customer reception is low. This means that you didn't mine the voice of the customer in a perfect manner. You may have gotten the wrong voice of the customer, or you may have gotten an incomplete voice of the customer. You may have gotten only what customers know how to say, and did not discover the hidden and unmet customer needs. You may have paid attention to minor concerns, and lost the big picture. You may have predicted the right voice of the customer at this moment, but failed to predict the voice of the customer in coming years. Most important, if you captured the wrong voice of the customer, no matter how well you do in product performance, quality, and reliability, you designed the wrong product and very few people will want it, unless you sell it at such a low price that it is almost free.

2. Product performance is inadequate. Assuming you mined the right voice of the customer, the low performance is caused by a flawed information transformation process and information and knowledge creation process. You did not translate the voice of the the customer into the best possible design solutions. Specifically, you can easily list the following possible flaws:

   a. You don't have a good enough bank of knowledge management and information; the design solutions that you pull are far from the best.

b. The design team is inadequately or incorrectly trained, so they don't know the necessary tools and methods, and do not apply best practices.

c. The design support tools and systems are poorly managed. People use all kinds of software, tools, and buzzwords, leading to a lot of incompatibilities and misunderstandings.

d. Poor information flow. Communications among people are poor, and meetings are unproductive.

e. Poor design architecture. Flawed design principles, excessive complexities, too many custom-made parts, and so on.

f. Poorly managed product development process. Everybody is busy, and there is a lot of fire-fighting, too many rushed, half-cooked projects.

g. Brain drain. People hop jobs too often; you need to retrain people, and reinvent wheels all the time.

h. Lengthy and ineffective quality improvement process. Too much money and time is spent in debugging the new products.

3. Excessive cost. As discussed earlier, there are three major cost components for the product:

a. Product development cost. Excessive cost in product development is usually caused by an inefficient product development process.

b. Production cost. Excessive product production cost is usually caused by poor product design and poorly run production systems.

c. Cost of running supporting business processes. Excessive business cost is usually caused by waste and an inefficient business process. Lean operation practices could be used to resolve this.

4. Excessive competition in the marketplace. You could design your product too similar to your competitors' product, so you have to endure brutal competition. You are struggling with either competing on cost or competing on performance. There is a product development strategy called the "blue ocean strategy," which calls for carefully examining your market position and your customer value proposition. This strategy has a roadmap of revising your product value position to break out of this brutal competition. I will discuss this blue ocean strategy in Chapter 3 of this book.

5. Inadequate quality and reliability. The common causes for inadequate quality and realiability include the following:

a. You did not debug the product adequately.

b. The product design is flawed or it is too complex.

c. There are too many immature technologies.

d. The product development process lacks discipline, and has too many holes.

e. There is poor quality control on suppliers.

f. The product design lacks integrity.

2.3.6.3 Deriving the Ideal Product Development Process I have discussed extensively the nature of the product development process, which consists of information mining, information transformation, and information and knowledge creation. The mining of information includes mining voice-of-the-customer information and mining technical information. The best mining process is to mine the gold! That is, you have to mine from the right customer, from the right source, obtain the right voice of the customer, and get adequate VOC information to populate the product requirements and figure out the core values of customers. The purpose of information transformation is to pull VOC information and technical information and effectively transform it into perfect product design solutions. So the best information transformation process is a process that has the highest speed and quality in pulling both the VOC and the technical information and transforming it into designs. The purpose of information and knowledge creation is bringing out innovative and breakthrough design solutions. The best such system should be able to bring out innovation and breakthrough solutions of the highest quality and at the highest speed. Within a product development system, there are many people and teams, so there are a lot of information flows. The purpose of information flow is to move information to the right people who need it. The best information flow is the one that delivers information accurately, smoothly, at the right time, and in the right amount when people need it.

The product development process is an information creation process. The consumer of this information is the product. If you design a high-end, complicated product such as commercial airplanes, it consumes a lot of information. You need an enormous amount of design, drafting, testing, simulation, and tooling design to generate enough information so that you know how to build the aircraft successfully. Axiom 2 of the axiomatic design principles says that the best design is the design that delivers all product functions and has the lowest possible information content. It makes a lot of sense here. For example, in an automobile design project, if you end up using a lot of specially-designed parts, extremely high part counts, and a lot of new and immature technologies, then you need to spend a lot of engineering hours to design, analyze, test, and troubleshoot these parts and subsystems. Overall, you need to generate a lot of information in these design tasks. So this design will consume a lot of information that was created by hard work. However, if you adopt the right design principle, you can use off-the-shelf parts to replace specially-designed parts, use value engineering and design for manufacturing practice to reduce a large number of parts, and reduce the number of immature technologies introduced into this product. If the product still can deliver the same functions as the previous design by doing all these, you reduce the information content in the design greatly, so you don't need to generate all that information any more, and you save a lot of engineering hours.

Therefore, the ideal product development process is such that it creates information and knowledge at the highest efficiency, speed, and quality, but the consumption of information for each good quality product is minimal. Obviously, if you generate information at the highest possible quality, efficiency, and speed, but the consumption of the information on each piece of good-quality product design is at a minimum, you will have the most effective product development process.

In recent years, the Toyota product development system (Liker 2005; Kennedy 2003) has been getting a lot of attention. Toyota is famous for its lean manufacturing practices. Some authors have called Toyota's product development process the *lean product development process*. In the next section, I will describe a value-based lean product development process, which is based on the combination of best practices in the product development process.

2.4 Customer-Value–Based Lean Product Development Process

At the end of the last section, I derived a very important result; that is, the ideal product development process is such that it creates information and knowledge at the highest speed, efficiency, and quality, but the consumption of information for each good quality product is at a minimum. In this section, I will discuss how to achieve this ideal product development process.

The lean operation technique is a very important tool in achieving the ideal product development process. I'll start by discussing the general lean operation in detail.

2.4.1 Lean Operation Principles

Lean manufacturing is a very effective manufacturing strategy first developed by Toyota. During a benchmarking study for the automobile industry in the late 1980s (Womack, Jones, and Roos 1990), Toyota clearly stood above their competitors around the world with the ability they developed to efficiently design, manufacture, market, and service the automobiles they produced. This ability made a significant contribution to both their company's profitability and growth, as consumers found their products to simultaneously exhibit both quality and value. The focus on recognizing and eliminating wasteful actions and utilizing a greater proportion of their company's resources to add value for the ultimate customer was found to be the key of their operating philosophy by the researchers. "Lean production" is first mentioned in this study, used to describe the efficient, less wasteful production system developed by Toyota, called the Toyota Production System. Lean production, in comparison to mass production, was shown to require one-half the time to develop new products, one-half the engineering hours to design, one-half the factory hours to produce, and one-half the investment in tools, facilities, and factory space (Monden 1993; Ohno 1990; Shingo 1989).

Although the lean manufacturing approach was originally developed in the traditional manufacturing industry, lean manufacturing mostly deals with production systems from an operational viewpoint, not from a hard-core technology viewpoint. It has been found that most lean manufacturing principles can be readily adopted in other types of processes, such as product development processes, office processes, and service factory processes.

The key objective of the lean operation is to eliminate all process wastes and maximize process efficiency. The key elements of lean operation include the following items:

- Waste elimination in process
- Value-stream mapping
- One-piece flow
- Pull-based production system

I will discuss each of these key elements in detail.

2.4.2 Waste Elimination in Process

In observing mass production, Tachii Ohno (Ohno 1990; Liker 2004), an engineering genius of Toyota, and the pioneer of the Toyota Production System, identified the following "seven wastes" in production systems:

1. Overproduction: Producing too much, too early
2. Waiting: Workers waiting for machines or parts
3. Unnecessary transport: Unnecessary transporting of moving parts
4. Overprocessing: Unnecessary processing steps
5. Excessive inventory: Semifinished parts between operations and excessive inventory of finished products

6. Unnecessary movement: Unnecessary worker movements

7. Defects: Parts need rework or are scrap

These seven wastes are called *muda*. Muda is a Japanese term for missed opportunities, or slack. These items are considered waste because in the eyes of customers, these activities do not add value to the products they desire.

In lean operation principles, the seven wastes can be identified mostly by a value-stream mapping method. The waste caused by overproduction can be reduced or eliminated by a pull-based production system. The waste caused by excessive inventory, waiting, unnecessary transport, and unnecessary movement can be greatly reduced by one-piece flow and work cells (cellular manufacturing).

2.4.3 Value-Stream Mapping

Value-stream mapping is a good method to chart a process and identify and quantify the waste in that process. Value-stream mapping was developed to map and analyze production processes, especially batch flow shops and flow shop processes. A *value stream* is all the activities (both value-added and non-value-added) required to bring a product through the main flows.

Figure 2.21 is a simplified value stream map for a production process.

Clearly, based on the definition of seven wastes, the staging, transportation, setup, and inspection are non-value-added steps; casting, machining, and assembly are value-added steps. In Figure 2.21, the horizontal length of each step is in proportion to the time required to perform the step.

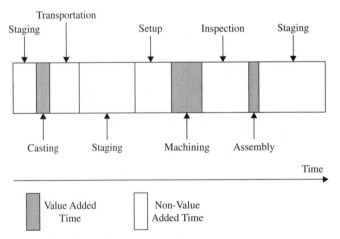

Figure 2.21 A Simplified Value-Stream Map for a Production Process

The total time duration from the beginning of the process to the end of the process is often called the *process lead time*. Clearly, in this example, the value-added time is a small portion of the total lead time.

You can see that this simple value-stream map identified and quantified waste in the process and provided the clue for process improvement. Clearly, the process can be improved if you can shorten the non-value-added time.

This kind of simple value-stream mapping can also be used to analyze a service process. Figure 2.22 shows a simplified value-stream map for a sales order process.

In Figure 2.22, the dark boxes are value-added steps; the light boxes are non-value-added steps. The first box, "search," does not really add value for customers, but it is an essential step, so for now, a color of yellow is used.

In many production systems, there are huge amounts of muda (seven types of waste) in the process. The ratio of value-added time over total lead time can be used as a measure of process efficiency, specifically:

$$\text{Process Efficiency} = \frac{\text{Value Added Time}}{\text{Total Lead Time}} \qquad (2\text{-}10)$$

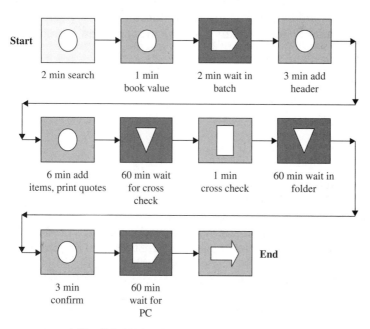

Figure 2.22   A Simplified Value-Stream Map for a Sales Order Processing Process

TABLE 2.3   Process Efficiency for Various Processes

Process Type	Typical Process Efficiency	World Class Process Efficiency
Machining	1%	20%
Fabrication	10%	25%
Assembly	15%	35%
Continuous manufacturing	30%	80%
Transactional business processes	10%	50%
Cognitive business processes	5%	25%

The major goal of a lean operation is to increase the process efficiency. A process that has high efficiency will have much less waste, shorter lead time, and lower cost. As a rule of thumb, a process is considered lean if the process efficiency is more than 25 percent. Based on research by Michael George (2003), the typical process efficiency and world class efficiency for many types of processes are summarized in Table 2.3.

Clearly, the process efficiency of typical processes is very low. A big proportion of process lead time is not used to do value-added work, but to do non-value-added work, that is, muda. Lean operation tries to redesign the process flow and layout so that the portion of process time in doing non-value-added work is greatly reduced.

The most frequently used techniques in lean operation include the following:

- One-piece flow
- Work cells (cellular manufacturing)
- Pull-based production

2.4.4   One-Piece Flow

There are several types of manufacturing processes, such as the job shop process, the batch flow shop process, and the line flow process. The job shop process is also called "machine village," which means that similar machines are grouped together. The job flow patterns of such production systems can be quite erratic and messy, as illustrated in Figure 2.23. A job shop process is featured by low utilization, long delays, high work in process inventory and long lead time. The advantage of the job shop process is that it can take on a large variety of tasks.

Many service processes are also in the form of job shop processes. For example, in most organizations, the departments are grouped together by function, such as the personnel department, accounting department, benefits department. If a new employee wants to finish all his or her

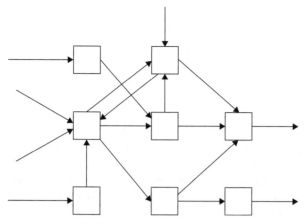

Figure 2.23 Typical Flow Pattern of a Job Shop

paperwork, he or she will go through all these departments. In many organizations, paperwork has to be approved by many departments, so each piece of paperwork will first go to one department, then it goes through interdepartmental mail to the next one. If there are any mistakes, the paperwork could be sent back for correction. It is quite usual for the documents to get lost or buried in the paper trail so that it takes a long time for them to be completed.

A batch flow process is better in its flow pattern, as illustrated by Figure 2.24.

However, there are still a lot of work-in-process inventories in a batch flow shop. Most of the lead time is spent on inventory waiting in the stock. It is better than the flow pattern of a job shop, but it is still inefficient.

One-piece flow, or single-piece flow, is the solution proposed by the lean operation principle. One-piece flow is illustrated by Figure 2.25. A one-piece flow means that the work piece is worked on one piece at a time, not one batch a time. This will eliminate the work-in-process inventory completely.

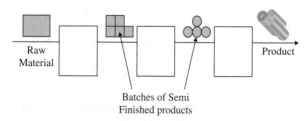

Figure 2.24 Flow Patterns of a Batch Flow Shop

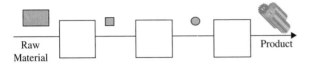

Figure 2.25   One-Piece Flow

On the other hand, in a single-piece line flow process, any error or defect in any process step will cause the whole line to stop. In traditional Western operation management, the work-in-process inventory, or buffer inventory, is used to temporarily "feed" the downstream process steps so the line will not stop. However, the Toyota production system believes that the buffer inventory has more disadvantages than benefits; buffer inventory ties up money, and hides hidden problems. In the Toyota production system, zero buffer inventory is used to expose all the hidden problems in the production process. It forces you to debug all hidden problems so that eventually, you will have a zero-defect production process.

2.4.5   Pull-Based Production

A pull-based production system means a "demand-driven" production system. The pull-based production system is modeled after supermarket shelf replenishment operation. On a supermarket shelf, there are lots of goods, such as milk, eggs, and orange juice, that are ready for customers to pick up. The customers "pull" the goods from the shelf, and then depending on how many items are taken away, the inventory person in the supermarket will refill the same amount of items by pulling them from the warehouse. Then the warehouse person will order roughly the same amount of items that are pulled from the warehouse.

Restaurant operation is a perfect example of pull-based production. The customer places the order, and then the kitchen produces exactly what the customer ordered. In general, the key feature for pull-based production is that the information flow direction is opposite to the material flow. The information flow means the production control order. In the restaurant case, the production control is the order from kitchen to cook; this order direction is clearly from the customer to the kitchen. On the other hand, the direction of material flow, which is the flow of food in the restaurant case, moves from the kitchen to the customer. Clearly, the information flow direction and material flow direction in the restaurant kitchen are opposite.

The opposite of pull-based production is "push-based" production. The key feature for push-based production is that the direction of information flow is the same as that of the material flow. In push-based production, each work stop will send the work downstream in the operation. That is, it pushes the work downstream, without considering whether the downstream can make use of it. Typically, activities are planned centrally but do not reflect actual conditions in terms of idle time, inventory, and queues.

Agricultural production is a typical push-based production. Because the production cycle is very long, there is no way that farmers can produce only the amount of food needed, based on real-time demand. The production plan is purely based on market forecast, and sometimes just based on last year's production. The production command will flow in the same direction as the workflow. It is well known that agricultural production often suffers from oversupply and market fluctuations. Clearly, pull-based production, whenever possible, will create much less overproduction, so the waste caused by overproduction can be reduced.

2.4.6 Lean Principles for Product Development

Lean operation practices have achieved a great deal of success in both the manufacturing industry and many service industries, such as banking, insurance, and health care. Can these lean operation principles achieve the same drastic results in the product development process? The answer to this question is positive. However, there are many distinct differences between the product development process and the manufacturing process, so the lean principles have to be modified to work well in the product development process.

The birthplace of lean manufacturing, Toyota, definitely has an edge in the product development process compared with North American automobile companies. Table 2.4 summarizes the performance differences between Toyota and North American automobile companies around 1990 (Womack 1990).

Toyota's product development system gained a lot of attention (Liker 2005; Kennedy 2003). However, there are many other best practices as well, such as Apple Computer's innovation approach and Samsung's Design for Six Sigma practices, and all of them have achieved great successes. In this chapter, we will outline a lean product development strategy

TABLE 2.4 Product Development Performance Comparisons

Measures	Toyota	North American Automobile Companies
Average engineering hours per new vehicle development (million hours)	1.7	3.1
Average development time (months)	46.2	60.4
Employees per team	485	903
Ratio of delayed project	1 in 6	1 in 2
Achieve normal quality After Launch	1.4	11

that combines many best practices and several sound design principles (Suh 1990; Nonaka 1995; Reinertsen 1997; Huthwaite 2004).

There are many significant differences between the manufacturing process and the product development process. For manufacturing processes, what you are going to produce is very clear in the beginning. The product that you produce has already been designed. For the product development process, you are very fuzzy about what the product will be like; and the outcome of the product development process is not a certainty. For a manufacturing process, the rework is treated as waste; for the product development process, iterative improvement on product design is quite common. Even the goal of lean operation is different between a manufacturing process and the product development process. For a manufacturing operation, the goal of lean operation is to minimize waste and increase speed; for the product development process, developing a top-notch product design that can lead to high sales and high profitability is the goal, so you are trying to maximize the value of the design, as well as reduce waste, and increase development speed.

Based on these differences, the following definition applies to this value-based lean product development process:

The value-based lean product development process aims to deliver more customer value by using fewer resources through

- Thoroughly capturing the voice of the customer and accurately understanding the customer value

- Effectively transforming the voice of the customer into high-quality design with high speed and low cost

- Relentlessly decreasing the wastes in the product development process

Based on this definition, you can see the difference between lean manufacturing and lean product development quite easily. While lean manufacturing focuses on reducing waste and increasing speed, the lean product development process focuses on both increasing the customer value and reducing waste and increasing speed.

Unlike the "seven wastes" in the manufacturing process, there are no universally agreed-upon waste categories for product development. However, the following waste categories are commonly found in the product development process:

- **Excessive design requirements**   Excessive tolerances, excessive material specifications, excessive operator requirements, and so on.

- **Excessive complexity in design**   The simplest design is the best design, assuming that it can deliver all the product functions.

- **Poor product architecture**   Poor product architecture often leads to redesign, mismatch, and performance problems.

- **Reinvention**   If someone else has already done this work, reinvention certainly is a waste of manpower and resources.

- **Mismatch of subsystems**   Many design rework problems happen in unexpected subsystem interactions.

- **Information loss and re-creation**   This happens a lot in most companies.

- **Miscommunication**   Miscommunication among product development team members often leads to doing the wrong work, and then having to redo it.

- **Searching for information, waiting for critical information**   This is certainly not a value-added activity.

- **Overburden on the people or resource**   Excessive workload and unrealistic deadlines often lead to half-cooked projects and bug-ridden designs, and eventually they lead to rework.

- **Unproductive meetings**   Meetings consume person-hours.

Here are some of the most important lean concepts in the product development process:

- **Value**   The capturing of true customer value.

- **Value stream**   The flow and transformation of the customer value and VOC information into completed product design, through the stages of product development phases.

- **Flow**   Flow of information and knowledge.

- **Pull**   Pull the right kind of VOC and customer value information from customers; pull the right kind of technical information and knowledge from the right sources, at the right time, in the right amount, by the right user.

- **Perfection**   No wastes, no information distortion and loss, perfect information flow, no reinvention.

After these thorough discussions on the lean principles for product development, I can define the value-based lean product development strategy. This strategy has five main components:

1. Mining the voice of the customer to capture value

2. Maximizing technical competence

3. Front-loading the product development process

4. Optimizing information transformation and flow

5. Creating a lean product

I will discuss these five main components in subsequent subsections.

2.4.7   Mining the Voice of the Customer to Capture Value

The ultimate success of the product depends on the customers' reception of the product. If you had magic powers so you could read your customers' minds completely and accurately, and were able to deliver what they want with low cost, then you would be guaranteed to get rich in no time! The same is true in discovering the real, accurate voice of the customer. You need to find the right sources for the voice of the customer, locate the sources, use the right method to extract good information from the source, and process the result into useful information for the product development team.

There are three objectives for the mining of VOC information:

1. Obtain sufficient information to develop product requirements. From the earlier discussions on axiomatic design, you know that product design is a sequence of mapping processes. The voice of the customer is the source for the product functional requirements. You need to get enough accurate information from customers to derive these product requirements. Chapter 3 of this book provides a discussion of what kind of voice of the customer we need to collect to serve this purpose. Chapters 4 and 5 give comprehensive descriptions of all methods and tools that are necessary to capture the voice of the customer. Chapter 6 discusses the technical aspects of how to refine and analyze the voice-of-the-customer data. Chapter 7 discusses the method of quality function deployment; this method can be used to transform VOC information into product requirements.

2. Obtain and analyze customer value information. Customer value is the value that the customer perceives for a given product. Customer value information is derived from the voice of the customer, because only the customers themselves can tell what kind of product benefits they really like, and what kind of liabilities they don't like. The kind of customer value that your product should have is a strategic decision in product development; I call it the customer value proposition. Customer value capturing and analysis is thoroughly discussed in Chapter 3. Chapter 8 and Chapter 9 discuss two effective methodologies to improve customer value and deploy that value into product development.

3. Explore alternative customer value propositions. Many marketplaces are flooded with competitive products, where the competition is brutal. Many successful innovative product developments offer

a completely new kind of product that has a different customer value proposition than its competitors. For example, in the late 1980s, when Chrysler introduced the SUV (sport utility vehicle) to the automobile marketplace, it was a completely new type of product with a new set of customer benefits, and thus, a new customer value proposition. Developing a new customer value proposition also needs voice-of-the-customer inputs. This strategy is called the "blue ocean strategy," and is discussed in Chapter 3.

A big portion of this book is about mining the voice of the customer and delivering customer value, so that these factors can serve as a technical manual for this value-based lean product development component.

2.4.8 Maximizing Technical Competence

Capturing the voice of the customer accurately is only a necessary condition for delivering valuable products. By capturing accurate VOC information, you will know what customers really want, but it does not tell you how to make this kind of product with low price, high performance, and high quality, and or how to bring this product to market quickly. Not only do you need the voice of the customer, but you also need strong technical competence. In Liker's book, *The Toyota Product Development System* (2006), he explained that one of the 13 Toyota lean product development principles is "develop towering technical competence in all engineers." Clearly, superior technical competence is an important pillar for a first-class product development team.

In the value-based lean product development process, you want to develop this capability in a cost-and time-effective manner. This strategy has the following key components:

- **An effective information and knowledge management system** This is a good bank of information and knowledge from which people can pull all the necessary information and explicit knowledge to create design solutions as needed. This information bank should be a live one that keeps improving and enriching itself.

- **Effective organizational knowledge creation** Adopt the knowledge-creation spirals mechanism for organizational knowledge creation, which was originally delineated in *The Knowledge-Creating Company* (Nonaka and Takeuchi 1995). We discussed this approach in detail earlier in this chapter.

- **Applying TRIZ to reduce reinvention**

- **Applying DFSS tools to support all stages of product development** Design for Six Sigma (DFSS) (Yang 2003) is an effective technically intensive approach to supporting the product development process. DFSS provides a roadmap and many effective tools to ensure product performance excellence and superior quality in every stage of the product development process. DFSS can go hand-in-hand with a lean product development process. DFSS is about high product performance, robustness and quality; the lean product development process can provide help in reducing waste and improving efficiency. DFSS and lean product development can work together very well in the product development process.

Table 2.5 summarizes many DFSS tools that support the first four stages of the product development process.

2.4.9   Front-Loading the Product Development Process

As we discussed earlier, the product development process is an information and knowledge generation process. Information has time value; you want the key information to be available earlier, rather than later. In Toyota's product development process, Principle 2 of the 13-principle Toyota lean product development process calls for "front-loading the product development process to explore thoroughly alternative solutions while there is maximum design space" (Morgan and Liker 2006).

TABLE 2.5   DFSS Tools and Product Development Stages

Product/Service Life Cycle Stages	What DFSS Tools Can Accomplish	DFSS Tools
Stage 0. Impetus/ideation	Ensure that new technology or ideas are robust for downstream development	Robust technology development
Stage 1. Customer and business requirements study	Ensure that the new product concept can come up with the right functional requirements that satisfy customer needs	QFD
Stage 2. Concept development	Ensure that the new concept can lead to sound design and is free of design vulnerabilities Ensure that the new concept is robust for downstream development	Taguchi method/robust design TRIZ Axiomatic design DOE Simulation/optimization Reliability-based design
Stage 3. Product/service design/prototyping	Ensure that the designed product (design parameters) can deliver desired product functions over its useful life Ensure that the product design is robust for variations from manufacturing, consumption, and disposal stages	Taguchi method/robust design DOE Simulation/optimization Reliability-based design/testing and estimation

The technical approach for this "front-loading" principle is the set-based design practice.

The *set-based design* approach is used in the concept design stage, and it works with modular design practice. *Modular design* is a design practice in which a product is broken into smaller subsystems. The subsystems are connected together via standard interfaces. In this case, the subsystems become "decoupled," that is, the designing of one subsystem is not dependent upon other subsystems. Therefore, the design work for each subsystem can be conducted in parallel.

For each subsystem, you start with the concept design. In regular design practice, you start with a small number of design concepts, then you select one seemingly good concept, and move into detailed design. After the design is ready, you perform an evaluation test. If the test shows the concept is acceptable, you move this concept to the parameter design and prototyping stage. If the test shows that this concept is not acceptable, you start another raw concept and do another round of development. You may iterate this process until an acceptable design is found. This regular design practice is illustrated in Figure 2.26.

On the other hand, set-based design simultaneously starts with several concepts. The initial sets of concepts are coming from

- Current knowledge

- New technology from recent research and development efforts

- Idea generation through brainstorming or TRIZ

This initial set of concepts should include at least one concept that is relatively mature and reliable. After the initial set is selected, you divide the engineers into teams, such that each team works on one concept. Each team will grow its concept by detailing, design evaluation, and tests. It is very important that the set-based design stay in the concept design stage; you can do some low-cost computer-aided design (CAD) simulations, alpha prototypes, small-scale lab tests, and so on. However, you don't launch high-cost prototype construction and validation tests at this time. This ensures that the set-based design approach will not be expensive and time-consuming. Figure 2.27 illustrates how set-based design works.

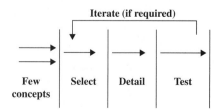

Figure 2.26 Regular Concept Design Process

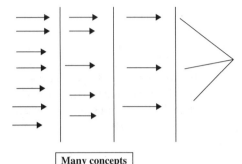

Many concepts
Each subsystem

Figure 2.27   Set-Based Concept Design

The whole set-based design will be subdivided into several mini-stages during its progress; the concepts are evaluated and tested, and weaker concepts are eliminated. The advantages of set-based design include discovering better concepts early in the design process, and gaining a lot of knowledge about what works well, and what doesn't. Cross-pollination of several good concepts may create an even better one. For example, in a university capstone design project, if you assign different projects to different teams of students, then every team will probably create an average, mediocre design. If you assign the same project to many teams, and each team works on it independently, then you will get many different designs. The best design will almost certainly be better than the average design, and by closely examining the top two or top three designs, you may end up with an even better concept by combining all the advantages of these designs, and eliminating the disadvantages. By sharing the results with all participants, everyone will gain a lot of knowledge in this subject.

2.4.10   Optimizing Information Transformation and Flow

Product development involves a lot of information transformation and information flow. *Information transformation* usually involves pulling design solutions to satisfy design needs. *Information flow* involves delivering information to needed people. The following is a list of types of waste that often occur in the information transformation and information flow processes:

- Reinvention
- Mismatch of subsystems
- Information loss and re-creation
- Miscommunication

- Searching for information, waiting for critical information
- Overburdening the people or resources
- Unproductive meetings

The optimized information transformation and information flow should be able to reduce those wastes to a minimum. There are several very effective approaches to improving information transformation and information flow process. However, in order to fully appreciate these approaches, we need to review some background knowledge in human factors and queueing theory.

2.4.10.1 Human Factors   People do the product development work. In order to find ways for people to work effectively, we need to study some basic principles of human factors:

- **Attention span**   *Attention span* is the amount of time a person can concentrate on a single activity. The ability to focus one's mental or other efforts on an object is generally considered to be of prime importance to the achievement of goals. When people are taking classes or training, or attending meetings, they can only focus for a certain period of time; after that period, their attention wanders. Some researchers state that a person's attention span is *10 + Your Age* minutes, and that anything taught after that is not taken in, but taking a 5- or 10-minute break after this time may help to replenish the person's attention span.

- **Mental focus and multitasking**   When people are doing some work, it takes some time to achieve mental focus on the job, and it takes some time to get even a subtask done. When people are constantly interrupted, the productivity will be very low. Also, when people handle several tasks, if they switch tasks very often, the progress on each task will usually be slower than when they are doing one thing at a time.

2.4.10.2 Queueing Theory   *Queueing theory* is the mathematical study of queues and waiting lines, called the queueing model. In the queueing model, jobs or customers are entering the system for service, the system has one or several servers, or even some complicated departments and sequences of service operations. The customers can be served in many ways, such as first-come first-serve, single queue, or multiple queues (multiple waiting times). There are several queueing performance metrics, such as average waiting time in the queue, queue length, and utilization of servers.

Queueing theory has many established results about how to improve queueing systems, such as how to improve the arriving process, and queueing the service process to improve job service times and reduce waiting times. These results can be readily used in helping the product development process as well. The product development process involves many projects. Each team or engineer will work on several projects during the whole product development process, so some projects have to wait in the queue until the current project is finished so the team or engineer can be free to work on them. In this case, how you sequence the jobs and how you assign the workload and timing will make a lot of difference in overall product development progress. Queueing theory can provide great help in reducing waiting time and improving throughput (number of projects finished/unit time).

There are several important results in queueing theory that are relevant to the product development process.

- **Batch queueing is inefficient**   *Batch queueing* means the jobs are coming into the queue in big groups, or batches. For example, if a planeload of international travellers arrives in an airport custom inspection and immigration inspection stop, a long queue will form immediately, and it takes a while for everyone to get cleared. If most people get out of work at about the same time, a traffic jam will form immediately and it takes a long time for everyone to get home. If the same number of people get out of work at different times, the traffic time for each person will be much shorter. The implication for product development practice is that if you load product development teams or engineers with big chunks of work, instead of staggering the work, your throughput will be low.

- **Nonlinear relationship between capacity utilization and queue length**   Queueing theory states that the relationship between server utilization and average waiting time is a nonlinear relationship, as illustated in Figure 2.27. *Capacity utilization* is defined as the percentage of time that the server is busy. What this relationship indicates is that when the server is partially loaded, say 50 percent, the waiting time will be very low. However, if you increase the capacity utilization by just 25 percent, queue length will grow to several times longer. When the server is 100 percent loaded, the waiting time will be extremely high. You can see this fact in your real-life experience. On a three-lane freeway that is not fully loaded, if a traffic accident suddenly occurs and just one lane is blocked, a long queue will form immediately, even though theoretically there is enough room to let every car go through. The implication for the product development process is that overburdening the product team or engineers will make the product development lead time much longer.

- **Constant job arrival rate vs. variable arrival rate**   Assume that you have two scenarios in the waiting queue system. The first one is such that every job arrives at exactly the same time interval, say exactly every 10 minutes. The second one is such that every job

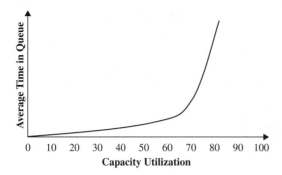

Figure 2.28 Queue Length vs. Capacity Utilization

arrives at variable time intervals, say one job arrives 2 minutes after the previous job, the next job 18 minutes later, and so on, but the average inter-arrival interval (for example, also 10 minutes) is the same as the first case. The average waiting time and queue length for the first queue system will be shorter than the second queue system. The implication for the product development process is that if you load jobs to engineers evenly, the throughput will be higher.

- **Uneven job size vs. similar job size**   Again, we are comparing two queue systems. In both queues, the arriving time patterns are the same. The first queue system is such that every incoming job takes about the same amount of time for the server to finish, for example, 10 minutes. The second queue system is such that every incoming job is a different size. Some jobs are big, and they take a longer time to finish. Some jobs are small, and they take a shorter time to finish, even if the average job processing time is the same as the first queue system (for example, also 10 minutes). Then the waiting time and queue length of the first queue system will be shorter than the second one. The implication for product development is that loading engineers with similar job sizes for each task will increase the job throughput.

After reviewing this background about human factor and queueing theory, I can now present several effective methods to greatly improve information transformation and flow.

2.4.10.3 Visible Knowledge   Visible knowledge is a very important component of Toyota's product development system. *Visible knowledge* is knowledge that has been captured and illustrated so that it is easier to share with other people within the organization. Here we will show two important visible knowledge tools, the A3 report and the planning wall.

A3 Report   An A3 report is a report that uses A3 size paper (11" × 17"). Each report will use exactly one piece of A3 size paper, which is equivalent to a two-page report for regular 8" × 11" size paper. Based on the objectives, there are primarily three kinds of A3 report:

- Knowledge sharing
- Problem solving
- Project status report

The contents in each A3 report for a knowledge-sharing objective usually include the following contents:

- Problem statement
- Current situation
- Prior research or work
- Root cause analysis
- Experimental methods
- Data analysis
- Recommendations

For other objectives, the contents will be different. An A3 report is primarily used in

- Communication before meetings
- Communication on the planning wall

In Toyota, before a meeting takes place, even a one-to-one meeting, the participants usually e-mail an A3 report to each other. Before the meeting starts, the participants can get enough information on the subject from each other so that the serious discussion can take place more quickly. In this way, meetings can be very productive. In many meetings, a lot of time is wasted in trying to understand what other people are really up to; it may take a long time just to get problems defined. After an hour or so, people are tired and stop paying attention, so the meeting is not productive.

The advantage of the A3 report is that it offers just about the right amount of information for people to digest. A three-line e-mail is too short and not enough information is provided. A 20-page report is too long and intimidating. An A3 report makes communication effective and the meeting productive.

Planning Wall   Planning walls are the walls in the project control room for many Japanese companies. Various A3 reports are displayed on its walls, giving team members an immediate "bigger picture" view of the project objectives and how they relate to overall corporate objectives,

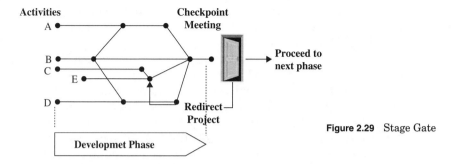

Figure 2.29   Stage Gate

visually oriented progress reports for all parts of the project (color-coded, to show which metrics are on target and which need immediate attention), and much more. Centralizing and distilling all of this project data in one location creates, in effect, a set of project management "dashboards" that team members can learn from, discuss, and collaborate around.

Clearly, the visible knowledge tools such as A3 reports and planning walls help to make information transformation and flow visible to all team members, so people will know what has been done, and what has not. They can also learn from other people's experience. This helps to reduce the waste caused by information searching, miscommunication, knowledge loss and re-creaction, and so on.

2.4.10.4 Stage Overlapping   The stage-gate approach is widely applied in many companies. The stage-gate approach brings a lot of discipline to the product development process and it plays a role in reducing quality problems during this process. However, if the stage gate is implemented in a very strict manner, no work in the next stage can start unless every check in the current stage is passed, as illustrated in Figure 2.29.

However, this approach is simply a batch queue, which is illustrated in Figure 2.30. In Figure 2.30, the projects are like customers in the queue. The strict stage-gate practice means that nobody can cross the

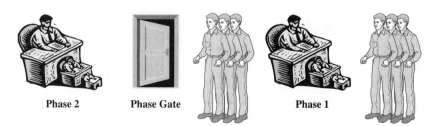

Phase 2   Phase Gate   Phase 1

Figure 2.30   Stage Gate as a Batch Queue

phase gate unless all the customers are served by the server in phase 1. The finished customers have to stay in phase 1 until the last customer is served, then all the customers flood into the next phase.

Obviously, this is a slow and ineffective approach. Clark and Fujimori (1991) discussed the stage-overlapping approach to overcome this deficiency. The idea of stage overlapping is illustrated in Figure 2.31.

In Figure 2.31, the product design and process design are two phases in the product development process. *Stage overlapping* means that the second phase does not need to wait until the first phase is completely finished. The second phase can start as soon as a necessary portion of the first phase is finished. In this way, the total duration needed to finish both phases will be much shorter.

2.4.10.5 Information Flow across Stages Clark and Fujimori (1991) also discussed the importance of letting information flow across different stages of the product development process, and among engineers. Clark and Fujimori defined information flow quality, as illustrated in Figure 2.32. The first row in Figure 2.32 indicates that a strictly phased and batched information flow has poor information flow quality; a fragmented information flow under overlapped stages has higher information flow quality. This is common sense, just as we eat bite by bite, rather than eating a big chunk at a time.

The second row of Figure 2.32 shows that the "throwing over the wall" type of batch document communication has poor information flow quality; a face-to-face, person-to-person communication has high information flow quality. The rest of Figure 2.32 is self-explanatory.

2.4.10.6 Reduction of Product Development System Complexity In Toyota's product development process, there are 13 lean product development principles. In this chapter, we have already discussed a few of these principles in earlier sections. If you can understand the importance of information flow and some basics of human factors and queueing theory, these principles are very easy to understand. Here we will discuss several more principles of Toyota's lean product development principles (Liker 2006).

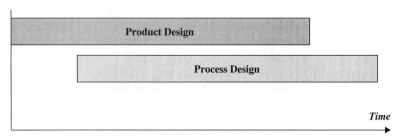

Figure 2.31 Stage Overlapping

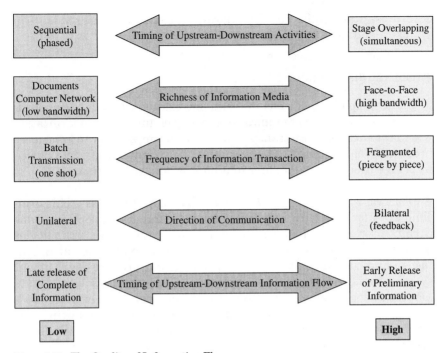

Figure 2.32   The Quality of Information Flow

Principle 3: Create Leveled Product Development Process Flow   This principle calls for synchronizing activities across different functional departments in a product development organization. It also calls for an evenly distributed workload to various departments and engineers; rather than having extremely busy days mixed with extremely idle days, you will want steady loads for all the people. This approach creates a steady workload and job flow so the tasks will flow through the organization smoothly and waiting line will be unlikely to occur. Even if you get a huge data set to be distributed to other department, you will cut the big data set into chunks and give the people one chunk at a time.

Overall, this principle stresses the importance of even job flow, avoidance of batch queues, and overloads the capacity of the queueing system. This principle makes perfect sense in queueing theory.

Principle 4: Utilizing Rigorous Standardization to Reduce Variation and Create Flexibility and Predictable Outcomes   This principle calls for applying the following four kinds of standardization across the product development organization:

1. Design standardization: Engineering checklist, standard architecture, sharing of common components

2. Process standardization: Standardizing common tasks, sequence of tasks, and task duration

3. Skill set standardization

4. Standardized skill inventories

This principle uses the fact that standardization will reduce complexities and confusion in communications among engineers. Standardization will make each job more transparent and uniform, so you can have more predictable outcomes. Standardization will also reduce the waste caused by reinvention, mismatch, information loss, and re-creation.

Principle 11: Adapt Technology to Fit People and Process   This principle calls for achieving a balance among technology, people, and the current product development process. Newer and fancier technologies may bring more capabilities, but they may also create mismatch, confusion, and communication problems on current product development systems and product development teams. You should try to integrate new technology seamlessly into existing technologies and existing product development systems before using it. The goal of introducing new technology is to improve and support the lean product development process, not to damage it.

Overall, this Toyota principle calls for streamlining and simplifying the overall system of people, technology, and process. Certainly it makes sense according to the information axiom of the axiomatic design; if the system can do the job, then the simpler system is a better system.

Principle 12: Align Your Organization Through Simple, Visual Communications   Principle 12 is the principle of visible knowledge.

Principle 13: Use Powerful Tools for Standardization and Organizational Learning   Principle 13 addresses the issue of organizational learning and knowledge creation.

2.4.11   Creating a Lean Product

Finally, recall the definition of the ideal product development process: The ideal product development process is such that it creates information and knowledge at the highest efficiency, speed, and quality, but the consumption of information for each good quality product is minimal.

The previous strategies all deal with creating information and knowledge at the highest efficiency, speed, and quality. We still need to develop a strategy to reduce information consumption for each good-quality product design. The information consumption is proportional to the complexity of the product design. So we need to trim all unnecessary complexities out of product design. A product design that is free of unnecessary complexities is called a lean product.

The complexity in engineering design is related to

- Number of functions and parts
- Complexity in product architecture (how different modules and design parameters are related to each other)
- Uncertainty (such as uncertainty caused by variation in quality and technical immaturity)
- Complex relationships between design parameters and product performance

Based on the work of Huthwaite (2004), the following approaches can be used to reduce the unnecessary complexities in the product design and create lean products:

- Reducing unnecessary product functions and parts
- Loosening up unreasonable tolerances
- Using standard or off-the-shelf parts
- Controlling technical immaturity
- Avoiding complicated user/operator requirements
- Avoiding complicated interface requirements

Customer Value and the Voice of the Customer

The bottom line for every company is not its short-term profitability but the value of its products in the eyes of potential customers. Short-term profitability reflects a company's recent history and past strengths, but without continuing enthusiasm from customers, profitability may not last. It is customers' opinions that will determine the price level, the size of the market, and the future trends for a product family.

Customers' opinions of the value of a product comprise the *customer value*. A product that has high customer value often also has increasing market share, increasing customer enthusiasm toward the product, word-of-mouth praise, increasing name recognition, a reasonable price, and a healthy profit margin for the company that produces it. Clearly, the ability to design and deliver service products that have high customer value is the key to success for service organizations.

There are plenty of books that discuss issues related to customer value. The famous book *Market Ownership* by William Sherden (1994) has an excellent chapter on customer value. Bradley Gale's book *Managing Customer Value* (1994) also presents workable methods for surveying customer value and using that information in product and service design.

The customer value for a given product or service may change over time, and a new product or business model that better fits the changing customer value could be a breakthrough product. Creating products or business models that fit customer value shifts is called a *blue ocean strategy* (Kim and Mauborgne 2005).

Overall, in this book, we will develop a comprehensive strategy that integrates several wonderful methods to create superior customer value for service products:

- Well-designed surveys are essential for obtaining customers' opinions, and there is some excellent literature on this topic. Customer survey design and analysis will be discussed in Chapter 4.

- The ethnographic method is a new and popular approach for collecting the voice of the customer. This method is rooted in anthropology and it is based on detailed observation on the customer's turf. This method can capture a lot of unarticulated and unspoken customer needs. The ethnographic method is discussed in Chapter 5.

- After collecting the voice of customer, the raw VOC data needs to be analyzed and processed into well-defined, quantitative quality metrics. VOC data analysis is discussed in Chapter 6.

- Quality function deployment, or QFD (Cohen 1985, 1989), is an excellent method developed in Japan, and it can be used to deploy customers' wants into product designs. QFD will be discussed in Chapter 7.

- Customer value is also highly related to brand recognition. Usually, customers are willing to pay more to buy a product with a well-established brand name than a similar product with no name recognition, so brand building should be an important strategic consideration in product design and customer value enhancement. Brand development will be discussed in Chapter 8.

- Value engineering is a technique that can systematically guide design engineers to develop high-value products with low cost. Value-engineering books (Park 1999) provide detailed definitions and methods for value analysis and cost reduction. The value-engineering technique will be thoroughly discussed in Chapter 9.

- Innovation and uniqueness are also huge factors for customer value. If the new product concept is right, customers may be willing to pay a premium price for a unique or "first of its kind" product. We will discuss the theory of inventive problem solving in Chapter 10.

This chapter will outline the customer value creation strategy and provide an overview for all these methods. Section 3.1 will formally define customer value and its components. Section 3.2 will discuss a practical analysis framework for collecting and analyzing customer value data, and Section 3.3 will discuss how to use the results in product development. Section 3.4 will discuss how customer value changes over time and how we can develop breakthrough products by capturing the changes in customer value. Section 3.5 discusses the relationship between customer value and the voice of the customer—specifically, what kind of VOC data we should collect in order to capture all aspects of customer value.

3.1   Customer Value and Its Elements

"Value" is a frequently used term, yet the concept is a confusing one. The nature of value has been extensively studied by many researchers, including value-engineering or value-analysis professionals. According to Park (2001), one of the leaders in the field of value engineering, "cost is a fact; it is a measure of the amount of money, time, labor, and any other expenses necessary to obtain a requirement. Value, on the other hand, is a matter of opinion of the buyer or customer as to what the product is worth, based on what it does to him/her. In addition, a person's measure of value is constantly changing to meet a specific situation." Dictionaries define value in terms of an item's equivalent in goods, services, or money; an item's market price; or its worth, usefulness, or importance to someone.

Value is related to worth, with worth being either a synonym for value or the quality that gives an item value. Carlos Fallon (1980) states that worth is a simple concept; it becomes value when it is related to cost. He further states that cost is a necessary component of value. Chris O'Brien (1982) further defines value as the ratio of worth and cost. In this definition, worth is an appraisal of the properties of a product—it is essentially an appraisal of the function of the product. In other words, value is the ratio of function to cost, with the function being what the product does for the customer.

Many value-engineering researchers and practitioners have developed precise definitions for value, such as the following:

Bryant   Value $V = \dfrac{\text{Wants} + \text{needs}}{\text{Resources}} = \dfrac{\text{SellFunctions} + \text{UseFunctions}}{\text{Dollar} + \text{People}}$

Harris $V = \dfrac{\text{Worth}}{\text{Effort}}$

Kaufman $V = \dfrac{\text{Functions}}{\text{Cost}}$

Wasserman $V = \dfrac{\text{Function}}{\text{Cost}} = \dfrac{\text{Utility}}{\text{Cost}} = \dfrac{\text{Performance}}{\text{Cost}}$

Fallon $V = \dfrac{\text{Objectives}}{\text{Cost}}$

In each of these definitions, the denominator is a unit that can be measured by dollars, effort, resources, manpower, and so on. In one way or another, they can all be converted to dollars. We can say that all these definitions converge to having the denominator be a measure of cost. The numerators converge to being a measure of function, or performance. Therefore, in value engineering, value is measured primarily as a "function-to-cost" ratio. A product with better functionality and lower cost has a higher value. If a product or method can accomplish a given function with a lower cost than all competitors, this product or method is the "best value."

Clearly, a higher function-to-cost ratio is important for increasing value in the eyes of the customer. However, in my opinion, the function-to-cost ratio alone is not sufficient to adequately measure customer value. Consider these two points:

- In value engineering the "best value" of a function is sometimes also defined as "the lowest cost to accomplish that function," measured in dollars. But this definition is difficult to reconcile with "value is function divided by cost." (Does this imply that function is measured in units of dollars squared?)

- There are many cases where people are willing to pay different prices for two products that have exactly the same functions. For example, the Toyota Corolla is exactly the same car as the Geo Prizm, but people are willing to pay $300 more to buy the Toyota Corolla. Similarly, the same item in a neighborhood convenience store will usually sell at a significantly higher price than in a large discount chain store. The function-to-cost ratio cannot explain the item's value in the eyes of customers adequately.

Sherden (1994) and Gale (1994) provided a broader definition for customer value. In their definition, customer value is perceived benefit (benefits) minus perceived cost (liabilities):

$$\text{Value} = \text{Benefits} - \text{Liabilities}$$

The benefits include the following categories:

- Functional benefits
  - Product functions, functional performance levels
  - Economic benefits, revenues (for investment services)
  - Reliability and durability
- Psychological benefits
  - Prestige and emotional factors, such as brand name reputation.
  - Perceived dependability (for example, people prefer a known brand rather than an unknown brand).

- Social and ethical reasons (for example, environmentally friendly brands)
- Psychological awe (many first-in-market products not only provide unique functions, but also give customers a tremendous thrill; for example, the first copy machine really impressed customers)
- Psychological effects of competition. For example, if there are many competitors producing a same or similar product, that usually will create a perception that this product is a commodity and it doesn't carry much value.

- Service and convenience benefits
  - Availability (how easy is it to access the product or service?)
  - Service (how easy is it to get service in case of product problems or failure?)

The liabilities include the following:

- Economic liabilities
  - Price
  - Acquisition cost (such as transportation and shipping costs, time and effort spent to obtain the service)
  - Usage cost (additional cost to use the product or service in addition to the purchasing price, such as installation)
  - Maintenance costs
  - Ownership costs
  - Disposal costs
- Psychological liabilities
  - Uncertainty about the dependability of the product or service
  - Self-esteem liability of using an unknown brand product
  - Psychological liability of poor performance of the product or service
- Service and convenience liability
  - Liability due to lack of service
  - Liability due to poor service
  - Liability due to poor availability (such as delivery time, distance to shop)

Clearly, this "customer value" definition by Sherden and Gale contains much more information than simply the product's function and cost, and it also is measured in dollars, or monetary worth. I will use this customer value definition throughout this book.

3.1.1 Value and Other Commonly Used Metrics

There are other product metrics that can easily be combined with the concept of value, including price, performance, cost, and quality. Let's look at the similarities and differences between value and these metrics.

3.1.1.1 Value and Price   Some economists define value as price, but as we saw earlier, price is only one factor that affects the value. Specifically, price is one important element in customer value (economic liabilities). In general, customers accept a high selling price for a product if they believe the product offers superior benefits.

A higher price may provide a higher profit margin for the company that sells the product, but the sales volume can be sustained only if customers think that this product will provide more customer benefits than costs; that is, if it offers high customer value (more benefits for a lower price). The gap between customer-perceived benefits versus liabilities determines the product's overall attractiveness to customers, and thus determines the size of the market, the sales volume, and the market share.

3.1.1.2 Value and Performance   Performance is also called function. The *function* of a product is what the product is supposed to do for customers, and it is only one component of the product's benefits; customer value is the benefits minus all costs. Therefore, more and better functions may not always give a better value. Also, a function can create value only if that function is something customers need or want. In 1979 the American Can Company thought that a product with more and better functions would always sell, so it designed a stronger paper towel called BOLT. It looked and performed like cloth, and it was sold at a higher price than regular paper towels. However, this product was a total failure, because customers did not perceive a benefit in a paper towel that could be washed. Actually, in many cases, customers enjoy a product change that adds more functions without increasing price, and often they are delighted to see a product change that reduces both functions and price. Adding and improving functions while also increasing the price is a risky strategy.

3.1.1.3 Value and Cost of Production   The costs of producing a product usually do not relate to customer value. A company could provide a product with a lot of features and high cost that customers do not appreciate. However, reducing cost will either increase the profit margin or create more room for price reduction.

3.1.1.4 Value, Quality, and Perceived Quality   Like value, quality is a tricky concept to define. Even quality gurus do not agree on a definition. The American Society for Quality (ASQ) defines quality as "a subjective term for which each person has his or her own definition. In technical usage, quality can have two meanings: 1. the characteristics of a product or service that bear on its ability to satisfy stated or implied needs. 2. a product or service free of deficiencies" (www.asq.org).

In ASQ's definition, the "characteristics of a product or service that bear on its ability to satisfy stated or implied needs" sounds a lot like "customer-preferred performance and function"; and "a product or service free of deficiencies" is definitely related to "dependability" and "reliability." In comparing quality with customer value, it is clear that value is a much broader concept. Quality is a part of value, not all of value.

Quality is subjective, as mentioned in the ASQ's definition, and it has a psychological component. For example, a brand-name drug may have exactly the same functionality and manufacturing quality as a generic drug. However, a substantial portion of consumers will still consider the drug with the brand name to have "higher quality." This overall customer opinion of the quality of a particular product or service is also called the *perceived quality*, and it is a better indicator of customer value than any objective measures of quality. There are primarily two components in perceived quality: the technical component that relates to performance, functionality, dependability, and level of defects; and the psychological component, such as brand image.

3.1.2   The Versatility and Dynamics of Value

Customer value can mean different things for different people—we call this the *versatility of value*. Value may also change over time with people's preferences and lifestyles—we call this the *dynamics of value*.

The versatility of value reflects the fact that the marketplace consists of different people, and it is difficult to find even two people who have exactly the same opinion. In marketing science, people can be divided into market segments for a particular kind of product or service; the customers within each segment display similar behavior and opinions. Some products or services can only find customers in a particular market segment; this is a niche market. For example, a good state-of-the-art computer engineering book can only be sold at university campuses and to computer engineers—this book is useless to an animal trainer. Some products and services, however, can find customers in mass market; for example, vegetables, fruits, pencils, and personal banking services have customers all over the social spectrum.

Given the versatility of value, it is important to know the types of markets that your product or service is in, and to understand what value your product or service can offer to customers. Even for a single product or service, its value can be broken down into several categories:

- **Use value**   Properties that make something work; this can also be called functional value.

- **Esteem value**   Properties that make something desirable to own; this can also be called emotional or psychological value.

- **Exchange value**   Properties that make it possible to exchange one thing for another.

For example, the use value of an airline ticket is the ability of a customer to take an airplane from point A to point B; even a coach-class ticket provides that use value. A business-class ticket provides a little more functional value, such as better seating and better food, but it also provides substantial esteem value — the feeling that "I am special, I have special status." Coach-class tickets are in a mass market; the business-class ticket is in a niche market.

If you want to enhance the value of a product or service so that it will be more successful in the marketplace, you need to determine what will make it more valuable. Because value is a matter of customer opinion, you need to understand what will motivate and excite people.

Abraham Maslow developed a simple scale to define people's psychological needs. He called this scale the "hierarchy of needs," as shown in Figure 3.1. Maslow said that people are motivated to do different things at different levels of psychological development, or at different levels of society. He divided these motivational factors into five basic needs. As each need is satisfied, other higher needs arise. Although the lower level needs may never disappear, they become weaker or less important. A person may have several needs at the same time but one need is dominant.

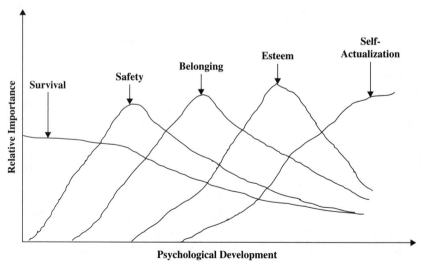

Figure 3.1   Value and Psychological Development

Maslow's theory provides a lot of insight into customers' buying motivation. For example, customers in developing countries usually prefer products that address basic needs, which are robust in harsh user conditions, and which are low in cost and without fancy features. This reflects the fact that survival and safety needs are predominant needs for these consumers. However, for many children in affluent countries, printing sports stars figures on cereal boxes might be a very effective way to increase sales because it addresses the "esteem" and "belonging" needs for such children.

Many factors can also change customers' preferences and the tradeoffs they're willing to make regarding products. This is the dynamic nature of customer value (changing over time). It can be affected by economic conditions, new technologies, and the changes in customers' psychological development.

For example, changes in oil prices will affect people's attitudes toward the types of cars they like. Similarly, bad economic times, fierce competition, and a tough job market can make discount chain stores a favorite shopping place. A strong competitor's emergence in a market segment will greatly change the expectations for a particular product or service.

3.2   Customer Value Analysis

In the last section, we established that customer value is the difference between benefits and liabilities:

$$\text{Value} = \text{Benefits} - \text{Liabilities}$$

where the benefits consist of the following elements:

- Functional benefits
- Psychological benefits
- Service and convenience benefits

and the liabilities consist of these elements:

- Economical liabilities (customer costs)
- Psychological liabilities
- Service and convenience liabilities

For any product or service, the company that provides more value to its customers than its competitors will eventually gain in sales and profitability. However, for each particular product or service, the profile of benefits and liabilities will be very different, and customers will give the benefits and liabilities different relative importance. When a product or service has several competitors, it is very important to do better in the areas most important to customers.

Bradley Gale (1994) developed a systematic approach to maximizing customer value in providing products and services. This approach consists of the following steps:

1. Conduct customer surveys to get information about the relative importance to customers of aspects of a particular product or service, and to get customers' ratings of competing products or services. Gale referred to this step as compiling the "market-perceived quality profile."
2. Collect information about the price or combined customer costs of the company's product or service, and the same information for competing products or services. Gale referred to this as compiling the "market-perceived price profile."
3. Complete a comprehensive customer value evaluation of the company's product or service compared with its competitors. Gale called this compiling the "customer value map."
4. Complete an area-to-area competitive analysis, to identify the critical areas in order to gain a competitive advantage.
5. Deploy the critical improvement into the product or service design.

In a customer value management approach, all non-cost-related attributes, such as functional benefits, psychological benefits, and service and convenience benefits, are considered to be components of market-perceived quality. If a product or service can offer higher market-perceived quality than all competitors, yet have its cost under control, this product will have a competitive edge.

3.2.1 Market-Perceived Quality Profile

A market-perceived quality profile is a detailed "quality scorecard" that provides quantitative quality ratings for a company's own product versus its competitors' products on all important non-cost-related attributes. According to Gale (1994), creating the market-perceived quality profile involves the following steps:

1. In forums such as focus groups, ask customers in the targeted market, including both your customers and competitors' customers, to list the factors other than price that are important in their purchase decisions.
2. Establish how the various non-price factors are weighted in the customer's decision, usually by simply asking customers to tell you how they weigh the various factors, distributing 100 points among them.
3. Ask well-informed customers, including both yours and your competitors, how you and your competitors perform on the various

quality attributes. Then for each attribute, divide the score of the product or service you are studying by the scores of competitors' products. That gives you the performance ratio on that attribute. Multiply each ratio by the weight of that attribute, and add the results to get an overall market-perceived quality score.

Table 3.1 shows an example of a market-perceived quality profile for frozen chicken (Gale 1994). The first column lists all of the important non-cost-related quality attributes in the chicken business: "yellow bird," "meat-to-bone," and so on. The second column lists the relative importance rating for the attributes listed in column 1; the relative importance ratings will add up to 100%. In this example, "yellow bird" accounts for 10% of relative importance, and "fresh" accounts for 15% of relative importance. These importance ratings are obtained from a specially-designed customer survey. The third column records the average customer rating of the product from "our business" on each quality attribute; the highest possible score is 10.0, and the lowest possible score is 1.0. These scores are also computed based on a specially-designed customer survey. The fourth column lists the average customer ratings of the competing products for each customer attribute.

The remaining columns are calculated from the preceding ones. The fifth column lists the ratio of column 3 to column 4; that is, the average customer rating of our product versus the average customer rating of the competitors' products. Clearly, if this ratio is less than 1, it means that in this quality attribute category our product performs worse than the competitor's average; if this ratio is greater than 1, it means that in this category our product performs better than the average competitor's. The last column lists the ratio from column 5 multiplied by the relative importance score in column 2. Clearly, if all the ratios are equal to 1, it means that our product is an average product in comparison with our competitors. In that case, the total score in column 6 would be equal to 100, and our market-perceived quality score would be 100. A product with a market-perceived quality score larger than 100 is considered to be a competitive product, and the higher the score, the more competitive the product is.

The information used to compile the market-perceived quality profile can be obtained by conducting a special kind of customer survey. A form for this kind of customer survey is illustrated in Table 3.2.

The survey population should include the customers of "our company," as well as all consumers who purchase this kind of product or service, including the customers of competitors. For example, if our company is McDonald's, then the survey population should be the customers for all fast food chains and should include the customer population of Burger King, Wendy's, and so on. Clearly, if we collected enough finished survey forms from customers, we would have all the information needed in a market-perceived quality profile study.

TABLE 3.1 Market-Perceived Quality Profile in the Chicken Business

| Quality Attributes | Customer's Weight of Attributes (Total = 100) | Industry Comparison | | Quality Scores | |
		Perdue (Our business) (1 = lowest, 10 = highest)	Average Competitors	Ratio (Ours/Competitor's) (Ratio>1.0 means "better than competitor")	Customer Weight x Ratio
Yellow Bird	10	8.1	7.2	1.13 = 8.1/7.2	11.3 = 1.13 × 10
Meat-to-bone	20	9.0	7.3	1.23	24.6 = 1.23 × 20
No Pinfeathers	20	9.2	6.5	1.42	28.4
Fresh	15	8.0	8.0	1.00	15.0
Availability	10	8.0	8.0	1.00	10.0
Brand Image	25	9.4	6.4	1.47	36.8
Total	**100**				**126.1 = market-perceived quality score**

TABLE 3.2 Survey Form for Customer Values

| Quality (Non-Price) Attributes | Importance Weights (Add up to 100) | Performance Scores 1 – 10, 1 = lowest, 10 = highest | | | |
		Company A (Our company)	Company B (Competitor 1)	Company C (Competitor 2)	Company D (Competitor 3)
1					
2					
3					
4					
5					
6					
7					
8					
9					
10					
Sum of Importance Weights = 100					
Price (perceived transaction price)					
Price Competitiveness					

Besides customer survey data, functional data can also be used in a market-perceived quality profile study. Table 3.3 gives such an example.

3.2.2   Market-Perceived Price Profile

For some industries, such as retail sales, the price for a particular item is paid only once, so it is very clearly understood. In this case, the price comparison simply involves comparing one dollar amount to another dollar amount. For many other businesses, the overall customer cost structure is rather complicated. For example, the cost-related factors in purchasing a car might involve a trade-in allowance, rebate, and finance rate, in addition to the purchase price of the car. In such a case, it is necessary to create a market-perceived price profile, because it will integrate all the cost factors and compile a combined price score.

The construction of the market-perceived price profile is very similar to creating a market-perceived quality profile. Customers are asked to list the factors that affect their perceptions of a product's cost. Table 3.4 shows an example of a market-perceived price profile in the luxury car market.

Because people consider lower prices to be better, however, having a higher customer satisfaction score in price level represent a better price is counterintuitive. In Table 3.4, Acura's market-perceived price score is 118.7—this is more than 100, which means that the Acura's overall price level is more attractive (lower) than other competitors. Using the inverse score (84.2) is more intuitive.

TABLE 3.3   Quality Profile Studies of Gallbladder Operations

		Endosurgery versus Traditional Surgery			
Quality Attributes	Customer's Weight of Attributes (Total = 100)	Industry Comparison			Quality Scores
		Endo	Traditional	Ratio	Customer Weight x Ratio
At Home Recovery Period	40	1–2 weeks	6–8 weeks	3.0	120
Hospital Stay	30	1–2 days	3–7 days	2.0	60
Complication Rate	10	0–5%	1–10%	1.5	15
Postoperative Scar	5	0.5–1 inch	3–5 inch	1.4	7
Operation time	15	0.5–1 hour	1–2 hours	2.0	30
Total	**100**				**232 = market-perceived quality score**

TABLE 3.4 Market-Perceived Price Profile: Luxury Cars

Price Satisfaction Attributes	Customer's Weight of Attributes (Total = 100)	Industry Comparison		Quality Scores	
		Acura (Our business) (1 = lowest, 10 = highest)	Average Competitors	Ratio (Ours/ Competitor) (Ratio>1.0 means "better than competitor")	Customer Weight x Ratio
Purchase price	60	9	7	1.29 = 9/7	77.4 = 1.29 × 60
Trade-in allowance	20	6	6	1.0	20 = 1.0 × 20
Resale price	10	9	8	1.13	11.3
Finance rates	10	7	7	1.00	10.0
Total	**100**				**118.7 = market-perceived price score**
Relative price ratio					**84.2 = (1/118.7) × 100**

If we just want to compare the sticker prices, a simple price ratio can be used. For example, if Acura's price is $35, 200, and the average competitor's price is $40,000, the price ratio is $35,200/$40,000 = 0.88. If a percentage score is used, the relative price ratio is 0.88 × 100 = 88%. If the relative price ratio of a product is less than 100, then its price level is lower than its competitors.

3.2.3 Customer Value Map

A customer value map is a very useful tool for identifying the competitive position of a particular product in comparison with other competitors' products. A product is competitive if it has high customer benefit and low customer cost. The customer benefit can be well represented by the market-perceived quality score, and the customer cost position can be represented by the relative price ratio.

A customer value map is a two-dimensional plot of the market-perceived quality score on the horizontal axis versus the relative price ratio on the vertical axis. Figure 3.2 shows what a customer value map looks like.

In this customer value map, each dot represents a particular product. The diagonal line in the customer value map represents the range where the market-perceived quality score is equal to the relative price ratio.

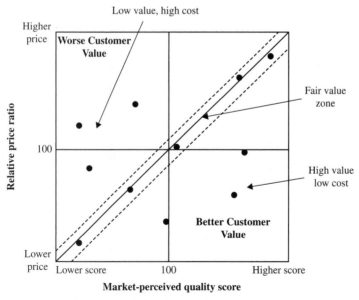

Figure 3.2 Customer Value Map

For example, if a product has a market-perceived quality score of 80, and its relative price ratio is also 80, the dot representing this product will be on the diagonal line; this product would have low value and low price. Similarly, if a product has a market-perceived quality score of 120, and relative price ratio of 120, it would also be on the diagonal line; it would be a high-price, high-value product.

Overall, the region around the diagonal line can be called the "fair-value zone," and products in this zone can be considered average products. The products in the lower-right corner of the customer value map feature lower relative price ratios and higher market-perceived quality scores. These are high-value, low-price products—they have superior competitive positions and are poised to gain market share. The products at the upper-left corner of the customer value map feature high relative price ratio and low market-perceived quality scores. These are low-value, high-price products that have inferior competitive positions in the marketplace and are vulnerable to losing market share.

Example 3.1 Toaster Customer Value Analysis   Table 3.5 lists 15 brands of toasters and their market-perceived quality scores and relative price ratios.

TABLE 3.5   Market-Perceived Quality Profile and Price Profile for Toasters

Name of Toaster	Market-Perceived Quality Score	Toaster Price	Relative Price Ratio
1. Cuisinart CPT-60	128	$70	215
2. Sunbeam	119	28	85
3. KitchenAid	117	77	237
4. Black & Decker	112	25	77
5. Cuisinart CPT-30	109	40	123
6. Breadman	107	35	108
7. Proctor-Silex 22425	104	15	46
8. Krups	101	32	98
9. Oster	94	45	138
10. Toastmaster B1021	91	16	49
11. Proctor-Silex 22415	87	35	108
12. Toastmaster B 1035	84	21	65
13. Betty Crocker	84	25	77
14. Proctor-Silex 22205	83	11	34
15. Rival	80	13	40
Average	**100**	**$33**	**100**

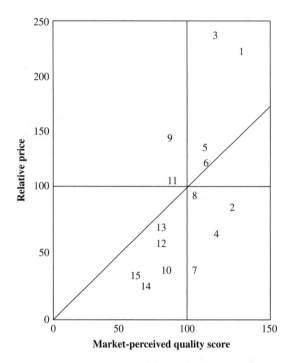

Figure 3.3 Customer Value Map of Toasters

By using the data from Table 3.5, we can draw the customer value map shown in Figure 3.3.

In the customer value map, the products above the diagonal line are the products that have low customer value:

- Products 1 and 3 (Cuisinart CPT-60 and KitchenAid) have good market-perceived quality scores but very high prices.

- Product 9 (Oster) has below average market-perceived quality but a high price.

- Products 5, 6, and 11 (Cuisinart CPT-30, Breadman, and Proctor-Silex 22415) are also relatively high in price and low in performance, but they are close to the fair value zone.

- Products 2, 4, and 7 (Sunbeam, Black & Decker, and Proctor-Silex) are low-price, high-performance products.

- Products 10, 14, 15, and 12 (Toastmaster B1021, Proctor-Silex 22205, Rival, and Toastmaster B 1035) are low in price and have reasonable performance.

- Products 8 and 13 (Krup and Betty Crocker) have better than average customer value but are in the fair value zone.

Overall, the products that are located in the lower right of the chart are "better customer value" products, and the further they deviate from the fair value line, the better customer value the product has. In Figure 3.3, products 2, 4, and 7 are in the lower-right portion and are the furthest from the fair value line, so they are the products with the best customer value. Product 2 has a higher price, so it is a best-value product at the higher price level. Product 7 has a lower price, so it is a best-value product at a low price level. Similarly, the products that are located at the upper-left corner of the chart are products with worse customer value, and the further they are from the fair value line, the worse customer value the product has.

3.2.4   Competitive Customer Value Analysis

A competitive customer-value analysis is a graph that compares products in important aspects of customer value. This analysis can show you the areas where you can improve your product most effectively.

Table 3.6 shows a market-perceived quality profile for two printers, printer A (our printer) and printer B (a competitor's printer), and Table 3.7 shows the market-perceived price profile.

TABLE 3.6   Market-Perceived Quality Profile of Two Printers

Quality Attributes	Customer's Weight of Attributes (Total = 100)	Industry Comparison		Quality Scores	
		Printer A (Our Printer) (1 = lowest, 10 = highest)	Printer B (Competitor's Printer)	Ratio (Ours/ Competitor) (Ratio>1.0 means "better than competitor")	Customer Weight x Ratio
Machine Up Time	25	8	7	1.14 = 8/7	$28.5 = 1.14 \times 25$
Print Speed	15	9	8	1.13	$17.0 = 1.13 \times 15$
Image Quality	15	7	8	0.88	13.2
Ease of Use	5	4	7	0.57	2.85
Service Response Time	15	5	7	0.71	10.65
Repair Time	15	5	6	0.83	12.45
Quality of Service	10	7	7	1.0	10.0
Total	**100**				**94.65 = market-perceived quality score**

TABLE 3.7 Market-Perceived Price Profile: Printers

Price Satisfaction Attributes	Customer's Weight of Attributes (Total = 100)	Industry Comparison		Quality Scores	
		Printer A (Our Printer)	Printer B (Competitor's Printer)	Relative Price Ratio	Customer Weight x Ratio
Purchase Price	40	$508	$585	0.87 = 508/585	34.8 = 0.87 × 40
Service and Repair	30	$60/year	$65/year	0.92 = 60/65	27.6 = 0.92 × 30
Toner	20	$235/year	$235/year	1.0	20
Paper	10	$124/year	$101/year	1.23	12.3
Total	**100**				**94.7 = relative price ratio**

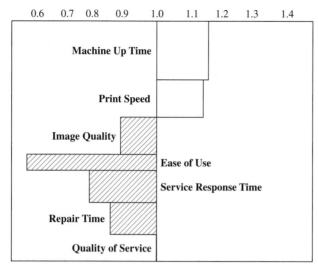

0.6   0.7   0.8   0.9   1.0   1.1   1.2   1.3   1.4

Machine Up Time

Print Speed

Image Quality

Ease of Use

Service Response Time

Repair Time

Quality of Service

Figure 3.4 Area-to-Area Customer Value Chart for Printers

Figure 3.4 shows an "area-to area" customer-value chart that compares printer A and printer B. Each bar represents a market-perceived quality characteristic. The horizontal dimension of the bar shows how much the product is better or worse than the competitor's, and the thickness of each bar is proportional to the relative importance of each characteristic. So the total white area represents our advantage, and the total shaded area represents our disadvantage. Our goal is to maximize the white area and minimize the shaded area in the most effective way.

Figure 3.5 shows a "head-to-head" market-perceived price ratio chart for printers, and each bar represents a customer cost component. Its horizontal dimension represents how much better or worse our product's price is compared with the competitors' price, and the bar's thickness represents the relative importance of that cost component in the eyes of customers. So the total white area minus the total shaded area represents our product's cost advantage. The larger the cost advantage, the more competitive our product is in price.

3.3   Customer Value Deployment

After we have completed our competitive customer-value analysis and relative price-ratio analysis, we need to find an effective way to overcome our disadvantages and strengthen our existing advantages to improve our customer value and win over the competition. To do that, we need to identify the critical areas of the company that are related to our key market-perceived quality factors and market-perceived customer price areas.

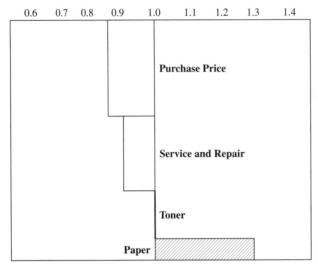

| 0.6 | 0.7 | 0.8 | 0.9 | 1.0 | 1.1 | 1.2 | 1.3 | 1.4 |

Figure 3.5   Area-to-Area Market-Perceived Price Ratio Chart for Printers

A quality function deployment (QFD) template can be very useful in deploying key customer values into our process improvements.

Table 3.8 shows the customer value deployment matrix for the printer example. In this matrix, correlation scores of 9, 3, 1, and 0 are used. A score of 9 means "very much related," a score of 3 means "related," a score of 1 means "slightly related," and a score of 0 means "not related."

TABLE 3.8   Customer Value Deployment Matrix for Printer Example

Quality Attributes	Design	Manufacturing	QC	Sales and Service	Distribution	Marketing
Machine Up Time	3	9	9	0	0	0
Print Speed	9	3	3	0	0	0
Image Quality	9	9	9	0	0	9
Ease of Use	9	0	0	9	0	3
Service Response Time	0	0	0	9	9	0
Repair Time	0	0	0	9	9	0
Quality of Service	0	0	0	9	0	0

For example, in the "machine up time" category, quality control and manufacturing are very critical in ensuring printer dependability; product design is also related to the dependability of the printer. In the "ease of use" category, design is very important in creating a printer that is easy to use. However, sometimes there is a gap between customer-perceived quality and the real quality level. For example, printer A may actually be easy to use, but because of poorly written customer instructions, poor service support, and poor marketing, significant numbers of customers may think that printer A is hard to use. To overcome this problem, redesigning the printer may not be the right solution—a comprehensive strategy that includes improving customer service, rewriting customer instructions, and developing a better marketing message might be the best approach.

3.4 Evolution of Customer Values—Blue Ocean Strategy

In today's marketplace, we are flooded by all kinds of products and services. With the development of information technology and a global economy, the pace of product development and cost cutting is accelerating. You could have a top-notch product with high customer value, and thus high customer appreciation, and your product could have a commanding lead in the marketplace this year. But very quickly, you will find competitors growing like mushrooms, and soon there will be many similar products in the marketplace with reasonably good quality, similar product characteristics, and cutthroat prices. You will be forced to enter a bloody competition, head-to-head, to compete on every item of the current customer values. You will be forced to make improvements on every item of the current customer values, but still cut prices. With this brutal competition, your profit margin will start to disappear.

This scenario is very common in today's marketplace in the airline industry, automotive industry, and many others. This situation is called the "red sea," as an analogy to fishermen along a crowded coastline competing for dwindling fish. In general, the red sea refers to a saturated market with fierce competition, crowded with companies providing similar services or goods.

The opposite of the "red sea" is the "blue ocean," which is an analogy to fishermen in a wide-open blue ocean, with plenty of fish and no competition. In general, the blue ocean refers to an untapped and uncontested market, offering little or no competition, since the market is not crowded. To create a blue ocean, you need to do something different from everyone else, produce something that no one has yet seen. The blue ocean strategy is a business strategy for capturing uncontested market space, making competition irrelevant (Kim and Mauborgne 2005).

Specifically, the blue ocean strategy enables a company to unilaterally alter the commonly adopted key customer value items to form a new set of customer values. Thus a new type of product or service is formed, without competition. Example 3.2 illustrates how the blue ocean strategy works.

Example 3.2 U.S. Wine Industry   The United States has the third largest consumption of wine worldwide, and this $20 billion wine industry is very competitive. California wines dominate the domestic market, capturing two-thirds of all U.S. wine sales. California wines compete head-to-head with imported wines from France, Italy, Spain, and South America. In recent years, there are also many emerging wine producers from such places as Oregon, Washington, and New York State. The number of wine brands keeps increasing, but overall market size remains stagnant, so the wine market becomes a typical "red sea" market. (Kim and Mauborgne 2005)

In the U.S. wine market, almost all wine makers compete based on the following seven factors:

- Price per bottle of wine
- An elite, refined image in packaging, including labels announcing the wine medals won and the use of esoteric enological terminology to stress the art and science of wine making
- Above-the-line marketing to raise consumer awareness in a crowded market and to encourage distributors and retailers to give prominence to a particular wine house
- Aging quality of wine
- The prestige of a wine's vineyard and its legacy (hence the appellations of estates and chateaux, and references to the historic age of the establishment)
- The complexity and sophistication of a wine's taste, including such things as tannins and oak
- A range of wines to cover all varieties of grapes and consumer preferences from Chardonnay to Merlot, and so on

Based on these seven factors, Kim and Mauborgne analyzed the wine market and drew the graph shown in Figure 3.6. The seven factors are shown on the horizontal axis, and the vertical heights indicate the level or values of each factor. For example, the premium wines have higher value in "price" than the budget wines, which means that the premium wines are higher in price. For other items, such as "aging quality," the higher value for premium wines means that premium wines have a

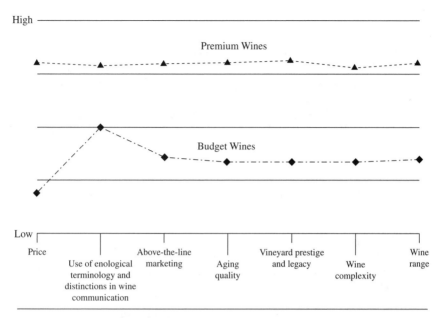

Figure 3.6   Value Curves of the U.S. Wine Market

higher aging quality than budget wines. Kim and Mauborgne (2005) call this graph the "value curves," or "strategy canvas."

We can easily see that this value curve has a lot of similarities with customer values. In Kim and Mauborgne's value curves, price is always the first item, and in a customer value analysis, price, and in the broader sense, the market-perceived price profile, is also always an important customer value metric. In Kim and Mauborgne's value curves, factors other than price are always key competitive factors—they are very similar to the market-perceived quality profile in the customer value. We can call all these factors the "value factors"—price is always one of the value factors, and the other value factors are quality-related factors.

One of the differences between Kim and Mauborgne's value curves and a customer value analysis is their graphing method. Figure 3.6 shows that although there are more than 1600 wineries in the U.S. wine industry, most wines fall into one of two categories: premium wines are high in price, but are also rated high in the remaining six key value factors, so premium wines are "high-quality, high-price" wines; budget wines are lower in price, but are also rated lower in the remaining six key value factors, so budget wines are "low-price, low-quality wines." The striking fact is that all wines are competing head-to-head on the same seven factors. The premium wines are trying to win by increasing the ratings of quality, which is indicated by the six value factors (other than price). Increasing these quality aspects requires investment, thus increasing the cost, and to recover the cost, the premium wines have to sell at a higher price. The budget wines are selling at a lower price, and to save production costs, the ratings of the six value factors (other than price) will have to be lower. When the overall market size is limited, competing either way is not very profitable, and the potential for growth is very poor.

The key for a blue ocean strategy is to systematically change these value factors:

- Which value factors that the industry takes for granted should be eliminated?

- Which value factors should be reduced well below the industry's standard?

- Which value factors should be raised well above the industry's standard?

- Which value factors should be created that the industry has never offered?

Casella Wines, an Australia winery, noticed that almost all wine makers focus on overdelivering on prestige, and on the traditionally accepted quality factors at their price points. However, Casella Wines noticed that the alternative drinks to wine, such as beer, spirits, and ready-to-drink cocktails, were a market three times as big as traditional wines. Casella also found that a lot of American consumers think traditional wines are too complex, confusing, and difficult to appreciate. These consumers think that beer and ready-to-drink cocktails are sweeter and easier to drink. Based on careful analysis, Casella Wines developed a brand new wine called Yellow Tail. Yellow Tail is a wine that has a new combination of wine characteristics that produced an uncomplicated wine structure. The wine is soft in taste and approachable like ready-to-drink cocktails and beers. The sweet fruitiness of the wine also keeps people's palate fresher, allowing them to enjoy another glass of wine without thinking about it. The result was an easy-to-drink wine that does not require years to make. In line with this fruity sweetness, Yellow Tail dramatically reduced or eliminated all the value factors that the wine industry has long competed on—tannins, oak, complexity, and aging. Figure 3.7 illustrates the value curves of this new wine.

We can easily see that Yellow Tail's value factors are very different from those of conventional wines. Three value factors for conventional wines— wine packaging, marketing, and aging quality of wine—are reduced to a minimum; another three value factors for conventional wines—vineyard prestige, wine complexity, and wine range—are also reduced; and three completely new value factors are added—easy drinking, ease of selection, and fun and adventure—and they are positioned at a very high level.

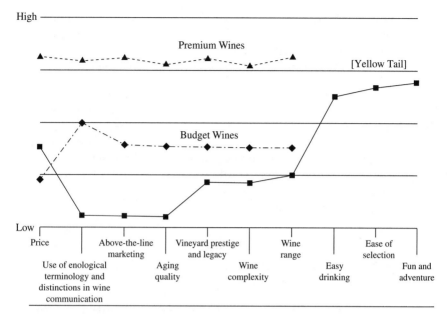

Figure 3.7 Value Curves of Yellow Tail

Yellow Tail is a great success. Two years after its launch, it became the fastest-growing brand in the history of both the U.S. and Australia wine industries, and it is the number one wine imported into the United States, surpassing the wines of France and Italy. By August 2003, it was the number one red wine in 750ml bottles sold in the United States. Moreover, this happened in the context of a global wine glut.

In summary, Yellow Tail's drastic change in value factors created a social drink for everyone: beer drinkers, cocktail drinkers, and non-wine-drinkers. This new wine works on different value factors and opened up a much broader market with very little competition. Thus it created a blue ocean.

We can derive the following observations:

1. In a red sea marketplace, every competitor delivers a similar product or service with an almost identical set of value factors; the competition is head-to-head; each competitor either delivers a high-price, high-quality product, or a low-cost, low-quality product; the market will soon be saturated; and profitability will be very marginal.

2. Customers' value systems may change over time, even for the same product or product class. The current customer value factors that the industry competes with may have been established long in the past, and customers may welcome a change in these customer value factors.

3. If a producer captures the right strategy in changing customer value factors, and develops a new set of customer value factors that resonate with customers' desires, the producer will create a blue ocean for its products with great market growth potential and little competition.

Clearly, creating a blue ocean is like hitting a jackpot or a gold mine. In the business literature, we can see many success stories resulting from bold changes in customer value factors, but there are also many blunders and failures due to bad moves. How can you formulate the right changes in customer value factors to create blue oceans, and are they even possible? Kim and Mauborgne proposed several ideas for systematically formulating blue ocean strategies by making the right changes in customer value factors.

3.4.1   Formulating a Blue Ocean Strategy

Kim and Mauborgne noticed that many companies tend to do the following in their business practices:

- Define their own industry as others define it, and fight head-to-head to be the best.

- Stick to their industry's accepted strategic groups, such as luxury automobiles or pickup trucks, and strive to stand out in their strategic group.

- Focus on the same buyer group, be it purchasers (for office equipment industry), or retail customers (in the clothing industry), or influencers (such as doctors for the pharmaceutical industry), even though there are usually many tiers of buyers.

- Define the scope of the products and services offered by their industry in much the same way as their competitors do.

- Adopt customer value factors similar to their competitors', and set similar priorities on functional and psychological benefits.

- Focus on the same point in time, and often on the current competitive threat.

These commonly adopted business practices are actually the main forces that lock businesses into red seas. In order to break out of red seas and into a blue ocean, you need to go beyond these business practices. However, breaking out of the currently accepted customer value factors in a given industry is also risky. You cannot arbitrarily abandon a well established set of customer values and replace them with unknowns.

Based on many years of research and observation, Kim and Mauborgne proposed formulating the blue ocean strategy by looking at the following six aspects:

- Alternative industries
- Strategic groups
- Buyer groups
- Complementary product and service offerings
- Functional-emotional orientation
- Time

3.4.1.1 Looking Across Alternative Industries   For a given industry, the alternative industries are the industries that provide different forms of products or services that serve the same purpose. For example, in one sense, restaurants and theaters serve very different kinds of products, but both restaurants and theaters give customers nice places to be and activities for their evening time. Another example is tax filing—you can hire a CPA, do it yourself, or buy tax-filing software. These are very different choices, but they serve the same purpose.

Different industries within the same alternative industries class may have quite different customer value factors; for example, the customer value factors for a restaurant are very different from those of theaters, and the customer value factors of a CPA are quite different from those of tax-filing software. However, these alternative industries serve the same customer groups for the same or similar purposes. In this case, some crossover of the features of these alternative industries can create new products or services that will be smashing successes. The case of Cirque du Soleil is illustrated in Example 3.3.

Example 3.3 Cirque du Soleil   Cirque du Soleil (French for "Circus of the Sun") is a successful entertainment enterprise based in Montreal, Quebec, Canada, and founded in Quebec in 1984. Though Cirque du Soleil is called a circus, it doesn't involve many animals as other traditional circuses do; it mainly depends on human performance. Cirque du Soleil not only inherited the tradition of circus, but it also integrated many elements of street performances and busking, opera, ballet, and rock music into its performance. Usually, the traditional circus emphasized stunts and skills in the performance without paying too much attention to story lines. Cirque de Soleil's performance is centered on story lines. Cirque du Soleil is also characterized by its beautiful colors, nice live music, and a mix of talents from all over the world.

From the customer value factors point of view, Cirque combined the customer value factors from several alternative industries: regular circus shows, opera, Broadway-style musical shows, Las Vegas style shows, and even Chinese acrobats. This cross-pollination created a unique entertainment style and a great success. Cirque du Soleil attracted customers for both the traditional circus—children—as well as customers who would usually not go to see a circus—corporate clients and adults.

Figure 3.8 shows the value curve for Cirque du Soleil. You can see that it eliminated four customer value factors that are commonly adopted by a regular circus—star performers, animal shows, aisle concessions, and multiple show arenas. Cirque du Soleil kept three customer value factors for a regular circus—fun and humor, thrills and danger, and a unique venue—and it introduced four new customer value factors— theme, refined watching environment, multiple productions, and artistic music and dance. These last four customer value factors are associated with musicals, Las Vegas shows, and others. Cirque du Soleil's new and unique customer value curve makes it a unique circus, and it has become so successful it is the largest cultural export of Canada.

3.4.1.2 Looking Across Strategic Groups Within Industries   In any industry, there are usually several strategic groups. A strategic group is usually a group of companies or entities within an industry that pursue a similar strategy. For example, within the automotive industry, there

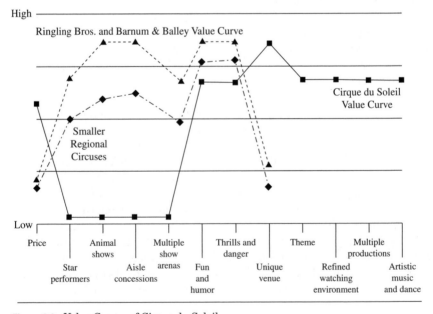

Figure 3.8 Value Curves of Cirque du Soleil

are several strategic groups, such as luxury cars, family cars, and low-price cars. Within the restaurant industry, there are fast food restaurants, family restaurants, and high-end restaurants.

Usually, strategic groups can be partitioned by price and performance; that is, high price and high performance, medium price and medium performance, and low price and low performance. The distinction between the luxury car, family car, and low-end car is one example of this kind of partition. Strategic groups can also be partitioned by functional variations within an industry. For example, in the automotive industry, there are strategic groups such as family sedans, trucks, and vans. Different strategic groups with an industry usually also have different customer value factors. Crossover of customer value factors among different strategic groups may lead to breakthrough products or services.

In the automotive industry, Toyota's Lexus combined the quality of high-end Mercedes, BMW, and Jaguar, with the price level of the Cadillac and Lincoln, thus creating a new market segment of "Lexus consumers." This is an example of crossover between two strategic groups within the automotive industry. Similarly, the Chrysler minivan was a crossover between family cars and regular vans—the minivan drives like a car, but it is more spacious and has more seats and cargo space like a van. Chrysler's minivan was a great success that led to the SUV (sport utility vehicle) revolution for American automobile makers.

3.4.1.3 Looking Across the Chain of Buyers   For most industries, there is usually a chain of buyers for the industry's products or services. For cloth manufacturers, their products may be bought first by retail chains; the products are then distributed to retail stores, and finally the products in retail stores are bought by individual shoppers. For the pharmaceutical industry, the products (medicines) are first bought by drugstore chains, and then they are distributed to individual drugstores. For prescription drugs, medical doctors will write the prescription, and the actual users will bring the prescription to the drugstore and get the medicine, although the medicine may be paid for by insurance companies.

The chain of buyers can be quite complicated, but generally there are three types of people in this chain of buyers: *purchasers* pay for the products or services; *users* actually use the products or services; and *influencers* play very important roles in the buying decisions. For example, doctors are the influencers in the pharmaceutical market place. These three types of people may overlap, but often they do not.

Though the products or services are used solely by users, the manufacturers usually pay the greatest attention to the most influential player in the buying decision-making process. For example, cloth manufacturers may pay a lot more attention to buyers from big chain stores, such as JCPenney, Marshall Fields, and Wal-Mart, because the bulk of their products will be sold to them. Pharmaceutical companies usually pay most of their attention to medical doctors, because it is doctors who write the prescriptions that lead to the sales of medicines. This convergence of attention on the most influential players in the purchasing decision usually leads to a fixed marketing and product strategy for the whole industry. Soon, every competitor in the industry plays the same game, and the whole marketplace is a red sea.

A path to business innovation is to look across the chain of buyers. Instead of focusing on the most influential player like most competitors, you can focus on other players in the chain of buyers. For example, Internet-based stores sell goods directly to users, instead of to retail chains, or stores. Another example is Novo Nordisk, a Danish insulin producer, discussed in Example 3.4.

Example 3.4 Novo Nordisk   Insulin is used to regulate the level of blood sugar for diabetes patients. Historically, the insulin industry focused its attention on the doctors, who are the key influencers in medicine-buying decisions. Before the 1980s, the insulin industry was competing on "who can produce the purest insulin." By 1980, the technology of producing insulin of the highest purity was developed and now many producers can do this. As a very common scenario, when many competitors have similar technological capabilities, the brutal price competition phase began in and the profit margin for each producer was quickly disappearing.

Novo Nordisk, a Danish insulin producer, who also had this insulin purification technology, thought about another strategy other than this price competition. Novo found that diabetes patents were supplied with insulin in vials. But self-application of these insulin vials involves the complex and unpleasant task of handling syringes, needles, and insulin, and of administering doses according to the individual's needs. Instead of competing on producing high-purity insulin vials to impress doctors, Novo developed a product called NovoPen in 1985, the first user-friendly insulin delivery solution, which was designed to remove the hassle and embarrassment of administering insulin. The NovoPen resembled a fountain pen; it contained an insulin cartridge that allowed the patient to easily carry, in one self-contained unit, roughly a week's worth of insulin. The pen had an integrated click mechanism, making it possible for even blind patients to control the dosing and administer insulin. Patients could take the pen with them and inject insulin with ease and convenience without the embarrassing complexity of syringes and needles.

This product is a hit, not because medical doctors liked it, but because many patients requested NovoPen for its easy and hassle-free application.

After NovoPen, Novo Nordisk introduced NovoLet, which is a prefilled disposable insulin injection pen with a dosing system that provided users with even greater convenience and ease of use. And in 1999 it brought out the Innovo, an integrated electronic memory and cartridge-based delivery system. Innovo was designed to manage the delivery of insulin through built-in memory and to display the dose, the last dose, and the elapsed time—information that is critical for reducing risk and eliminating worries about missing a dose.

In this case, Novo Nordisk shifted its attention from medical doctors to the patients themselves. Novo Nordisk developed a product that fits the customer value of diabetes patients rather than that of medical doctors. This product made Novo Nordisk the dominant player among insulin producers.

3.4.1.4 Looking Across Complementary Product and Service Offerings

Most products or services are not consumed alone—they are consumed with other products or services. In the automobile industry, cars and trucks are the main products, but without auto insurance, auto repairs, auto parts suppliers, and so on, consumers would not be able to use cars and trucks in the long run. In this case, insurance, auto repairs, and parts are complementary products and services for the automobile industry. For most industries, there are many hidden untapped values in complementary products and services. To open up the market boundaries and create blue oceans, we should not only focus on the original industry itself, but also on these complementary products and services.

For many utility industries, such as electrical power generation, producing and selling power generation products, such as turbines, is the main focus. However, there are many competitors in this market, and power-generation equipment usually has very long life cycles, as much as 30 to 40 years. The number of units that each company can sell fluctuates greatly. If equipment producers only count on their revenue from equipment sales, they will be in a very precarious position. However, the demand for complementary products and services, such as parts, upgrades, and maintenance, is very steady and these items can be cash cows. As a result, most equipment manufacturers will not only provide equipment, but will also be very actively involved in providing these complementary products and services.

Many successful enterprises look at both the core products and services, and at complementary products and services to create highly successful business models. Starbucks not only provides coffee, but it also provides other services associated with coffee consumption, such as meeting and chatting places, Internet connections, and nice settings. Compared with traditional bookstores, where only books are sold, Borders and Barnes & Noble redefined the scope of the services they offer. They transformed the product they sell from the book itself, into the pleasure of reading and intellectual exploration, adding lounges, knowledgeable staff, and coffee bars to create an environment that celebrates reading and learning. Both Borders and Barnes & Noble are very successful.

3.4.1.5   Looking Across Functional or Emotional Appeals to Customers

As mentioned earlier in this chapter (Section 3.1), for every product or service, the customer value always has functional and psychological benefits. Usually one of the most important psychological benefits of a product or service is the emotional benefit or emotional appeal of the product. For example, a very successful businessman may like to drive a luxury car. From a functional point of view, a car provides a means of transportation, and to serve this functional purpose, an economy car is sufficient. However, a luxury car is also a status symbol, and the businessman who drives it shows that he and his business are both successful.

In a given industry, some products compete mostly on price and function, and others compete largely on emotional appeal. When a product adopts its unique combination of functional and emotional benefits, it tends to stick with them. The producer of the product gets familiar with the strategy, and often adopts the mentality of "if it's not broken, don't fix it." On the other hand, customers of the product also get used to it and develop a fixed "customer image" for the product. When several products are adopted with similar combinations of customer value components, you get the "red sea" type of competition. To break out this market boundary, you can think about redefining the combination of functional and emotional benefits to create a unique product and a blue ocean market.

For example, in the airline industry, almost all airlines adopted similar customer value strategies and competed on the same things, such as getting high-end business travelers. This meant that many airlines built business lounges, supplied good meals, and so on. This approach offers a fair combination of functional and emotional benefits. However, Southwest Airlines adopted a very different combination of customer values: it reduced a lot of the emotional benefits, reducing the quality of meals, not offering luxury lounges, and so on. But Southwest drastically increased the convenience of boarding airplanes and offered frequent and flexible departures, and it offered this more functional service with much lower cost. The bulk of customers are willing to take a very basic, but reliable and safe airline service, with low cost. They don't care much about in-flight meals, business lounges, and so on. Southwest achieved a lot of success by adopting a low-cost, purely functional customer value.

There are also many success stories that work the other way around, that is, adding more emotional benefits on top of functional benefits. For example, regular coffee shops only provide coffee, and if you adopted a function-based customer value combination, you might just try to improve the quality of the drink by using high-grade coffee beans, making fancy coffee drinks, and selling them at a higher price. Starbucks Coffee's strategy can be seen as a fair combination of functional benefits and emotional benefits—it provides reasonable quality coffee drinks, but on top of that, Starbucks provides a social setting, a place to take a break or to have brief business meetings, and so on. It adds a lot of emotional benefits.

3.4.1.6 Looking Across Time All industries are subject to external trends that will profoundly affect them over time. These trends include technological developments, life-style evolution, climate change, global economic changes, and so on. For example, the rapid development of computational technologies, data storage technologies, and microelectronics will greatly affect the way people communicate and interact with each other. When technology or lifestyle undergoes a "quantum leap," it greatly affects customers' attitudes and value systems.

Many companies only pay attention to the needs of the marketplace in the very short term, and fail to pay attention to these trends. Companies that pay attention to these trends and catch them ahead of time may be able to develop breakthrough products much earlier than their competitors and capture a huge share of the market. One such example is the development of the Toyota Prius hybrid automobile. One fundamental trend in the world is that nonrenewable resources will get scarcer and scarcer. When Toyota put in a lot of effort to develop the Prius in the late 1990s, petroleum was still quite cheap. If Toyota had just paid attention to the short-term market, it would not have made sense to develop this high fuel-efficiency, but still immature technology. However, the drastic oil price hikes in the early 2000s made fuel efficiency one of the most important factors in car-purchase decision making. The reputation of Toyota's fuel efficiency and of the Prius, in particular, greatly increased Toyota's market share worldwide.

3.5 Customer Value and the Voice of the Customer

So far in this chapter, we have thoroughly considered the concept of customer value. Customer value consists of key factors that determine how well customers will appreciate a given product or service. Investigating and analyzing customer value and planning with it in mind will help you determine the strategic positions for products or services in the

marketplace. In a broader product or service development process, however, more needs to be done.

Product development is an information mining, transformation, and
creation process, as illustrated in Figure 3.9. For a customer-centric product development, the first step is to mine the information about what
customers really want. This information is the "customer attributes." Even
for a product based on new technology, knowing what customers like is
still very important. Mining the customer attributes is often referred to
as capturing the voice of the customer. As you can see in Figure 3.9, complete information about what customers really want is the foundation for
a good product development process because it is both the first step and
the source of all information regarding product development.

The first "mapping" in the product development process is to transform customer attributes into functional requirements for the product
or service. Functional requirements are a minimum set of independent requirements that completely characterize the functional needs
of customers with regard to the product or service that they purchase.
Specifically, products or services are designed and produced to deliver
functions, and customers purchase or use products for their functions.
The functional requirements define those functions.

For example, customers purchase and use pens to "make marks" on
paper, so "making marks" is one of the main functions of a pen. For
complex products, such as an automobile, the functions that it delivers
are usually more complex and hierarchical. Automobiles can deliver
many functions such as

- Moving people from A to B

- Providing a nice riding environment

- Providing a nice image for passengers

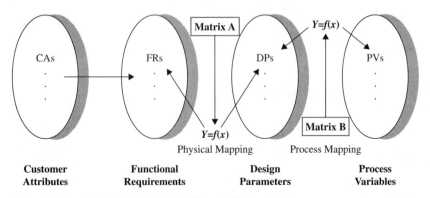

Figure 3.9 Product Development as a Mapping Process

For each of these main functions, such as "moving people from A to B," there are many details, such as "move with varying speed," "change directions," "stop," and "park"—these are the second-level functions. And for each second-level function, such as "move with varying speed," there are many third-level functions, and so on.

The second mapping in the product development process is the transformation from functional requirements to design parameters. Design parameters are the elements of the design solution in the physical domain that are chosen to satisfy the specified functional requirements. For example, the power-train system in an automobile is the design parameter that provides power to move the car around. Clearly, design parameters can be very complex and hierarchical, taking the form of systems, subsystems, components, and parts.

The last mapping is the transformation from design parameters to process variables. Process variables characterize the manufacturing process that satisfies the design parameters. Essentially, this transformation involves figuring out how to design the manufacturing process and equipment to make the physical products. The process variables are usually complex and hierarchical as well.

From Figure 3.9, you can see that the nature of product development is information mining, creation, and transformation. If we do not capture the VOC data accurately in the first stage of product development, we may develop products that customers do not like. If we do not capture sufficient amounts of VOC data, we will not be able to link every product requirement to what customers really like when we develop functional requirements, and we would have to fill in those customer requirements arbitrarily, or by guesswork. This could lead us to develop products with the wrong performance specifications.

The transformation from functional requirements to design parameters also depends on the accuracy and completeness of the VOC data we capture at the beginning of the product development process; if we derive the wrong functional requirements, we will end up transforming these wrong requirements into wrong product design parameters, and we will develop the wrong product. On the other hand, the transformation from functional requirements to design parameters requires a good understanding of technology, science, engineering, and economics, and it also requires knowledgeable product development team members, as well as a good product development process. Similarly, the transformation from design parameters to process variables also depends on the accuracy and completeness of the VOC data captured, as well as on knowledge and a good product development system.

We can conclude the following:

■ **Importance of VOC data**   Clearly, customer value defines the strategic position of the product or service in the marketplace.

Information on the customer value of our product can only come from the voice of the customer. In detailed product development, the voice of the customer is also the source of information on the product development process; we need to get enough information so that the three mappings in the product development process can be carried out flawlessly. Therefore, the VOC is the source of information for both the strategically important customer value proposition, and all the building blocks in the product design.

- **Accuracy of VOC data**   Because of the importance of the voice of the customer, we need to capture VOC data accurately. Only with accurate VOC data can we develop an accurate customer value proposition and product performance specification. There are many methods for capturing the voice of the customer, but arbitrary use of these methods will not lead to capturing the voice of the customer accurately. We will discuss how to accurately capture the voice of the customer in Section 3.6.

- **Completeness of VOC data capturing**   Not only do we need to accurately capture the voice of the customer, but we also need to capture a sufficient amount of VOC information to define customer value and benchmark our customer value position with our key competitors. Sometimes, we need to collect enough VOC information to figure out ways to redesign our customer value position to create breakthrough new products. The voice of the customer is also the source of information for product development, and we need to capture sufficient VOC data to drive the product development process. Specifically, we need to get sufficient VOC information to drive the transformations from customer attributes to functional requirements, from functional requirements to design parameters, and from design parameters to process variables.

3.6   Capturing the Voice of the Customer

Quality can be defined as meeting customer needs reliably and providing superior value, and meeting customer needs requires that those needs be understood. "The voice of the customer" is the term used to describe the stated and unstated customer needs or requirements. The voice of the customer can be captured in a variety of ways, including customer interviews, customer surveys, focus groups, ethnographic studies, and so on.

The purposes of capturing VOC data are usually the following:

1. Product development
2. Market and product planning

3. Customer feedback and product improvement

4. Customer-centric innovation and market opportunity search

For different purposes, the objectives of capturing VOC data will be different, as will the strategy and methods used. In general, though, the objective of VOC capturing is to capture all relevant customer attributes so we will have enough information to create detailed functional requirements, design parameter specifications, and manufacturing process variables for all details.

3.6.1 Plan for Capturing the Voice of the Customer

Once our customer value proposition and product plan are established, the next step is to plan how to capture the customer's needs. This includes determining who our target customers are, which customers to contact in order to capture their needs, who will do VOC data collection, what methods to use to collect the data, and scheduling and estimating resources required to capture the VOC data.

3.6.1.1 Who Are the Target Customers   There is no single voice of the customer. Even for the same product, there are diverse voices from customers about what they think the product should be like. For example, in the automobile market, some customers would prefer larger sizes and more power, while some customers would prefer low gas mileage, low cost, easy maintenance, and so on. We usually can find that there are several major segments in this market, and within a market segment, the customers' opinion are similar; however, there are significant differences among different market segments. In this case, we may want to capture the voice of the customer from all market segments and develop a portfolio of products to satisfy different needs. In order to do this, we should target a representative set of customers from each market segment to get complete information on customers' needs.

For many products, there might be multiple voices from each buying unit. When buying a car, a husband and wife may have different opinions. When making a purchasing decision about commercial equipment in an organization, the voice of the direct user, the voice of the purchasing department, and the voice of support personnel might be very different. In this case, all kinds of voices should be considered and the corresponding product design should take into account all these voices.

For a given product, we also need to capture the voice of the following types of customers:

1. Current customers. These are customers who are buying our product; we will try to know what they want in order to keep them.

2. Competitors' customers. These are customers who need this kind of product, but they are not buying our product, so we will try to determine what they want in order to improve our product to capture more market share.

3. Potential customers. These customers are not buying our product or competitors' products. They are not customers of our industry; we will try to know why they are not and integrate their VOC data into our product design, and hopefully they can be our customers in the future.

4. Lead customers. These are either our current customers or competitors' customers. Lead customers are those customers who are the most advanced users of the product, customers who are pushing the product to its limits, or customers who are adapting an existing product to new uses. The voice of lead customers is important to catch the future market trend and develop a new generation of product.

How many customers should we talk to? The number depends on the complexity of the product, diversity of the market, product use, and the sophistication of customers. The goal is to get to the 90–95% level in capturing customer needs. Research for a range of products indicates that, on average, this is 20 customers.

3.6.1.2   Who Should Collect VOC Data   Traditionally, marketing departments have had responsibility for defining customer needs and product requirements, but this approach has many flaws. First of all, design is a mapping process—the voice of the customer will have to be transformed into functional requirements, and usually it is the engineering people who actually perform this transformation. If engineers are out of the loop, and marketing people do all the work in capturing the voice of the customer and compiling market study reports, and then throw it all over the wall to the engineers, severe distortions will happen as marketing language is translated into engineering language. The accuracy of the voice of the customer will be severely compromised.

Second, marketing people are not engineers; they don't do engineering design. It is up to engineers to define all the necessary functional requirements, and to derive design parameters in fine detail. If the marketing people have not captured sufficient amounts of the right VOC data, the engineers will not be able to derive accurate functional requirements for the product. For example, in a power saw design project, the marketing people might tell design engineers that customers would like a "wider handle" for the saw. The design engineers might then be puzzled by why a wider handle is needed, and how wide is wide enough.

Therefore, product development personnel need to be directly involved in researching customer needs. This may involve visiting or meeting with customers, observing customers using or maintaining products, participating in focus groups, or rotating development personnel through marketing, sales, or customer support functions. This direct involvement provides them with a better understanding of customer needs, the customer environment, and product use; it develops greater empathy on the part of product development personnel, minimizes hidden knowledge, overcomes technical arrogance, and provides a better perspective for development decisions. These practices have resulted in fundamental insights, such as engineers of highly technical products recognizing the importance to customers of ease of use and durability as opposed to the latest technology.

3.6.1.3 What Kind of VOC Data Should Be Collected? As we saw earlier, the purpose of collecting VOC data is to get sufficient and accurate information to finish the product development process. Most importantly, we need to get accurate and sufficient data to derive complete functional requirements for the product from VOC data. This clear purpose will determine what kind of VOC data should be collected.

According to Ulwick (2005), the following four kinds of VOC data are not suitable for product development purposes: solutions, specifications, needs, and benefits. However, three kinds of VOC data are good choices: jobs to be done, desired outcomes, and constraints.

- **Solutions** Many customers offer their own solutions for a particular product design when they are asked what they want in a product. For example, in a customer survey study about laptop computers, many customers said that they wanted the laptop computer to have a larger screen. However, in the product development process, technical solutions are actually part of the design parameters, which are usually derived in a later phase of the product development process. Most customers are not engineers, technical specialists, or scientists, and they do not always have the best solutions; their solutions may also interfere with other product requirements. In the laptop computer example, one of the manufacturers actually adopted the customer's solution and produced a laptop with larger screens, but they found that very few customers wanted to buy this kind of laptop, because it was too big to carry around. Later the computer manufacturer found that customers actually want a clear image on the screen, which can be produced with a screen the same size or smaller, but with higher resolution. This example emphasizes that the first mapping in the product development process is to transform the voice of the customer into functional requirements, which should be stated in a solution-neutral environment. Functional requirements should not be tied to any particular solutions. Once all the relevant functional requirements are defined, we can determine the best set of technical solutions from our design process. Customer-provided solutions simply confuse the process.

■ **Specifications**   Customers often give their own opinions on product specifications, such as size, weight, color, shape, and so on. They may want something like a "wider handle," "sleeker look," or "lighter weight." Again, specifications are part of the design parameters, and design parameters should be derived after functional requirements. Customers are not engineers, and accepting their specifications may limit other design solutions. For example, when customers want wider handles, they may actually want a better grip, which could be accomplished by many technical means. If we adopt a "wider handle," we may design a tool that is too big for many customers.

■ **Needs**   Needs are often expressed as very vague, high-level descriptions of the overall quality of products or services. They are typically stated as adjectives and do not imply a specific benefit, and they give no information about "how" and "how much." For example, customers often say that they want a product to be more reliable, dependable, robust, or effective. These needs statements give some information about what customers might care for, but they are very imprecise statements; they do not provide enough information to derive any meaningful functional requirements. For example, if customers say that they want to have "more reliable" products, how reliable is reliable enough? The specific reliability requirement will still be left to guesswork by design engineers.

■ **Benefits**   Customers often use benefit statements to describe what values they would like a new product or service to deliver. They often use words like "easy to use," "faster," or "better." Again, these statements are imprecise and they do not provide sufficient information to derive functional requirements.

Clearly, solutions, specifications, needs, and benefits do not provide sufficient information to derive functional requirements. Now let's look at the three types of VOC data that Ulwick recommended be collected.

■ **Jobs to be done**   The jobs to be done are actually the functions that a product or service is supposed to deliver. For example, the jobs to be done for a razor include: "remove hair," "prevent bleeding," "prevent dry skin," "cause little or no pain," and so on. The jobs to be done for an energy drink may include "quench thirst," "help people focus," and "recover from exercise." Clearly if we can collect a complete list of jobs to be done by a product, it will be easy to derive functional requirements. At least we will know what functions a product should deliver.

- **Desired outcomes**   Desired outcomes are specific descriptions of the end result for each job to be performed. In a power saw product development project by Bosch (Ulwick 2005, 26–145), Bosch collected many customers' statements about their desired outcomes for the power saw:

  - Minimize the amount of kick that occurs when starting the saw
  - Increase the likelihood that the blade will begin cutting precisely on the cut line
  - Minimize the amount of time the cut line is blocked from view when making a cut
  - Minimize the amount of pressure that must be exerted to keep the saw flat on the cutting surface
  - Minimize the likelihood that the blade guard will snag the material
  - Minimize the likelihood of the cut going off track when approaching the end of the material/cut
  - Minimize the frequency with which the cord gets in the way of the cut path
  - Minimize the frequency with which the cord (end plug) gets caught on the material
  - Minimize the frequency with which the blade binds
  - Increase the number of cuts that can be made with a single blade
  - Minimize the amount of damage a worn or dirty table inflicts on the material when a cut is being made
  - Minimize the amount of dust/debris that is generated by the saw

  These are just a few of the 80 or so outcomes they captured from customers. For most jobs, even those that may seem somewhat trivial, there are typically 50 to 150 or more desired outcomes, not just a handful. Companies must capture information about *all* the desired outcomes in order to have sufficient information to derive correct functional requirements and subsequent design parameters.

- **Constraints**   Constraints are physical, regulatory, and environmental "road blocks" that prevent the product from doing the job. Identifying the constraints often provides potential chances for developing breakthrough products. For example, one constraint for diabetes patients to the use of home test kits is that the patients often have shaky hands and blurred vision, and the current device needs precise hand positioning. Obviously, this constraint prevents home testing from being done successfully. This constraint, however, provided an unique opportunity for improvement—Roche developed a new device that does not need precise hand positioning, and this design improvement enabled Roche's product to increase its market share from 20% to 45% (Ulwick 2005).

As you can see, VOC data in the form of jobs to be done, desirable outcomes, and constraints will provide excellent information for deriving functional requirements. They are a good type of VOC data.

3.6.1.4 Organize VOC Data   Once data about customer needs has been gathered, it must be organized. The mass of interview notes, requirements documents, market research, and customer data needs to be distilled into a handful of statements that express key customer needs. Affinity diagramming is a useful tool to assist with this effort. Brief statements that capture key customer needs are transcribed onto cards, and a data dictionary that describes these statements of need is prepared to avoid any misinterpretation. These cards are organized into logical groupings or related needs. This makes it easier to identify any redundancy and serves as a basis for organizing the customer needs.

In addition to stated or spoken customer needs, unstated or unspoken needs or opportunities should be identified. Needs that are assumed by customers and therefore are not verbalized can be identified by preparing a function tree. Excitement opportunities (new capabilities or unspoken needs that will cause customer excitement) are identified through the voice of the engineer, marketing, or customer support representative. These can also be identified by observing customers use or maintain products and recognizing opportunities for improvement.

4

Customer Survey Design, Administration, and Analysis

One of the most important success factors in designing superior service products is to know your customers. For a world-class company, the design decision should be driven by the voice of the customer. Obtaining detailed customer data is even more important for the survival of a service institution because maintaining customers' loyalty is extremely vital. Customer surveys are essential tools that can yield information about customer expectations, customer satisfaction, and strategies for improvement.

Commercial enterprises use customer survey findings to formulate marketing strategies and design their products. Television and radio programs are evaluated and scheduled largely based on the results of consumer surveys. Libraries, restaurants, financial institutions, and recreational facilities use customer surveys to gather information about customers and their desires.

Customer surveys use a sample of customers to learn about the whole population of a customer base. In a customer survey, information is collected from the sample and analyzed. The sample may contain as few as 30 people (a small sample) to as many as 30,000 or more people (a very large sample). Survey methods are based on sound statistical principles, and over 70 years of modern survey practices show that surveys based on a relatively small sample are usually quite accurate.

4.1 Customer Survey Types

There are three customer survey methods: mail-out, in-person, and telephone.

4.1.1 Mail-Out Surveys

The mail-out survey involves mailing printed questionnaires to a sample of predesignated potential respondents. Respondents are asked to complete the questionnaire and mail it back to the survey researcher.

The advantages of the mail-out survey include the following:

- **Low cost**   Other survey techniques require trained interviewers, and may incur high travel costs.

- **Convenience**   The questionnaire can be completed at the respondent's convenience.

- **Privacy**   Because there is no personal contact in mail-out surveys, respondents may feel more comfortable, and their privacy is preserved.

- **Lack of time pressure**   Respondents can take their time to complete the questionnaire and consult their personal records if necessary.

The disadvantages of mail-out questionnaires include the following:

- **Lower response rate than other methods**   The mail-out survey response rate usually ranges from 20–30 percent in the worst case to 85–90 percent in the best case. Therefore, adequate numbers of questionnaires need to be mailed out, and many follow-ups are generally needed to achieve the desired sample size.

- **Comparatively slow**   It usually takes a few weeks for questionnaires to be returned for a mail-out survey.

- **Self-selection**   The survey researcher only receives the returned questionnaires, and the people who choose to fill out and return the questionnaires may not adequately represent the population. For example, people who have low reading and writing proficiency may never return their questionnaires, and thus will not be represented in the sample.

4.1.2 In-Person Interviews

In-person interviews involve an interviewer directly talking to respondents to get the information.

The advantages of in-person interviews include the following:

- **High response rate**   The response rate of in-person interviews is much higher than that of mail-out surveys.

- **Ability to contact hard-to-reach populations**   Certain groups are difficult to reach by mail or telephone, and in-person interviews are the only way to reach them.

- **Greater complexity**   Because of the direct interaction between interviewer and respondents, complex questions can be asked, and the questions can be explained.

- **Comparatively quick**   It takes less time to finish the survey compared to mail-out surveys.

In-person interviews have the following disadvantages:

- **High cost**   In-person interviews can be very costly due to personnel requirements, travel expenses, interviewer training, and so on.

- **Interviewer bias**   Interviewers may subconsciously introduce personal bias and affect respondents' choices in answering the questions.

- **Greater stress**   The interviewing process is usually stressful for both the interviewer and respondent.

4.1.3 Telephone Surveys

Telephone surveys collect information through telephone interviews between trained interviewers and respondents.
Telephone surveys have the following advantages:

- **Rapid response**   Telephone interviews usually get results more quickly than in-person interviews and mail-out surveys.

- **Lower cost**   The cost of a telephone survey is usually much less than using in-person interviews, and can be less costly than mail-out surveys.

- **Privacy**   Respondents' privacy is better preserved than when using in-person interviews.

Telephone surveys also have the following disadvantages:

- **Less control**   The interviewer has less control over the interview process than when using in-person interviews—the respondent can hang up the phone easily.

- **Greater stress**   Receiving a stranger's call may be very annoying for many respondents.

- **Lack of visual material**   Telephone interviews cannot use visual aids, such as maps, charts or pictures to explain the survey questions.

4.1.4 Other Methods of Gathering Information

Besides customer surveys, there are several methods that can be used to gather customer information, such as secondary customer data research, direct data measurement, and direct product usage involvement.

■ **Secondary customer data research**   The customer information may already be available somewhere, such as in libraries, from government agencies, or on the Internet, and secondary research is the process of retrieving this information. By using data mining techniques (Berry and Linoff 2000, Edelstein 1999), researchers can extract important information from these potentially huge data sources and produce valuable clues that can guide sales and promotion efforts. Data mining is primarily used by companies with a strong customer focus, such as retail, financial, communication, and marketing organizations. It enables these companies to determine relationships among internal factors such as price, product positioning, or staff skills, and external factors such as economic indicators, competition, and customer demographics. Data mining enables companies to determine the impact of these factors on sales, customer satisfaction, and corporate profitability, and to develop marketing and sales strategies to enhance corporate performance and cut losses.

■ **Direct data measurement**   This technique involves directly counting, testing, or measuring data without interviewing customers. Testing cholesterol levels, monitoring customers' arrival time and time spent in a service institution, and recording and counting the type and number of errors in insurance claims are typical examples of direct measurement.

■ **Direct product usage involvement**   The key idea in this technique is to ask the product design leaders to play the role of the consumer and practice using the product. This technique is practiced by Toyota. One story (Liker 2004) described a Japanese design leader who had never been to the U.S., but was assigned to design a car for the North American market. To overcome his lack of knowledge of the North American market, he landed in the U.S., rented a car, and drove through all the continental United States and the Canadian provinces to experience the actual driving and car usage conditions in North America. As a result of this first-hand experience, he made some very good changes in the car design.

4.2   Stages of the Customer Survey

Customer survey research is well-established, and a step-by-step procedure is available to guide the whole survey process. These are the stages of a typical customer survey:

Stage 1: Establish goals and objectives of the survey.
Stage 2: Set the survey schedule and budget.
Stage 3: Establish an information base.
Stage 4: Determine population and sampling frame.

Stage 5: Determine sample size and selection procedure.

Stage 6: Design the survey instrument.

Stage 7: Pretest the survey instrument.

Stage 8: Select and train survey interviewers.

Stage 9: Implement the survey.

Stage 10: Analyze the data and report.

The following sections give a brief overview of each stage.

4.2.1   Stage 1: Establish Goals and Objectives of the Survey

Determining the goals and objectives of the survey is always the first stage in a customer survey study. The company that uses the survey information has to do this itself, even if the actual survey job will be contracted out, because only the company itself can define its own needs.

The following aspects are essential in establishing the goals and objectives of your survey.

1. Determine survey objectives. This is the most critical part in customer survey design. The survey objectives will determine what kind of questions will be in the survey. For example, if the objective in the customer survey is to gain the customer requirements for a product, as I described in Chapter 2, then you should design a survey that provides appropriate questions to extract accurate and sufficient information to derive product functional requirements and design parameters. If the objective in the customer survey is to study the key factors in customer value proposition, as I described in Chapter 3, then you need to design the questions to extract the customer rating on these factors. The survey objectives should be spelled out clearly and very specifically.

2. Identify the users of the survey results. The users of the survey results are the customers of the survey study. For example, in the product development process, design engineers are often the users of the survey results. It is very common that the people who do the survey study and the people who use the survey results are from different backgrounds, speak different buzzwords, and having different mindsets. For example, if the survey study is to be conducted by the marketing department, but the users are engineers, many times their interpretations of the same word, such as "operation," are quite different. Many true meanings in the original customer statements will be "lost in translation" in this case. In some big corporations, the VOC information will be transmitted through a hierarchy of different departments, and when it reaches the final user, the VOC information has become so distorted that it becomes useless. So, it

is necessary to identify all significant survey results users and let them participate in all stages of customer survey study.

3. Identify what specific information is needed. After identifying the users of the survey, you need to determine what specific information the users really need. For example, in the product development process, you need to consult with design engineers, maybe even manufacturing engineers, to find out what kind of information they need to determine the design specifications and product requirements. In this case, good cooperation between the marketing and engineering departments is really essential; the marketing department may know the marketing trends and how to deal with customers, and the engineers know what information, how much information, and the level of detail in the information they need in order to do their design and production job.

Another example is the restaurant business. The professional marketing people know how to deal with customers; however, if the customers don't like the taste of the food, the marketing people will not know how to fix it. If the marketing people only tell the chef that "your food tastes bad," it won't help very much. So the chef and the marketing person should work together to figure out ways to get the right information so the chef can produce the food that customers really want.

4.2.2   Stage 2: Set the Survey Schedule and Budget

After the goals and objectives of the customer survey are determined, you need to develop a budget and timeline for this survey project.

If the survey study will be contracted out, then a major cost item in the budget will include the contract cost. If the survey study will be done by internal resources, then manpower from several departments will be needed, and some trade-offs between manpower needs in this project and other routine jobs may need to be made.

The timeline of the survey project will depend on when you will need this information. In product development, the availability of the survey information should fit the product development schedule.

The timetable should be flexible enough to accommodate unforeseen delays.

4.2.3   Stage 3: Establish an Information Base

Before developing the survey questionnaire, you need to collect detailed information to populate the survey questions. This should be a joint effort by the following people:

- People who will design and administer the survey

■ Survey information users

■ Representatives of customers

A focus group meeting is a popular approach to bring all these people together in order to reach consensus about the information needs in the survey. The size of a typical focus group ranges from 5 to 12. In this focus group meeting, based on the goals and objectives of the survey established in Stage 1, an exhaustive list of "raw" questions will be developed and they will serve as a basis to develop a survey instrument.

4.2.4   Stage 4: Determine the Population and Sampling Frame

The population is the entire set of people, organizations, households, etc., that are addressed by your survey research. For example, for a fast-food chain, the relevant population will be "fast-food eaters." For a suburban hospital, the population will be nearby residents.

The portion of the population that can be identified for interviews is called the sampling frame. For example, the population of fast-food eaters may include all people except homeless and sick people. But if a telephone interview is to be conducted, only the people with known telephone numbers can be reached, so "people with telephone numbers" will be the sampling frame for fast-food eaters. The concepts of population and sampling frame will be further discussed in Section 4.4.

4.2.5   Stage 5: Determine Sample Size and Selection Procedure

The survey researcher will have to select a sample that adequately represents the population under study. In general, larger samples will yield greater accuracy than small samples in terms of analysis results. The sample size is usually determined by balancing the analysis accuracy against the increased cost and time required for surveying a larger sample size. Once the sample size is determined, the method of sampling will be chosen. Commonly used sampling methods include random sampling, stratified random sampling, and cluster sampling. How sample sizes and sampling methods are chosen will be discussed in Section 4.5.

4.2.6   Stage 6: Design the Survey Instrument

The survey instrument, or questionnaire, is a key tool in the customer survey process. At this stage, the survey researcher must populate the right questions in order to obtain the information needs in the survey.

It is recommended that the following people should participate again in the development of survey questions:

- People who will design and administer the survey
- Survey information users
- Representatives of customers

The development of the questionnaire can be a time-consuming process, requiring great attention to detail. The best-designed questionnaire is short and concise with well-worded questions. Long and wordy questionnaires often result in lower response rates and higher survey costs. The design of the survey instrument is discussed in Section 4.3.

4.2.7   Stage 7: Pretest the Survey Instrument

Once a draft questionnaire has been created, it must be pretested with a small group of respondents. During this pretest, poorly worded questions can be identified and refined, improving the quality of the questionnaire.

4.2.8   Stage 8: Select and Train Survey Interviewers

For telephone and in-person interviews, trained interviewers are required. Interviewers can be college students, part-time workers, and so on. Interviewers should be familiar with the questionnaire and know how to handle uncooperative respondents.

4.2.9   Stage 9: Implement the Survey

The process of administering the survey instrument is a crucial stage of a customer survey. It is very important to follow the sampling procedure to ensure the validity of the survey. It is also very important to stick to the time schedule. Care must also be taken to ensure the privacy of the respondents and minimize their inconvenience. The implementation of surveys is further discussed in Section 4.4.

4.2.10   Stage 10: Analyze the Data and Report

The data from returned questionnaires will be summarized and analyzed with statistical methods, and the findings will be reported. Section 4.5 discusses the details of data analysis and report.

4.3 Survey Instrument Design

One of the key tasks in surveying customers is developing the questionnaire. The key concerns in questionnaire development are the types of questions, the wording of questions, the sequence of questions, and the overall length of the questionnaire.

There are two types of questions that can be used in customer survey: close-ended questions and open-ended questions. Close-ended questions allow respondents to select an answer from a list of choices, and open-ended questions allow respondents to answer using their own words.

4.3.1 Close-Ended Questions

Close-ended questions offer a fixed list of responses and ask respondents to select one or more options to best answer the question. Close-ended questions have many different formats.

4.3.1.1 Multiple-Choice Questions Multiple-choice questions have a list of answers and usually ask for a factual response. In multiple-choice questions, only one answer is supposed to be applicable. Here is an example of a multiple-choice question:

What department do you work in?

a. Sales

b. Marketing

c. Manufacturing

d. Research

When you construct this kind of question, you must be careful to make sure the list of answers is exhaustive and that the answers are mutually exclusive.

In the preceding example, if someone were working in the personnel department, he or she would not find an appropriate answer. In such a case, the question has an inadequate number of answers. However, the total number of answers in a multiple-choice question should be no more than 10 or 12.

4.3.1.2 Checklist or Inventory Questions Checklist or inventory questions ask respondents which items on a list are applicable. Here is an example of a checklist question:

Please indicate what sources you use to get information about new music and movies. Check all that apply.

___Radio	___Television	___Internet	_____Newspapers
___Friends	___Magazine	___Other (Please specify)	

Clearly, in this example, it is possible that more than one answer may apply.

4.3.1.3 Rating Questions

Rating questions ask respondents to use a given scale to judge something. Here is an example of a rating question:

Please rate your teacher's performance in the following categories on a scale of 1 to 5, where 1 is "very poor," 3 is "average," and 5 is "excellent."

Rating	Category
_____	Knowledge of course contents
_____	Instruction
_____	Office hours

4.3.1.4 The Pros and Cons of Close-Ended Questions

Close-ended questions have several advantages:

- The set of answers is uniform, so it is easy to compare the differences among respondents.

- The uniformity in the set of answers for each question will make computer data entry easier.

- The fixed list of answers tends to make the question clearer to the respondent.

Close-ended questions have the following disadvantages:

- Close-ended questions will compel respondents to choose the answer that most closely represents their actual response, which may deviate from their true opinions.

- When respondents are unsure which is the best answer, they may choose a random answer, which will lead to errors.

4.3.1.5 Level of Measurements in Close-Ended Questions

In analyzing the results of a survey, the survey data will be organized as variables. A variable is a specific characteristic of the population, such as age, sex, preference, rating, and so on. Depending on the design of the questions and answers, the variables used in the survey will have different measurement properties, referred to as levels of measurement or measurement scales. The commonly used measurement scales include the nominal scale, the ordinal scale, and the interval scale. In survey design, there is also a specially designed interval scale called the Likert scale. We will now look at these scales in detail.

Nominal Scale   The nominal level of measurement simply divides survey answers into categories. For example, a variable such as political party preference in the United States can be categorized using three classes: Democrat, Republican, or Independent. When using the nominal scale, survey data is placed into categories and the frequency of occurrences is counted. There is no ranking or ordering for the categories.

Ordinal Scale The ordinal level of measurement goes one step beyond the nominal scale—it ranks the categories by certain criteria. For example, people's education levels can be classified using the following categories: high school graduate or lower, two-year college degree, B.A./B.S. degree, M.A./M.S. degree, and Ph.D. degree. Clearly, we can rank these education levels: Ph.D. is higher than M.A., for example. But it is difficult to define a numerical difference between these educational achievements.

The Interval Scale The interval level of measurement gives the greatest amount of information about the variables. It labels and orders them and also uses numerical measurements to indicate the exact value of each category. For example, variables such as income, age, and weight use interval scales.

Likert Scale The Likert scale (named after Rensis Likert) is used to measure attitudes and opinions. A Likert scale may contain several options such as strongly agree, agree, neutral, disagree, and strongly disagree. Here is an example:

Firefox is easier to use than Internet Explorer:

1. *Strongly disagree*
2. *Disagree*
3. *Neutral*
4. *Agree*
5. *Strongly agree*

Sometimes, a numerical scale is explicitly displayed in the questionnaire:

What is your general impression of how the city government affects your business?

Highly Negative				*Highly Positive*
1	*2*	*3*	*4*	*5*
___	___	___	___	___

Likert items are ordinally scaled. It is not assumed that the difference between "strongly agree" and "agree" is the same size as the difference between "agree" and "neutral." However, in survey data analysis, it is a general practice to treat the Likert scale as an interval scale. For example, in college course evaluations, there are many Likert scale questions about a professor's teaching abilities, and the scores for each question from all students are averaged as the evaluation score. Clearly this treatment assumes that the Likert scale is an interval scale.

4.3.2   Open-Ended Questions

Open-ended questions allow respondents to provide longer, more complex answers than closed-ended questions do. Here are some examples:

*What is your favorite place to go for your summer vacation?*_____
How long have you and your family lived in your current location?
_____

What is the first foreign language you learned? _____

There are several disadvantages of open-ended questions:

- They will elicit some irrelevant answers.
- The number of distinct answers from the respondents may be very high and answers could be messy and confusing.
- They are difficult to analyze with statistical methods, because statistical methods require some degree of data standardization.

Overall, it is highly recommended that most survey questions be in the form of close-ended questions to ensure higher response rates, shorter questionnaire completion times, and easier data analysis.

4.3.3   The Wording of Survey Questions

The wording of survey questions is very important for a successful customer survey. Good survey questions should be:

- Clear and easily understandable, stated directly and straightforwardly
- Specific and precisely stated so that the respondent knows exactly what is being asked
- Unambiguous and unequivocal so there is only one way to understand or interpret what the question is asking
- Simple and brief, rather than complicated, cluttered, and long-winded
- Stated in terms that your respondents are likely to be familiar and comfortable with, without using complex technical terminology, jargon, or overly sophisticated wording

Good survey questions should *not*

- Lead the respondent toward a specific answer or make some answers clearly unattractive or undesirable
- Ask about two or more things together in the same question
- Make the respondent uncomfortable, or put the respondent in a difficult or compromising position

Example 4.1 Multipurpose Question   Here is an example of a multipurpose question:

Do you believe the development of the I-696 freeway entrance will affect the image and property value for our whole subdivision?

     Yes_____   *No_____*

This question might be difficult to answer, because a Yes or No answer suggests that the image and property value will move in the same direction. However, it is possible that the image of the community will go up because of the better freeway access, but the property value will go down due to increased traffic, noise, and commercial development.

Example 4.2 Ambiguous Question   If a survey question about income is worded as "What is your income?" it will generate all kind of answers, such as annual income, hourly pay rate, monthly income, or total household income. A better wording for the question would be as follows:

What is your total annual income before taxes?

a. Below $20,000
b. Between $20,001 and $40,000
c. Between $40,001 and $60,000
d. Between $60,001 and $80,000
e. Between $80,001 and $100,000
f. Over $100,000

4.3.4 Order of Questions in Surveys

The order in which the questions are presented can affect the overall customer survey significantly. A poorly organized questionnaire can confuse respondents, bias their responses, and jeopardize the quality of the survey.

It is a good idea to start the survey with some easy introductory questions. Here are some examples:

Do you own a car? Yes____   No____
Do you have an e-mail address? Yes____   No____
How long have you lived at your current address? _____(years)

You should save more complicated questions that may require some careful thought for later, after you warm up the respondents with introductory questions.

There are several typical organizational patterns for survey questions:

- **Chronological order**   The questions are presented in sequential or temporal order, for instance, from most recent to least recent.
- **Funnel pattern**   The questions are ordered by topic from broad to specific.
- **Inverted funnel**   The questions are ordered by topic from specific to general.
- **Tree pattern**   The questions branch out in different directions, depending on the respondent's answers to earlier questions.

The choice of organizational pattern for the questions should serve the goals and objectives of the customer survey. Usually, topically related questions should be grouped together.

Example 4.3 Grouping Survey Questions   The following is an example of good grouping of questions:

1. *How would you describe the current relationship between labor and management?*

    *Good*_____   *Fair*_____   *Poor*_____

2. *During the past five years, do you think this labor-management relationship has:*

    *Improved*____ *Remained the same*_____   *Worsened*_____

3. *In what way do you think the labor-management relationship can be improved?*

    _____

The preceding three questions are related and are presented in a logical order. The following three questions are confusing when grouped together:

1. *Do you or your coworker participate in a company-sponsored suggestion program?*

    *Yes*_____   *No*_____

2. *During the last five years, do you think this labor-management relationship has:*

    *Improved*____ *Remained the same*_____   *Worsened*_____

3. Would you be interested in a job training program?

   *Yes*_____   *No*_____

The respondents will be able to answer the questions, but they will get disoriented after answering 10 or more of these misplaced questions.

4.3.5   Questionnaire Length

The questionnaire should be as concise as possible while covering the subjects determined by the goals and objectives of the survey. Practice has shown that when the survey becomes too long, the response rate and the quality of the answers goes down significantly.

As a general guideline, telephone interviews should be less than 20 minutes, and should ideally take between 10 and 15 minutes. Mail-out questionnaires should not take more than 30 minutes to answer, and ideally take about 15 minutes. In-person interviews should be limited to less than an hour, and ideally to less than 30 minutes.

4.4   Administering the Survey

Once the survey instrument is designed, pretested, and revised, it is time to administer it. For different survey methods—mail-out survey, telephone survey, and in-person survey—the method of administering the survey will be different. In this section, we will look at how these different types of surveys are administered.

4.4.1   Administering Mail-Out Surveys

In mail-out surveys, the questionnaire should be designed in the form of a booklet in order to ensure a professional appearance. Any resemblance to an advertisement brochure should be strictly avoided. The professional appearance of the questionnaire is very important for ensuring a satisfactory response rate. There should be adequate spacing between questions, and the questionnaire should be designed so that it is convenient for the respondent to mail it back. A good cover letter explaining the purpose of the survey is very important.

There are two ways to present questionnaires to respondents: personal delivery and direct mailing. Delivering the questionnaires directly to respondents is more costly in terms of time and effort, but it is likely to result in a higher response rate, more rapid responses, a higher percentage of completed questions in the questionnaire, and perhaps more valid and accurate responses. Direct mailing will usually result in a lower response rate, although this can be remedied somewhat by follow-up mailings or follow-up phone calls. Usually these follow-ups should be done three to four weeks after the questionnaires are sent by mail. Direct mailing plus follow-up usually will achieve a 50–60 percent response rate. Additional follow-ups may raise the response rate to over 70 percent.

4.4.2 Administering Telephone Surveys

The telephone survey is less complex to implement than the mail-out survey. The most important aspect of the telephone survey is the selection and training of telephone interviewers. A good source of possible interviewers is university students, especially graduate students. Training usually involves the interviewers learning the questionnaire by themselves, learning about pretest results and potential "tough issues" in the questionnaire, and learning about general ethical issues. Interviewers should not introduce any bias in the interview process, and should not express any opinions in response to the answers from the respondents.

Many companies that conduct telephone interviews use computer-assisted telephone systems where the interviewers sit at a computer that dials the telephone and displays the question to be asked so the interviewer can read it to the respondent. The software that manages this can take care of the data entry and coding, as the interviewer uses the computer's keyboard or mouse to indicate the respondent's answer to the questions.

4.4.3 Administering In-Person Surveys

In-person interviews are the most expensive type of survey, in terms of both time and money, and the most intrusive of all the methods. However, a major strength of the in-person interview is that you can deal with complex topics. Because you can see how the respondent reacts to the questions as you ask them, you will have a better idea of how well they understand the question and what confuses them. You will also have the opportunity to resolve any glitches in the interview process.

Taking care in selecting and training in-person interviewers is even more important than for telephone interviewers. The methods for selecting and training interviewers are almost the same as for telephone interviewers; the additional abilities that an in-person interviewer should have include politeness; nice smile; good eye contact; the ability to observe customers' subtle gestures, body language and language tones, and make sense out of them; and skills to alleviate customers' stress.

4.5 Survey Sampling Method and Sample Size

The main goal of a customer survey is to produce an accurate picture of the population based on information drawn from a scientifically selected subset of that population. Sampling is necessary because it is impractical to seek information from every member of the population.

In this sampling process, we first need to determine what our population is. Then we need to define a sampling frame, which is a list of elements in the population that may be selected in the sample. Elements in the population could be people, business entities, organizations, and so on. Third, we need to identify a sampling method—a method for selecting a subset from the sampling frame.

Before we do the sampling, we need to identify an adequate sample size for the survey to ensure the credibility of the data analysis. In this section, we will look at these issues.

4.5.1 Population and Sampling Frame

The first consideration in survey sampling is the *unit of analysis*: the individual, object, institution, or group that is relevant to the survey. For example, for a fast-food chain owner, the individual consumer could be the unit of analysis; for a mortgage lending organization, each household could be the unit of analysis; for a medical equipment supplier, each hospital or clinic could be the unit of analysis.

The *population* is defined as the collection of units of analysis that the findings of the survey will apply to. For example, the population of "fast-food chain customers" is the collection of all potential individual customers that the chain can reach; the population of customers for a mortgage lending operation is the collection of all households that the lending operation could do business with; the population of customers for a medical equipment supplier will be the collection of all potential hospitals and clinics that could do business with this supplier.

Usually, in any population, not all of the units of analysis can be identified and reached. For example, if a population is all the people living within a metropolitan area, the unit of analysis will be each resident. From a practical point of view, it is unlikely that all the residents of this metropolitan area can be identified and reached. People are born and die, people move in and out. Some people do not have telephones or stable living places. Usually only a portion of a population is identifiable and reachable, and this portion of the population is often called the *working population*.

From the working population, it is possible to develop a list of units of analysis that can be readily reached in a customer survey. This list is called the *sampling frame*. For example, if the population is "all the residents in a metropolitan area," and the working population is "all residents that can be reached by phone," the sampling frame could be the residents listed in the local telephone directory. Other possible sources of sampling frames include the voters' list, utility (gas, electric, water) customer lists, motor vehicle registrants, and magazine or newspaper subscriber lists.

With most sampling frames, you will have to deal with some of these problems:

- **Missing elements**   Some legitimate members of the population will not be included in the sampling frame. For example, in some polls of U.S. elections in 2004, only traditional phone users were polled, and people who only had cell phones were not included in the poll list. Therefore, a sizeable proportion of young professionals was left out.

- **Foreign elements**   Some people's names will be listed in the sample frame, but they are actually no longer in the population. For example, people could move away but their names would still be in the phone directory.

- **Duplicate elements**   Population members may be listed more than once in the sample frame.

In all these situations you need to determine how many missing, foreign, and duplicate elements are in the sampling frame, and how big a proportion these elements are as a percentage of the whole sampling frame. If this proportion is large and it will affect the accuracy of the poll, you should consider using a different sampling frame. For example, in some opinion polls for the U.S. 2004 election, people with only cell phones were excluded in the opinion poll sampling frame. If these people were a sizeable proportion of voters and their opinions were significantly different than those of traditional phone users, the opinion poll might be unreliable.

4.5.2   Sampling Methods

Sampling methods can be classified into *probability sampling* and *nonprobability sampling*. Probability sampling is used when you would like to draw conclusions for the whole population based on the data you collect in the sample. If your goal is just to learn something about the sample, and you do not intend to draw conclusions for the whole population, you can use nonprobability sampling.

There are two characteristics of probability sampling:

- The probability of selection is equal for all elements of the sampling frame at all stages of the sampling process.

- The selection of one element from the sampling frame is independent of selection of any other element.

For example, consider a sampling frame of 1000 people whose names are written on equal-sized pieces of papers. The pieces of paper are thoroughly mixed and selected one by one without looking at them. If 1000 people are selected in this sample, the probability of selecting any person in the first draw is 1/1000; the probability in the second draw is 1/999; and so on. Finally, the chance of selecting any person in the hundredth draw is 1/901. Though the probability of selecting a particular person is slightly different in each draw, the probability of selection for all the available people within each draw is the same. This is in keeping with the first rule of probability sampling—that the probability of selection is equal for all elements of the sampling frame at all stages of the sampling process. Also, the probability of selecting a particular person in any drawing is clearly independent of previous drawings, so this sampling practice is an example of probability sampling.

There are several methods of probability sampling: random sampling, systematic sampling, stratified random sampling, and cluster sampling. The following sections will discuss these probability sampling methods as well as nonprobability sampling in detail.

4.5.2.1 Random Sampling The best known probability sampling method is random sampling. In random sampling, each unit in the sampling frame is assigned a distinct number, and then units are chosen at random by a process that does not favor certain numbers or patterns of numbers. The chosen units become the sample.

A commonly used method to randomly choose units from a sampling frame is to use a table of random numbers, as shown in Table 4.1. Suppose that there are 1000 people in the sampling frame, and we want to select a random sample of 30 people. Each person will be assigned a number ranging from 000 to 999. Using Table 4.1, we can choose to use any three digits from the five-digit numbers given. For example, we could choose to use the last three digits, so we would select the people with number 073, 849, 761, 622, 905, 276, 837, ..., 033.

For large samples, using a random number table is tedious and time-consuming, so computer-generated random numbers can be used to select a random sample.

4.5.2.2 Systematic Sampling Systematic sampling is an adaptation of the random sampling method that can be used if the sampling frame is already randomly distributed. It is useful when the sampling frame is quite large and the sampling units cannot be easily numbered. For example, if a sampling frame has 3 million people and a sample of 1,500 people is

TABLE 4.1   A Portion of a Table of Random Numbers

77073	51849	15761	85622	38905	72276
20837	95047	50724	16922	04405	30858
37504	15645	36630	28216	10056	97628
40392	58557	60446	11553	60013	38037
53408	14205	33152	70651	17314	93033

required, a random sampling approach might be unrealistic—numbering 3 million people is already a big task.

If the original list of these 3 million is randomly distributed, we can select sample units by selecting them from the list at fixed intervals (every nth entry; for example, every 20th car on the highway, every 50th customer in a store). If we are selecting 1,500 people from 3 million, we can select 1 out of every 2,000 people in the sampling frame (3,000,000/1,500 = 2,000). If we start with a random starting point, and we select every 2,000th person, we will create a random sample of 1,500 people.

4.5.2.3 Stratified Random Sampling   Stratified sampling assumes that the sampling frame consists of several mutually exclusive groups, called *strata*. In stratified random sampling, the total number of samples are divided among strata by a predetermined proportion. Then, random samples are taken from each stratum.

For example, in a community, assume that 60% of the population is white, 15% is black, 15% is Hispanic, and 10% is Asian. If a sample of 1000 people is needed, the stratified sampling method will divide these 1000 people into four ethnic groups based on their proportion in the population. So 600 samples will be allocated to whites, 150 samples will be allocated to blacks, 150 samples will be allocated to Hispanics, and 100 samples will be allocated to Asians. Then the 600 people in the White strata will be randomly selected from the White sampling frame; 150 people in the Black strata will be randomly selected from the Black sampling frame, and so on.

4.5.2.4 Cluster Sampling   Cluster sampling is used when there is a hierarchy of sampling units. The primary sampling unit is a group (or cluster), such as counties, cities, schools, subdivisions, and so on. The secondary sampling units are the individual elements within these clusters from which the information is to be collected.

For example, suppose we want to study the needs of first and second graders. It is difficult to directly locate the sampling frame from a "raw population list," such as a telephone directory, but it is easy to identify the clusters, such as public and private schools, in which there are first-grade and second-grade classrooms. After we select a subset of classrooms, we can randomly select sample units from these classrooms.

4.5.2.5 Nonprobability Sampling   In nonprobability sampling, the probability that a particular unit will be selected is unknown. In this case, we cannot generalize the findings from within the sample to the overall population. We can't assume any valid statistical relationship between the sample and the population, so we can't use such useful probability distribution models as normal distribution.

However, nonprobability sampling can still be helpful in getting a feel for what a portion of customers may think, and selecting a sample is much easier. For example, nonprobability sampling can be used to quickly select a small sample of respondents (say 30 people) to pretest a survey instrument. The conclusions from these 30 people cannot be generalized to the general population, but a lot of shortcomings in the survey instrument can be identified.

There are a couple of commonly used nonprobability sampling methods. The most common is the "sidewalk" survey. For example, for a general population of shoppers, an interviewer might interview passersby near a shopping center. In this approach, the sampling frame is not explicitly identified and numbered, and the probability of selecting any particular passerby is unknown. The advantage of this approach is the ability to get a lot of information quickly.

The other commonly used nonprobability sampling technique is "snowball sampling." This method is particularly beneficial when it is difficult to identify potential respondents. Once a few respondents are identified and interviewed, they are asked to identify others who might qualify as respondents. The list of respondents will be increased quickly.

4.5.3   Sample Size Determination

One critical question in a survey project is how many units are needed in a sample for the results to be generalized to the whole population. The answer to this question depends on two key factors:

- **Accuracy**   The greater the level of accuracy required in the study, the larger the sample size must be.

- **Money and time**   Larger sample sizes will certainly mean higher costs and more time will be required to complete the survey.

Therefore, the sample size is mostly determined by the trade-off between the desired level of accuracy on the one hand, and the budget and time available on the other.

4.5.3.1   Determining Sample Size for Variables Expressed in Proportions   In survey data analysis, many variables are expressed in terms of proportions. For example, we might ask customers a question like this:

Do you like the service of ABC Bank? Yes____No____

The proportion of people in the survey sample answering "yes," which is often called the sample proportion, $\hat{p}$, is often used as the statistical estimate of the population proportion, p, where p is the real proportion of customers who like ABC Bank's service. Of course, we would like $\hat{p}$ to be as close to p as possible. From the properties of normal distribution, and if a random sampling method is used, the probability distribution of $\hat{p}$ is:

$$\hat{p} \sim N\left(p, \frac{p(1-p)}{n}\right) \qquad (4\text{-}1)$$

The $100(1-\alpha)\%$ confidence interval for p is:

$$\hat{p} \pm Z_{\frac{\alpha}{2}}\sqrt{\frac{p(1-p)}{n}} \approx \hat{p} \pm Z_{\frac{\alpha}{2}}\sqrt{\frac{\hat{p}(1-\hat{p})}{n}} \qquad (4\text{-}2)$$

We can use $\Delta_p = Z_{\frac{\alpha}{2}}\sqrt{\frac{p(1-p)}{n}} \approx Z_{\frac{\alpha}{2}}\sqrt{\frac{\hat{p}(1-\hat{p})}{n}}$ to represent the half width of the confidence interval for p. The magnitude of Δ_p represents the accuracy of $\hat{p}$ as an estimator of p, because:

$$P(\hat{p} - \Delta_p \le p \le \hat{p} + \Delta_p) = (1-\alpha)100\% \qquad (4\text{-}3)$$

Δ_p is also called the margin of error.

Example 4.4 Calculating Sample Size with the Margin of Error = 3%   In a customer satisfaction survey, if the preliminary results indicate that the proportion of unsatisfied customers is very close to the proportion of satisfied customers, what sample size is needed if we want the accuracy of the survey to be within $\pm 3\%$ of the true proportion, with 95% confidence?

In this case, clearly $p \approx 50\%$, from the statement of the problem. We want

$$\Delta_p = Z_{\frac{\alpha}{2}}\sqrt{\frac{p(1-p)}{n}} = 3\%$$

Therefore,

$$n = \left(\frac{Z_{\frac{\alpha}{2}}\sqrt{p(1-p)}}{\Delta_p}\right)^2 \qquad (4\text{-}4)$$

is the sample size formula for this case. Specifically,

$$n = \left(\frac{1.96 \times \sqrt{0.5(1-0.5)}}{0.03}\right)^2 = 1067$$

for this example, where $Z_{0.025} = 1.96$.

So a sample of 1067 or more people is needed to ensure the accuracy of $\pm 3\%$.

TABLE 4.2 Minimum Sample Sizes for Proportions

Confidence Interval (Margin of Error %)	Sample Size	
	95% Confidence	99% Confidence
±1	9,604	16,590
±2	2,401	4,148
±3	1,067	1,844
±4	601	1,037
±5	385	664
±6	267	461
±7	196	339
±8	151	260
±9	119	205
±10	97	166

Table 4.2 lists the relationship between sample size, margin of error, and confidence level.

4.5.3.2 Determining Sample Size for Variables Expressed in Proportions for Small Populations

The sample size rule specified by equation (4-4) is based on the assumption that the population size is infinite or very large. In some survey studies, however, the population size is rather limited. For example, the customer base for a medical equipment supplier will consist of a number of hospitals and clinics—the population size will be in the hundreds at most. If the population size, N, is known, then according to Rea and Parker (1992), the sample size n can be calculated as follows:

$$n = \frac{Z_{\frac{\alpha}{2}}^2 [p(1-p)]N}{Z_{\frac{\alpha}{2}}^2 [p(1-p)] + (N-1)\Delta_p^2} \qquad (4\text{-}5)$$

Example 4.5. Calculating Sample Size for Proportional Variables and a Small Population   In a customer satisfaction survey, the preliminary results indicate that the proportion of unsatisfied customers is very close to the proportion of satisfied customers, and the population size is N = 2500. What sample size is needed if we want the accuracy of the survey to be within ± 3% of the true proportion, with 95% confidence?

By using equation (4-5),

$$n = \frac{Z_{\frac{\alpha}{2}}^2 [p(1-p)]N}{Z_{\frac{\alpha}{2}}^2 [p(1-p)] + (N-1)\Delta_p^2} = \frac{1.96^2 \times (0.5 \bullet 0.5)(2500)}{1.96^2 \times (0.5 \bullet 0.5) + 2499 \times (0.03)^2} = 749$$

This sample size is smaller than the one calculated in Example 4.4.

4.5.3.3 Determining Sample Size for Interval Scale Variables   In survey analysis, some variables are interval scale variables. For example, personal income, age, and evaluation scores based on the Likert scale are all interval scale variables. The population means of these interval scale variables, μ, are usually our interest. The sample mean of the interval scale variable, $\bar{x}$, is often used as the statistical estimate of population mean, μ. We would like $\bar{x}$ to be as close to μ as possible.

From the properties of normal distribution, and if a random sampling method is used, the probability distribution of $\bar{x}$ is

$$\bar{x} \sim N\left(\mu, \frac{\sigma^2}{n}\right) \tag{4-6}$$

The $100(1-\alpha)\%$ confidence interval for μ is

$$\bar{x} \pm Z_{\frac{\alpha}{2}} \frac{\sigma}{\sqrt{n}} = \bar{x} \pm \Delta_\mu \tag{4-7}$$

where Δ_μ is the margin of error for μ.

By using the relationship

$$\Delta_\mu = Z_{\frac{\alpha}{2}} \frac{\sigma}{\sqrt{n}} \tag{4-8}$$

we can derive the sample size rule:

$$n = \frac{Z_\alpha^2 \sigma^2}{\Delta_\mu^2} \tag{4-9}$$

Example 4.6 Calculating Sample Size for Interval Scale Variables   In a survey study of household incomes for County Y, the preliminary estimate of average household income is \$40,000, and the standard deviation is estimated to be \$6,000. If we would like to determine a survey sample size so that the margin of error for the average household income is no more than \$1,000, what is the minimum sample size, if a confidence level of 95% is desired?

By using Equation (4-9):

$$n = \frac{Z_\alpha^2 \sigma^2}{\Delta_\mu^2} = \frac{1.96^2 \times 6000^2}{1000^2} = 139$$

Therefore, a minimum sample size of 139 households is required.

4.5.3.4 Determining Sample Size for Interval Scale Variables for Small Populations   The sample size rule specified by equation (4-9) is based on the assumption that the population size is infinite or very large. Sometimes, however, the population size is rather limited. If the population size, N, is known, then according to Rea and Parker (1992), the sample size n can be calculated as follows:

$$n = \frac{Z_{\frac{\alpha}{2}}^2 \sigma^2}{\Delta_\mu^2 + Z_{\frac{\alpha}{2}}^2 \sigma^2 \big/ (N-1)} \qquad (4\text{-}10)$$

Example 4.7 Calculating Sample Size for Interval Scale Variables and a Small Population   In a survey study of household incomes for County Y, the preliminary estimate of average household income is $40,000, and the standard deviation is estimated to be $6,000. If we would like to determine a survey sample size so that the margin of error for the average household income is no more than $1,000, and it is known that the total number of households in County Y is 5,000, what is the minimum sample size, if a confidence level of 95% is desired?
By using Equation (4-10),

$$n = \frac{Z_{\frac{\alpha}{2}}^2 \sigma^2}{\Delta_\mu^2 + Z_{\frac{\alpha}{2}}^2 \sigma^2 \big/ (N-1)} = \frac{1.96^2 \times 6000^2}{1000^2 + \dfrac{1.96^2 \times 6000^2}{4999}} = 135$$

4.6   Internet Surveys

Since the late 1980s, the Internet has emerged as a popular medium for spreading, transmitting, and collecting information. The Internet reaches more and more people every day, and in the not-too-distant future it will reach an overwhelming proportion of the world population. E-mail has gradually become a preferred form of communication. The Internet and e-mail are more convenient, more interactive, more flexible, cheaper, and faster than paper-based information media.

The emergence of the Internet has also affected the way that businesses conduct customer surveys. In the late 1980s and early 1990s, e-mail surveys were becoming popular, because e-mail can instantaneously transmit many surveys at little or no cost. However, at that time, e-mail surveys could only be text-based, and the user interface was poor and prone to response errors. In the mid-1990s, the web-based survey was gaining the upper hand because its user interface was much better and it could be highly interactive. After the mid-1990s, e-mailing respondents with a hyperlink to a web survey became one of the preferred forms for surveys.

In this section, we will discuss all major aspects of Internet-based surveys, and compare Internet-based surveys with traditional forms of surveys.

4.6.1 Drawing People to the Internet-Based Survey

The first step in an Internet survey is to draw a sample of individuals to participate in the survey. Specifically, a group of individuals must be identified, selected, and contacted for the survey. This group should fit the goals and objectives of this Internet survey. For example, if Amazon.com would like to know what its customers think about the quality of Amazon's online purchasing service, this group should consist of current Amazon.com customers.

The first stage of an Internet survey should involve the following steps:

1. Specify the target population.

2. Develop a sampling frame.

3. Choose a sampling method.

4. Determine the size of the sample.

5. Contact the selected sample group.

Let's look at these steps in detail.

4.6.1.1 Specify the Target Population   As in all surveys, the target population is the people relevant to the purpose of the survey. For example, suppose Amazon.com would like to know what its customers think about Amazon's online shopping process in order to improve its processes and make customers happier. In this case, the target population should be all Amazon.com online customers. If Amazon wants to improve its online shopping process to attract more customers, the target population should be all Internet customers, which would include not only all of Amazon's current customers, but also the online shoppers of other companies, such as eBay's customers, and potential future online shoppers.

4.6.1.2 Develop a Sampling Frame   The sampling frame is the subset of people in the target population who are readily reachable by the survey researcher. For example, for the target population of "all Amazon.com online customers," the sampling frame could be "all registered Amazon.com online customers." In this case, survey researchers could send online surveys by e-mail to all registered customers because they are readily reachable. However, if the target population is "all online customers," the relevant sampling frame will be much harder to come by, because there is no single database for all online customers. The best that survey researchers can do might be to acquire customer lists from major online shops.

Usually, a sampling frame can be developed in one or both of the following ways:

- **E-mail list**   Surveying people by e-mail is very similar to telephone surveys or mail-based surveys. You send e-mails to the people in the e-mail list and hope to get responses. However, getting the right e-mail list to start with may not be an easy task. Many businesses have their own customer e-mail lists, but these may not provide an adequate sampling frame, because in many survey studies the researchers want to reach competitors' customers and other potential customers. In such cases, the survey researchers need to acquire e-mail lists from the relevant e-mail list holders, for example, from other companies, or other sources. Another difficulty in using e-mail lists is that e-mail addresses in the lists may not be accurate and are subject to change.

- **Web users**   Survey researchers can include links to the survey in relevant websites. While e-mail is the virtual equivalent of postal mail, communicating with customers through the web is comparable to advertising through billboards. Considering that there are billions of web pages, the number of people attracted by a particular web page is often quite low. However, putting advertisements in popular search engines such as google.com, or on highly popular websites, could attract a sizeable number of web users.

4.6.1.3   Choose a Sampling Method   After the sampling frame is determined, the members of the sampling frame will be selected and contacted. As with telephone and mail-based surveys, there are two basic approaches to the sampling method: probabilistic and nonprobabilistic. These two sampling methods were discussed in the "Sampling Methods" section earlier in this chapter.

The sampling method chosen depends on the purpose of the survey. If the survey is designed to make inferences to or predictions about the target population, a probabilistic sampling method is required. However, if the purpose is only to describe the population, a nonprobabilistic sampling method can be used.

4.6.1.4   Determine the Size of the Sample   After establishing the sampling frame, researchers must determine the number of people within it to solicit for the survey study. The size of the sample will determine how accurate the survey study will be. Obtaining sufficient sample size is not usually a problem in Internet surveys—the marginal cost of contacting an additional individual is very low, and the transmissions can be done instantaneously, so adding respondents does not generally increase the cost and time required for the study.

However, there are two common problems with Internet surveys. One is the extent of invalid contact information; a sizeable proportion of e-mail addresses might be wrong, or outdated. The other problem is the response rate—people usually get a lot of e-mails every day, and e-mail solicitations for surveys may get ignored or misclassified as junk mail and may never reach the intended customers.

4.6.1.5   Contact the Selected Sample Group   The next step is to contact the members of the sampling frame and solicit their participation. In current Internet survey practice, two techniques are used most commonly: e-mail-based solicitation and website-based solicitation.

E-mail-Based Solicitation   Survey participants can be contacted by e-mail. The sampling frame may be an e-mail list from a particular organization, or e-mail lists from several organizations. The samples obtained from these kinds of e-mail lists are usually probabilistic samples. Alternatively, the survey participants may be solicited from public mailing lists, or from paid mailing lists, such as http://lists.indymedia. org or http://www.infousa.com. In this case, because members of the target population are usually not equally likely to be included in the lists, samples from these lists may not be probabilistic.

After securing a list of e-mail addresses, the survey recruitment e-mails can be sent to the e-mail addresses. However, survey researchers should be mindful of the threat posed by spam. Spam is the name given to unsolicited bulk commercial e-mail, and most e-mail systems have spam filters that redirect those bulk mailings to a junk mail folder. Researchers must construct messages that will not only elude spam filters but can also overcome the readers' suspicion—many users ignore or delete such messages either because they look like spam or because they fear computer viruses. A very appealing e-mail heading is essential to overcome these fears.

An e-mail heading is the virtual equivalent of the address information on a postal envelope. It generally contains six text fields:

- **From**   The e-mail address of the sender
- **To**   The e-mail address of the recipient
- **Subject**   The topic of the e-mail
- **CC**   The e-mail addresses of users receiving a disclosed "carbon copy" of the e-mail
- **BCC**   The e-mail addresses of users receiving a "blind carbon copy" of the e-mail
- **Attachments**   Any files to be added to the e-mai.

Survey researchers must be careful about what information is entered on these lines, because e-mail users frequently decide whether or not to open a message based on these fields.

The "From" field should contain the e-mail address of someone familiar to the recipient if at all possible, because spam software often filters out unknown senders. The "To" field should contain a single recipient. Though most e-mail software permits multiple names to be placed in the "To" field, spam software frequently screens for bulk e-mails, particularly ones containing unfamiliar addresses.

The "Subject" field should contain a brief, precise phrase or sentence inviting the user to participate in a research study. Survey researchers should avoid words commonly used to market products, such as "free," "money," or "offer," because spam software often filters on such text. As a result, incentives are best not mentioned in the subject field. The focus of the subject line should be on legitimizing the e-mail, either by referencing the survey researcher's home institution or the subject of study.

The "cc" and "bcc" fields should be left blank, because spam software will often filter out bulk e-mails, especially with unfamiliar e-mail addresses in the "cc" or "bcc" fields. Also, e-mail receivers are usually very skeptical about e-mails with large "cc" lists—long cc lists are usually a sign of "junk mail." The "attachment" field should also be left blank. Some people will not open e-mails containing attachments for fear of being infected with a computer virus.

The body of the message in the e-mail invitation should explain the objectives, procedures, and expectations of the survey, as well as how the individual's name and e-mail address were obtained. Spam software also screens the body of the message for keywords, hyperlinks, or images, so survey researchers should avoid using words and phrases commonly found in product ads.

Website-Based Solicitation   The web is also a platform for recruiting large, diverse, nonprobabilistic samples of the general population because hundreds of millions of people use the web every day. In website-based solicitations, advertisements should be posted on popular web pages. Interested viewers can simply click through the ad and be immediately directed to the web survey site, where they can be formally recruited. Survey researchers can also list their website with a search engine, such as google.com. Search engines direct viewers to these websites based on keywords the viewer enters in the search engine. The survey websites can also be listed in the advertisement section of search engine pages.

The content of the advertisement is very important in attracting viewers to the survey website. Menon and Khan (2002) found that customers are more attracted to colorful, image-laden websites than to monotone, simple websites.

4.6.2 Administering a Survey on the Internet

After recruiting sufficient survey participants, the second stage of the Internet survey is to administer the survey. Survey researchers need to construct the survey instrument and distribute it to all participants. To do this, the researchers need to

1. Select a communication mode

2. Control access

3. Format the survey instrument

4. Help participants finish the survey

In the following sections we will look at these steps in detail.

4.6.2.1   Selecting a Communication Mode   Survey administration starts with selecting a communication mode that transmits the survey instrument to survey participants. Currently, there are three popular methods: e-mailing the survey instrument directly to survey participants; posting the survey instrument on the web; and e-mailing participants with a message and link to the survey website.

E-mailing the Survey Instrument Directly to Survey Participants   The advantage of directly e-mailing the survey is its convenience and efficiency. You can send survey instruments to all participants in no time.

There are two ways to transmit the survey instrument through e-mail. One is to embed the survey instrument in the e-mail message. However, some e-mail programs cannot display anything but text e-mails, or for security reasons are set to only display text, and completing surveys in text format is not a very convenient process. The other option is to send the survey instrument as an e-mail attachment. This approach has several disadvantages as well: Some people are afraid of opening e-mail attachments for fear of computer viruses, and participants may not have software that can open the file type of the survey instrument.

Posting the Survey Instrument on the Web   Posting the survey instrument on a website has quite a few advantages. Websites have a lot of design flexibility—they can integrate multimedia and interactive elements in visually appealing formats. However, web posting is not the right tool for contacting many participants directly and quickly.

E-mailing Participants with a Message and the Survey Website Hyperlink   This option is a hybrid of the previous two approaches. In this approach, e-mails are sent to participants, and a hyperlink to the web page hosting the survey instrument is provided in the body of the e-mail message. This approach can contact a large number of participants quickly, and the hyperlink will conveniently guide the participants to the survey website where a user-friendly and highly interactive survey instrument can be hosted. However, this approach might require too much computer knowledge for some people, such as senior citizens who may have less experience with computers and the Internet.

4.6.2.2 Controlling Access   Once the survey researchers have selected a transmission mode, they have to develop and implement procedures to control access to the survey instrument. The purpose of controlling access is twofold: You need to make sure that a participant can only fill out one survey instrument, and you need to make sure that the right participants get the right survey instruments. (The second reason is particularly important if the researchers are conducting more than one survey, or if the website hosts multiple survey instruments.)

For e-mail solicitation, controlling access is relatively easy; the survey researcher will send one e-mail for each participant. If the e-mail contains a hyperlink, the researcher can simply include the correct hyperlink for the appropriate survey instrument. Usually, researchers can also assign each participant a password, and the password can be set in such a way that it only works for the right type (or version) of survey instrument, and so that it only works once for that survey instrument. For web-based solicitation, the website can also be designed in such a way that each user will be assigned one password.

4.6.2.3 Formatting the Survey Instrument   The next stage of the administration process involves formatting the survey instrument for presentation. The formatting of the survey instrument has two important aspects: questionnaire design and style design.

Questionnaire Design   The following list describes some best practices in Internet survey questionnaire design.

- **List only a few questions per screen**   The survey designer should present only one, or very few, questions per screen so that participants do not have to scroll down to answer the questions. Excessive scrolling may stress participants, causing them to abandon the survey.

- **Eliminate unnecessary questions**   The survey designer should avoid unnecessary questions, such as the current date (the computer can provide this), participant's name (which may already be available from another source), and so on.

- **Limit the length of the survey to 15 minutes**   Research (MacElroy 2000) indicates that the dropout rate (the proportion of participants who abandon the survey) is proportional to the survey length. The dropout rate is only 9% if the survey can be finished in 10 minutes;

however, the dropout rate jumps to 24% when the survey takes 15 minutes to finish, and 35% when the survey takes 20 minutes to finish.

- **Avoid content that slows down computers**   Some computer graphics and fancy animation might take a long time to download and thus may slow down the computer. This will add to frustration for the participants.

- **Use very clear language; avoid ambiguous questions and answers**   As in any survey, the questions should be worded very clearly in such a way that different readers will interpret each question identically.

- **Provide some indication of survey progress**   With a mail survey, participants can easily flip through the pages to see how much of the survey has been completed so far. Without an indicator showing how much of the survey has been completed, participants will feel that the survey is endless, leading to stress. Having a progress indicator will help participants to budget their time.

- **Allow participants to interrupt and then re-enter the survey**   Survey participants should be offered an option of leaving the survey, saving the survey results, and going back to the same survey at the same point, especially for long surveys.

Style Design   The style design of Internet survey instruments involves the following aspects:

- **Display**   The survey researchers must decide how to visually display the survey instrument. There are three important factors: monitor resolution, viewable screen area, and the size of the text.

  - **Monitor resolution**   This is the number of pixels visible on the screen, with a pixel being a single point projected onto a monitor. For example, an 800×600 pixel screen can display 800 pixels on each of 600 lines. The popular monitor configurations are 640×480, 800×600, 1024×768, and 1280×1024 pixels. Creating a large survey instrument (as measured in pixels) will make the page wider than some computer screens, forcing participants to scroll horizontally to read it.

  - **Screen area**   Different types of desktop and laptop computers have different screen sizes. Currently, most monitor sizes range between 14 and 21 inches measured along the main diagonal. A 14-inch monitor has a 12-inch horizontal screen, and a 17-inch monitor has a 14-inch horizontal screen. Because survey instruments could be viewed on all kinds of computers, it is a common practice to put survey instruments into a 10-inch horizontal space in order to minimize scrolling.

- **Text size**   The most popular font sizes in Internet surveys are 12 or 14 points.

- **Color**   One of the attractive features of the Internet is the ability to apply color to the survey instrument. Color can be used to liven up text, provide contrast, and make images more eye-catching. In Internet surveys, it is desirable to have a high contrast between the background color and the text. Black on white, blue on yellow, and red on green are commonly used as color pairs for the text and background. Also, research has indicated that dark text on a light background is easier on eyes than light text on a dark background.

- **Graphics, multimedia stimuli**   Internet surveys can add a lot of features that are not possible in paper-based surveys. Paper-based surveys usually are printed in black and white in order to reduce costs, and explanations and special help are minimized to save space. Internet surveys don't have those limitations. They can include the following:

  - Hyperlinks to other documents can be used to provide "help" screens with detailed definitions and explanations.

  - Graphics and animations can be added to illustrate the survey contents.

  - Videos, voice, and sounds can be added to clarify the contents and stimulate the participants so the survey is not boring.

However, these multimedia stimuli should be used sparingly because they may slow down the survey process, requiring files to be downloaded. Participants also may not have the software required to view these advanced features.

4.6.2.4   Helping Participants to Finish the Surveys   Survey researchers can't just send the survey instruments to participants and hope everything will be fine. Researchers should offer help and incentives to participants in order to get a higher response rate and ensure the quality of the survey.

Providing Instruction   Survey researchers should give clear instructions to participants on how to access, complete, and submit the survey instruments. Survey researchers should not assume that any task is too simple or obvious. Instructions should be designed to inform all targeted participants, regardless of their education, technical skills, or online experience, on how to perform each task in the survey.

The instructions should explain how to get access to the survey instrument, including the hyperlink to the website, how to get a password, and so on. The instructions should also include a good introductory statement that describes the objective of the survey and the major contents in the survey. The instructions should also tell participants how to enter their responses, and how to submit the survey when finished.

Ensuring Response Rate   The response rate is the proportion of survey participants who complete and submit the survey to those who were contacted by the survey researcher. Survey researchers can use these techniques to improve on the response rate:

- **Personalized contact**   Personalized contact emphasizes how important individual participants are to researchers and stimulates participants to complete the survey. It is helpful to inform the participants how researchers acquired their name and contact information, and to emphasize the importance of the survey and how much the participants are appreciated.

- **Financial incentives**   Another option for inducing participation is to offer a financial incentive either when participants agree to participate or once they have completed the instrument.

- **Sending reminders**   Sending e-mail reminders to participants once the administration of the survey instrument is under way can help remind participants about the survey.

4.6.3   Comparing Paper-Based Surveys with Internet Surveys

We have now seen how to conduct Internet surveys. For survey researchers, though, the real question is which type of survey is better, the paper-based survey or the Internet survey? And which type of survey should be used? To answer these questions, we need to compare the following aspects of paper-based surveys with Internet surveys:

- Response rate
- Cost
- Lead time
- Data quality

4.6.3.1   Response Rate   Researchers (Schonlau et al. 2002) have shown that the response rate of e-mail surveys ranges from 6% to 68%, and the response rate for web-based surveys ranges from 7% to 44%. These response rates are somewhat lower than those for mail surveys (21% to 78%). The response rate of Internet surveys is definitely lower than that of telephone surveys.

4.6.3.2   Cost   The cost of Internet surveys consists of the following components: the survey instrument design and survey website development (this is a one-time charge), the cost of contacting participants, and the cost of compiling the survey results. On the other hand, mail-based surveys have the following cost component: the survey instrument design (this is a one-time cost), the cost for printing and mailing surveys, and the cost of data entry for survey responses and compiling survey results.

Based on the study of Schonlau (2002), for Internet surveys the website design incurs significant cost, but the cost of contacting participants is much cheaper for Internet surveys than for mail surveys. Internet surveys also do not require data entry when compiling survey results, because the participants have already entered their results in electronic form.

Overall, for a small-scale survey (a survey of a few hundred up to a thousand), the mail-based survey is cheaper, because of the high cost of website development. For larger-scale surveys (say 1,000 participants or more), the Internet survey will be cheaper, because website design is a one-time charge, and the cost of contacting additional people and the cost of compiling survey results are lower for Internet surveys.

4.6.3.3 Lead Time The time between the "kickoff" of the survey to the completion of survey data analysis is the survey lead time. This includes the time required to contact participants and for them to respond, and the time for following up and compiling the data. Compared with mail-based surveys, Internet surveys can cut down the time required to contact participants and compile survey results. But Internet surveys can't cut down the time required for participants to respond. Overall, Internet surveys are somewhat faster, but they may not be as fast as many people hope for.

4.6.3.4 Data Quality Survey data quality usually refers to the quality of submitted surveys. The data quality is usually measured by the number of participants who missed at least one question, or by the percentage of missed items on questionnaires. For open-ended questions, longer answers are usually considered to be more informative and of higher quality. Research (Schonlau 2002) has shown that compared with mail-based surveys, more participants will miss survey questions in Internet surveys. However, for open-ended questions, Internet survey participants usually give longer answers.

In summary, there are pros and cons for both paper-based surveys and e-mail/Internet-based surveys. Overall, because more and more people use e-mail/Internet as their dominant way of communication, e-mail/Internet surveys are gaining ground. However, face-to-face interviews and telephone interviews can provide more direct personal interaction between survey researchers and customers, and the quality of the survey data is also better than that from both paper-based surveys and e-mail/Internet surveys, so they are still indispensable forms of customer surveys.

5

Proactive Customer Information Gathering—Ethnographic Methods

In developing a new product, one of the key questions is "What do customers really want?" In other words, "What is the real voice of the customer?" In the last chapter, we discussed methods for collecting VOC data, such as surveys and focus groups. These are important and dominant methods of collecting VOC data, and they have been in use for many decades. The common feature of these methods is that they are based on "what customers *say*," either in groups or individually, in person or over the phone, on paper or on the Internet. However, if product developers rely on methods that are based on what customers say, rather than on what they *do* and how they use the product, they could end up with significant errors in capturing the real voice of the customer.

In one focus group study conducted by Microsoft, customers were asked to use Microsoft software for a few hours and then answer questions interactively. It went something like this:

Question: Did you like the product?
Answer: Yup!
Question: Any features you do not like or want to add?
Answer: Nope!

Based on these answers, it might appear that Microsoft had a winning product. But when Microsoft developers began recording keystrokes and videotaping customers' behavior, they discovered a wide range of negative customer reactions—grimaces, hesitations, etc. This simple example, and numerous other cases, demonstrate that traditional VOC collection methods based on what customers say are not adequate. There are many problems with simply asking customers what they think about a product:

- What customers say is highly dependent upon how they are asked. Customers usually will not talk about problems that are not related to the interview or survey questions. However, survey and interview questions are usually designed by the product developers or their business partners, so they are usually highly subjective and are determined by the mindset of product developers.

- Customers usually are not technical experts. They generally cannot express the precise problems they are having with a product just by talking about it. They often have difficulty describing what type of product they need or want.

- When customers are asked about things that happened in the past, their answers are subject to many kinds of errors. Customers tend to remember events in a way that conforms to what they expect; they are highly influenced by the way questions are asked.

- Customers' ability to describe a product's problems usually depend on how familiar they are with the product. When they are not familiar with the product and struggle to learn how to use it, they notice difficulties in using it. When they have used a product repeatedly, their usage becomes routine and they have a hard time noticing details.

- Customers are often reluctant to state the truth to avoid sounding ignorant or exposing their perceived weaknesses. They see themselves as possessing qualities that they do not really possess, so they answer questions for their "ideal" selves rather than their real selves. They are embarrassed to admit that they have problems with products because they fear those problems might reflect poorly on their abilities.

- Customers have all sorts of agendas that interfere with strict accuracy. They develop opinions about what interviewers want to hear and consciously or unconsciously tailor their statements based on these opinions.

Research indicates that what people say does not exactly reflect what people do. Wicker (1969) found that there is only a weak correlation between speech and behavior. Loftus and Wells (1984) showed that people's descriptions of past events are highly fallible. Clearly, by using traditional VOC collection methods alone, we cannot get accurate information on what customers really want. We need some other method to supplement the traditional techniques discussed in Chapter 4, so we can capture what customers really want rather than just what customers say. Ethnographic methods can be used to capture this extra information, and this chapter will look at them in detail.

5.1 What Are Ethnographic Methods?

Ethnographic methods are research methods used by anthropologists to understand people's behavior in context. Ethnographic methods are used in social science to study people's social behavior, such as family and community relationships, cultural beliefs, and values in their native habitat. Ethnographic methods are used in business and product development to study people's behavior when using products in the actual environments of use, such as using a recreational vehicle (RV) on highways and in trailer parks, using detergent in coin laundries or at home, and so on. Even if the product is not yet available, ethnographic researchers will make every effort to learn how users will use something similar to the product, such as a prototype of the new product, a competitor's product, or an old generation of that product. It is important to observe people's behavior in realistic environments, because each usage environment is unique. Without knowing the specific usage environment and how this environment will affect the usage of the product, it is difficult to develop a top-notch product that works perfectly with consumers.

In Jeff Liker's popular book, *The Toyota Way* (Liker 2004), he tells the story of the redesign of the 2004 Toyota Sienna. In this redesign effort, Yuji Yokoya was assigned to be the chief engineer. Yokoya had worked on Japanese and European projects, but he had never been to North America, and the North American market is the "native habitat" for the users of the 2004 Sienna. Yokoya didn't feel he understood the North American market, so to overcome this deficiency, he flew to North America and drove an older version of the Toyota Sienna to all 48 contiguous states in the U.S., 13 Canadian provinces, and all parts of Mexico. On this extended trip, he found quite a few "surprises" and made many design changes that would make no sense to a Japanese engineer living in Japan. For example:

- The roads in Canada have a higher crown than in America (they are more bowed up in the middle), perhaps because of the amount of snow in Canada. He learned that controlling the "drift" of the minivan is very important for drivers in Canada.

- When Yokoya was driving on a bridge over the Mississippi River, a gust of wind blew him very hard, and Yokoya realized that sidewind stability was very important. Driving through the crosswinds of Ontario, he was alarmed at how easy it was for trucks to blow the minivan aside. Based on this observation, the newer Sienna design improved its ability to handle crosswinds.

- When he was driving the narrow streets of Santa Fe, Yokoya found that it was hard to turn the corner with his older version of the Sienna. Based on this observation, Toyota improved the turning radius of the newer Sienna by 3 feet. This is a huge improvement, since the 2004 model was also significantly larger.

- By practically living in the Sienna for all his driving trips, Yokoya learned the value of cup holders. In Japan, driving distances are usually shorter, and while you can buy drinks in cans, it is more common to drink them outside the car. In America, distances between stops can be much longer, and you need space to put drinks in the car—not just coffee, but other kinds of drinks as well. Based on this experience, there are 14 sturdy cup holders in the 2004 Sienna, so each passenger can have 2 to 3 cup holders.

- Yokoya also noticed the American custom of eating in cars rather than taking the time to stop and eat. In Japan, the roads are narrower and there are many trucks around, so driving is more stressful and requires high concentration—driving while eating is not a good idea. In America, the roads are wider and less crowded and driving is more relaxing, but trip distances are longer. So it makes sense to eat in the car. Yokoya learned the value of having a place for hamburgers and fries, so he added a flip-up tray accessible from the driver's position.

Besides the results of Yokoya's experience, Toyota's global R&D head, Dr. Akihiko Saito, also found that Americans buy larger items, such as 4' × 8' sheets of plywood, and put them into the back of their pickup trucks and minivans. Dr. Saito noticed that the older version of the Toyota Sienna did not have this capacity, so he approved the change of design for the new Sienna so that it could accommodate a 4' × 8' sheet of plywood. As this story demonstrates, it is important in the product development process to study consumers' behavior in the actual product usage environment.

In general, ethnographic research is the direct, first-hand observation of daily behavior where the researcher may even participate as a participant observer. It is a research method based entirely on fieldwork, and it seeks to observe phenomena as they occur in real time. In other words, ethnographic researchers have to study behavior in natural settings, sometimes getting their hands and pants dirty. Another principle of ethnography is that you cannot understand people's behavior fully if you don't understand the symbolic world of the people you are studying—you have to see the world through their eyes and use their shared meanings. This involves learning the language in use, the dialect, any buzzwords or jargon, special words, and so on. Ethnographic research involves staying in the field—you have to walk miles in a customer's shoes to develop a real understanding of their experiences.

Unlike regular customer survey studies, in which the sample size is usually large enough to make statistically sound inferences, ethnographic research uses small but "good" samples of "informants." Informants are people who may have special knowledge of the area under study. When ethnographic research is used to study indigenous tribes, the "informants" are usually influential and knowledgeable people in the tribe; when ethnographic research is used in customer research and product development, the informants are usually "lead users" (usually "heavy users" or "state-of-the-art" users) of the products. For example, in an ethnographic research study for computer-aided design (CAD) software, the "lead users" could be research and development scientists or engineers who use CAD software very heavily in advanced research, who often take their use of the software to the edge, and who often develop driver programs or customize the software. These people really understand the product and they often can provide valuable insight on it. Ethnographic researchers can work with informants to gain in-depth knowledge in the area.

5.1.1 Frequently Used Ethnographic Methods

There are several frequently used ethnographic methods: participant observation, nonparticipant observation, formal interviews, informal interviews and casual conversation, and informant diaries.

5.1.1.1 Participant Observation   Participant observation is an approach in which "you are directly involved in community life, observing and talking with people as you learn from them their view of reality" (Agar 1996). In ethnographic research, there are four levels of involvement in observation: the complete participant, the participant-as-observer, the observer-as-participant, and the complete observer (Gold 1958). However, as Hammersly and Atkinson (1983) pointed out, the quality of information that can be gathered by people at either of the two endpoints is severely limited, and the difference between the middle two is highly questionable. The ideal approach is to minimize the effect of the researcher on the researched, and to maximize the depth of information that is obtained. The actual approach can range from a hidden or disguised voyeur to an active participant who acts as a member (not as a researcher), as long as these roles will not alter the flow of interaction in an unnatural way.

5.1.1.2 Nonparticipant Observation   In some cases, it is undesirable to conduct customer research as a participant observer because the introduction of an outsider may disturb or destroy the natural customer behavior. In this circumstance, nonparticipant observation is preferred. One typical example of this is when gathering data about how parents and children interact, as described by Rust (1993). In this study, researchers posed as shoppers in supermarkets and toy shops, waiting in aisles for a parent-child group. On seeing a shopping party enter the aisle, the researchers would estimate the child's age, record some basic information about the shoppers, and then take notes on what the parent and child said and did. As soon as the party left the aisle, the researchers would finish up the notes and wait for the next party to appear. In this study, 200 records were gathered and used to develop a number of marketing strategies.

5.1.1.3 Formal Interviews Formal interviews are important tools for data collection in ethnographic research. A key feature of the ethnographic interview is the use of nondirective questions. These questions are designed as triggers that stimulate the interviewee into talking about a particular broad area. The focus of the interview is not to draw conclusions on a given set of hypotheses, but to uncover as many unknown details about the research subjects as possible.

5.1.1.4 Informal Interviews and Casual Conversations Much of the rich data that ethnography can get comes from the whole realm of informal talk between researchers and informants. The main feature of informal interviews is that the researcher does not have a written list of questions, but rather a set of question-asking strategies to let the conversation flow. These interviews take place in all kinds of circumstances, such as when working with informants, taking a coffee break, relaxing in front of the TV, and so on. Informants have all kinds of freedom in responding to questions—they can criticize a question, correct it, and answer in any way they want.

5.1.1.5 Informant Diaries Informants can be asked to keep diaries relevant to the research area. Because these diaries are created by the informants themselves, they provide some "untainted" information.

5.1.2 Data Recording Methods

There are a couple of common methods of recording data in ethnographic research, which include making field notes and taking photographs or videos.

5.1.2.1 Field Notes In the ethnographic tradition, field notes play a very important role. Field notes are records of social activity written up by the researcher as soon as possible after the event. According to Spradley (1980), a field note should provide the following information:

- **Space** The physical place or places
- **Actor** The person or people involved
- **Object** The physical things that are present
- **Act** A single action that the actors perform

- **Activity**   A set of related acts that the actors perform
- **Event**   A set of related activities that the actors carry out
- **Goal**   What the actors are trying to accomplish
- **Time**   The sequence of acts, activities, and events over time
- **Feeling**   The emotions felt and expressed.

5.1.2.2   Photographs and Videos   Visual data can be extremely useful in interpreting behavior, including the temporal flow of events, culturally significant moments, and human-object interactions. They can also be used to stimulate discussion with informants.

5.1.3   Types of Ethnographic Research Used in Product Development

In product development, there are three kinds of VOC research in which ethnographic methods play important roles: discovery, definition, and evaluation.

5.1.3.1   Discovery Research   Discovery research is an open-ended exploratory effort to learn about consumer culture. It is useful for developing original product and service ideas, or for finding new applications for existing and emerging technologies. In discovery research, ethnographic researchers collect and analyze a combination of verbal, observational, and contextual information to identify what people say and do in their natural environment. The inconsistencies in what customers say and do, and the "work-arounds" people use, can help to identify unarticulated, unrecognized needs and gaps.

The ethnographic methods and other VOC data collection methods used in discovery research include:

- Contextual observations
- Participant observations
- Video recording
- Open-ended interviews
- Focus groups
- Content analysis (text, audio, and video)

One good example of discovery research is described by Squires (2002). He describes an ethnographic case study about family morning routines and breakfast behavior. In this case study, researchers visited families as early as 6:30 in the morning, and observed details about how people chose their breakfasts. Overall, they found that people want good quality, good portions, and nutritional meals; however, the time pressure made American families start their days earlier and earlier, when many kids and teenagers were barely over their sleeping cycles. So what was actually eaten at breakfast time was very far from parents' expectations, ranging from eating very little to skipping breakfast entirely. Based on these observations, Yoplait developed Go-Gurt, which is squeezed out of a tube directly into the mouth, instead of being scooped with a spoon. The sale of Go-Gurt took off quickly, and the key to the success of this product was that it filled unarticulated needs: kids liked it because it was fun and mobile, and parents liked it because it was nutritious food for their children.

5.1.3.2 Definition Research   Definition research is used when there is already a product concept—it helps developers take a product concept and turn it into a meaningful form. Definition research can be used to determine what the product should do, how it should be used, and how it can be promoted to consumers. Therefore, definition research focuses on identifying the use and use features of a product, and on firming up what a fully functional product prototype should be like.

The methods used in definition research are different from those of discovery research because we already know the product concept, the key parameters of the product, and something about the consumers. Because of this knowledge, we don't need to start from a "clean slate"—we can zoom in on the user and the product. This research can compare existing products in context and determine how the product concept and design could be changed to fit users' daily use. The ethnographical and other VOC data collection methods used in definition research include:

- Contextual observations
- Directed and semistructured interviews
- Ratings, rankings, and comparisons of the relevant products
- Participant observation
- Scenario analysis

Squires (2002) offers a good example of definition research in his discussion of a project for a personal care product. Personal care products include tissues, lotions, toothbrushes, and so on. Usually people use them in their homes, but some professionals, such as front office secretaries or salespeople, need to use them during long working days. In this case, the product concept exists, but the personal care product needs to be modified to fit into the front office environment. In this type of product development, it is very important to know the culture, the work surroundings, and the user context for potential customers—the office workers. Ethnographic researchers need to understand the culturally appreciated practices for using these products. Researchers found that while it is important to maintain one's personal image by using these personal care products during workdays, their usage should be kept low key. In this case, the researcher found that packaging these products to fit in with the office environment was the most important feature.

5.1.3.3 Evaluation Research   Evaluation research deals with situations where there is already a working product or prototype. Evaluation research helps to validate the product, test customer usability, and refine or fix feature details. Usually companies perform evaluation research at the end of product development. The ethnographic methods and other VOC data collection methods used in evaluation research include:

- Surveys
- Focus groups
- Usability tests
- Product simulation
- User-experience interviews

5.1.4 Key Winning Factors for Ethnographic Methods

In recent years, ethnographic-based VOC methods have been on the rise. So what are the key advantages of ethnographic VOC methods over traditional VOC methods? The following sections outline the key winning factors for ethnographic methods that set them apart.

5.1.4.1 Objective and First-Hand Information   Just as with any anthropology practices, the ethnographic researchers conduct their investigation in a "beginner's mind." That is, they should not bring any pre-existing judgments into the study, but observe and record all the fine details during the fieldwork as they happen. On the other hand, in the traditional VOC data collection methods, such as customer survey and focus group interview, the designing of the survey questions is very dependent on pre-existing judgments.

5.1.4.2 Detailed and More Accurate Information   In ethnographic research, ethnographic researchers observe how customers use the product in real time and in the actual usage environment. They can observe and record what goes right with the product, and what goes wrong in great detail. These observations could reveal unexpected product usage, product inconvenience, and failure modes. On the other hand, the completeness and accuracy of the traditional VOC methods are severely limited by the product manufacturer's subjective assumptions about how the product is used.

5.1.4.3 Audio and Video Documentation   Audio and video recording is a part of standard practices in ethnographic research. Audio and video recording often captures some information that cannot be captured in traditional VOC methods, such as facial expression, hand gestures, reluctance, and other body language. This information can reveal many important hidden facts about customers' true opinions, doubts, and true appreciation for the product.

5.1.4.4 Cultural and Contextual Meaning Capture   One of the basic principles in ethnographic research is to understand the customers' worldview, cultural values, and the meaning of the product in their life. Ethnographic researchers do this by observing and even living with customers to study their schedules, their lifestyles, their work environments, and their buzzwords. By putting all these puzzles together, the ethnographic researchers can gain a very deep understanding of customers' values and opinions on the product.

5.2 Ethnographic Research Project Planning

The first step in any ethnographic research project is planning. There are several steps in the planning:

- **Research objectives**   Determine what is to be accomplished with the project.
- **Research design**   Decide what kind of informants, or customers, the project will work with.
- **Research methods**   Determine what kind of ethnographic methods will be used in the project, such as participant observation, video recording, and so on.
- **Research team and rules**   Develop the ethnographic research team and set up the ground rules for the project fieldwork.

5.2.1 Determining Research Objectives

Ethnographic methods can be applied to many stages of the product development process, as was mentioned in Section 5.1. We can generally classify ethnographic research projects into three types: discovery research, definition research, and evaluation research. In an ethnographic research project, the objective should be specific enough to give direction but broad enough so that we are not limited to any predetermined narrow framework. In most cases, we can roughly define three types of ethnographic projects, as illustrated in Figure 5.1.

When we deal with "fuzzy front end" problems, we don't have a product concept defined yet, and we are researching product concepts, discovering market opportunities, and looking for new product development opportunities. This situation calls for a discovery research project. When we have already developed a product concept, but are not sure about product details, or how to satisfy specific needs from customers, or market-needs details for the product designs, we should go conduct a definition research project. When we already have developed a detailed product design, and we want to launch the product and develop the best marketing strategy and packaging and test the product, we should set up an evaluation research project.

Example 5.1. Lipstick Study   One lipstick company that wanted to maintain and enhance their product's market position wanted to develop new lipstick products and brand extensions. Ethnography was chosen as a research method because it would allow researchers to better understand lipstick products and usage in context. When determining research objectives, they decided to focus solely on women in the context of their lipstick usage. They had two objectives:

- To understand how, when, and where women used lipstick (specific lipstick usage)

- To understand the needs and drivers for women who used lipstick (contextual behavior)

Product Development Process		
Fuzzy Front End	**Concept Design, Product Design**	**Launch**
• Determine customer needs • Uncover unmet needs • Find problem to solve • Generate new product	• Learn how customer use product • Discover functional emotional benefits • Determine product satisfaction and dissatisfaction • Determine product performance targets	• Test new product • Determine packaging
Ethnographic Research Project		
Discovery Research Project	**Definition Research Project**	**Evaluation Research Project**

Figure 5.1   Types of ethnographic research projects

Clearly, this project was a definition research project, because the product concept—lipstick—was not new (Wellner 2002).

5.2.2 Recruiting Informants

After the research objectives have been determined, it is important to determine who the best "informants" are, or which key customers should be interviewed or observed, and how to recruit them. The research objectives determine what types of informants, or customers, are needed for the study. For example, for a definition research project, the product concept has already been developed, and existing customers can be studied for this kind of product. However, for a discovery research project, the type of informants should be considerably different.

Example 5.2. Lipstick Study (Continued)   A lipstick manufacturer wants to beat the competition and promote its own product. This calls for a definition research project. The product concept (the lipstick) had already been developed, but this project is not benchmarking or evaluating the product—it is researching customers' needs so that the manufacturer could improve the product or the marketing. The ethnographic team selected the following four types of customers:

- Loyal lipstick users
- Occasional lipstick users
- Those who rejected the usage of lipstick
- Those who were aware of lipstick but did not use it

Clearly, this selection includes a very broad base of current customers and people who have opinions on the product.

Example 5.3 Breakfast Study   An ethnographic research project was conducted to study the breakfast habits of Americans. The sponsor of this research, a large breakfast food company, was interested in discovering new product development ideas. This project is discovery research. The team didn't have a specific product concept to evaluate, so the target informants needed to include more than the sponsor's existing customers, or just packaged food consumers. In this study, the target customers were all the different types of Americans who consume breakfast, such as men, women, kids of different ages, and senior citizens, people with different income status, education levels, and so on.

5.2.2.1 Psychographics, Lifestyles, and Sample Size Besides the basic informants' qualifications, which are based on research objectives, psychographics is also worth considering when recruiting informants— psychographics focuses on the attitudes people have about themselves, their lives, and their futures. Psychographics are criteria for segmenting customers by attitude, personality, value system, behavior, and experience.

Determining informants' psychographics profiles is valuable and important in the following respects:

■ It is important to choose informants with psychographics profiles that match the objectives of the ethnographic research project. For example, if you are doing discovery research on a high-end product, you should recruit informants who are very self- confident and have upbeat personalities and mostly successful experiences.

■ When you try to make a case based on an informant's interviews and observations, you have to pay attention to the informant's psycho- graphics profile. For example, if the informant is a successful profes- sional who has a high income, a stable job, a very positive outlook on the future, and identifies him/herself with prestigious brands, and who is very positive about the product concept, but the product is really designed for the low-end mass market, you should be careful about making inferences about the product's market potential.

Lifestyle—the way a person lives his/her life and spends his/her money—is also an important factor in recruiting informants. For exam- ple, does he spend most of his time and energy making his home beau- tiful? Does she put a lot of her money into travel? Does your audience consist of young people who spend most of their money on clothes? Lifestyle profile is important when you are dealing with a product or service that expresses personal identity.

The sample size is the number of informants who should be recruited in the ethnographic research project. Based on the experience of ethno- graphic researchers (Abrams 2000), if the sample size is equal or fewer than 15, it is likely that just a few eccentric people will throw off the conclusion of the study. If the sample size is 50 or more, the project will often be too long and costly. A sample size between 20 and 30 is very common in ethnographic research projects, and if enough attention is paid to interviewing and screening informants, this sample size is good enough to extract a rich profile of information.

5.2.2.2 Recruiting and Screening of Informants The recruitment of in- formants consists of two tasks: generating the initial pool of candidates and them screening them. The initial pool of candidates can come from the following sources:

- Known customers or friends of friends: This type of informant may work very well because there is mutual trust between them and ethnographic researchers. However, it is necessary to prevent and reduce the possible bias from this kind of customer.

- Customers from past focus groups: Interesting and articulate focus group participants may be good candidates for ethnographic research projects. They may have already been screened, and be known to be articulate and responsive participants.

- Ads in newspapers, newsletters, postings, and online bulletin boards: This recruiting method may provide a large fraction of unsuitable candidates, so careful screening is essential.

The screening of candidates is usually done by a screening survey. This survey contains many questions that screen the candidates based on demographic, psychographic, and lifestyle considerations. The following are key ingredients for a good screening survey:

- Start with broad questions and move to more specific questions

- Ask nonleading, multiple-choice questions

- Screen the person for past participation in market research and competitive employment.

Example 5.4 shows a complete screening survey.

Example 5.4 A Screening Survey for an Ethnographic Research Project   An ethnographic research project is launched for a food company that developed a new product prototype consisting of noodles and a flavored coating for chicken. Here is the complete screening survey:

Ethnographic Study Participants Survey Questions
Hello. My name is _____ from _____, a research company in _____. We're conducting a study on meals and would like to include your opinion. Do you have time to answer a few questions?

1a. Do you or anyone in your household work for:

    (   ) a market research company or advertising agency? (TERMINATE)

(   ) a food processor, retailer, or wholesaler or any related business? (TERMINATE)
(   ) none of the above (CONTINUE)

1b. Have you participated in a market research study, focus group, or telephone survey concerning any food product during the last six months?

(   ) Yes. (TERMINATE)
(   ) No. (CONTINUE)

2. Do you shop and prepare meals for your family?

(   ) No. (TERMINATE)
(   ) Yes. (CONTINUE)

3. Does your family include at least one child between the ages of:

(   ) 6 and 10? (CONTINUE)
(   ) 10 and 13? (CONTINUE)
(   ) 13 and 16? (CONTINUE)
(   ) None of the above (TERMINATE)

4. Is your spouse living with you at present?

(   ) Yes
(   ) No
At least 15 families must include both husband and wife.

5. What category best describes your total annual household income?

(   ) Under $30,000 (TERMINATE)
(   ) $30,000–$40,000 (CONTINUE)
(   ) $40,000–$50,000 (CONTINUE)
(   ) $50,000–$60,000 (CONTINUE)
(   ) Over $60,000 (CONTINUE)
Respondents' incomes should represent a spread.

6a. Do you sometimes prepare meals with one or more of the following side dishes?

(   ) Prepackaged, flavored rice? (CONTINUE)
(   ) Prepackaged, flavored, dried noodles? (CONTINUE)
(   ) Prepackaged, flavored, dried pasta? (CONTINUE)
(   ) None of the above (TERMINATE)

6b. How frequently do you use these kinds of side dishes?

(   ) Less than twice a month (TERMINATE)
(   ) At least twice a month (CONTINUE)

7. **I'm going to read the names of various brands to you. Please tell me if you've ever used them, if you've used them in the past three months, or if you've used them in the past month:**

Product	Ever	Past Three Months	Past Month
Kraft Rice & Cheese	( )	( )	( )
Noodle Roni	( )	( )	( )
Uncle Ben's Long Grain & Wild Rice	( )	( )	( )
Rice-A-Roni	( )	( )	( )
Savory Classics	( )	( )	( )
Uncle Ben's Country Inn	( )	( )	( )
Kraft Pasta & Cheese	( )	( )	( )
Golden Saute	( )	( )	( )
Lipton Rice & Sauce	( )	( )	( )
Lipton Noodles & Sauce	( )	( )	( )
Lipton Pasta & Sauce	( )	( )	( )

If respondent has not used any one of the above brands during the past month, TERMINATE.

If respondent has used at least one brand during the past month, CONTINUE.

8a. **Have you used Hamburger Helper at any time during the past six months?**

( ) No
( ) Yes

Respondents must include at least five Hamburger Helper users.

8b. **Have you used Shake 'N' Bake at any time during the past six months?**

( ) No
( ) Yes

Respondents must include at least five Shake 'N' Bake users.

9. **How frequently do you serve your family chicken?**

( ) Less than once a week (TERMINATE)
( ) Once a week or more (CONTINUE)

10a. Have you served pasta or noodles with an alfredo flavor sometime in the past six months?

( ) No
( ) Yes

10b. Would you consider serving pasta or noodles with an alfredo flavor?

( ) No

( ) Yes

If answer is No to questions 10a and 10b, TERMINATE.
If one answer is yes, CONTINUE.

A well-known food company would like to invite you to participate in a very unusual market research study. It would involve an interview in your home as you prepare and serve a meal to your immediate family. The interview would be conducted by two women and would last an hour to an hour and a half. The interview would be videotaped—but only for market research purposes. Your privacy would be completely respected. You would receive seventy-five dollars for your time and your participation, plus the cost of main ingredients other than those we would supply.

Respondent's name: _____
Address: _____
City: _____ State: _____ ZIP: _____
Home phone: _____ Office phone: _____ _____

5.2.3 Selecting Research and Data Collection Methods

Based on the specific objectives of the ethnographic research project, the appropriate research methods and data collection methods have to be determined. The most frequently used research methods are participant observation and customer interviews.

- **Participant observation** means to dive deep into the environment where the product is being used. In a participant observation study, detailed notes are taken about observations on product users, usage environment, and all activities associated with the usage. These notes will be analyzed to discover possible new insights about the product.

- **Interviewing** is the other part of the equation. The research team explores and probes customers' actions and beliefs in real time. This allows the research team to check their assumptions, which leads to new and robust interpretations. Interviews are generally one-on-one and in-depth, but sometimes the interviews are done in groups, if appropriate; for example, research that is conducted with a teen best-friend pair, family unit, or an adult group of friends who share a passion or influence each other. The interviews are open-ended, which allows customers to cover all the relevant topics based on their own way of seeing the world. Subsequent interviews are typically more structured, as knowledge and relevant questions are developed that need testing.

The amount of time spent with the customer and the number of visits depend on the research objectives, but some general rules are as follows:

- **Longitudinal versus short study**   A longitudinal study is defined as a study with more than one cycle of fieldwork, and a short study is one cycle of fieldwork. If a subject topic is intimate or of a private nature, such as feminine hygiene or finances, a longitudinal study is appropriate. In this case, it is important to build rapport with the interviewee over time before delving into more personal topics; thus multiple visits to a site would be required. A short ethnographic study is appropriate for everyday products that are not private and do not require a lot of cognitive thought.

- **Number of site visits or interviews**   Ethnographic sampling is very different from sampling methods used in quantitative work. An ethnographic sample is quite small and broadly representative of a constituency. The goal is to understand and gain insight, not the kind of generalization that comes from more quantitative and larger samples. Like a detective, ethnography looks for patterns and clues that get closer to the reality of others. As more time is spent with the customer, more insight is gained. It generally takes about 9 to 20 in-depth site visits to gain the most insight from the participants.

In ethnographic research projects, there are primarily four data collection methods:

- **Field journals or field notes**   A field journal is a written record of the site visit. The research team takes verbatim shorthand notes during the site visits and then revisits those notes to fill in any gaps and highlight the most important points. Notes are critical to the team's ability to debrief after their field experience and create an archival record. This method should be used in most ethnography studies.

- **Audio recordings**   Audio recordings of the site visit allow for an in-depth objective review of the content. Choose high-quality equipment that is as unobtrusive as possible.

- **Photographs**   Photography provides a visual picture of the environment of the site. This method combined with others allows other people to get a feel for the context of the entire site. Equipment varies for this method, but again, unobtrusiveness is important. One benefit of photography is that it can be relatively inexpensive, depending on the type of camera, yet still provide a visual record of the site visit. Digital photographs can easily be sent with the research to others.

■ **Videos** Video can provide a visual record of conversations, behavior, and the environment of the site visit. Video combines a visual and verbal record of the conversation and gives the most realistic perspective for those not involved in the research. This method can be more obtrusive and expensive and needs to be weighed against the goals of the research and comfort of the consumer. Coding, editing, and transcribing the tapes can be time-consuming, but if done correctly, it can provide a wonderful overview of the research.

The selection of data collection methods should fit the objectives of the ethnographic research project and should be practical in the field.

Example 5.5 A Complete Ethnographic Research Project Plan This is a project plan for a manufacturer of plastic bags for trash and garbage disposal.

Objectives

- To understand how consumers store, sort, and dispose of their trash
- To uncover problems and frustrations involved in handling trash— particularly those not currently addressed by available products
- To help open potential areas for new trash-handling product development or current product repositioning.

Methodology

- House Calls will videotape and interview consumers in their homes in two or three separate suburban areas to discover where and how they store various kinds of trash, how they sort their trash, and how they dispose of it.

Detail

- There are various kinds of common household trash:
  - Garbage generally developed in the kitchen
  - Wastebasket trash (usually accumulated in bathrooms, bedrooms, home offices, and so on)
  - Newspapers and magazines
  - Dead leaves and grasses (probably seasonal)
  - Heavy-duty cleaning trash (possibly seasonal)

In addition, trash can be categorized as wet or dry, organic or nonorganic, cans, bottles, plastic, or metal.

House Calls will seek to understand whether consumers in various parts of the country make these distinctions and whether they treat different kinds of trash in different ways.

- Half the recruitment will take place in areas with strong recycling mandates and half in areas with no strong recycling laws. Respondents will be told not to empty their garbage or wastebaskets before the interview.

- House Calls will perform a visual inventory of all the trash in each household, encouraging respondents to talk about the origins and frequency of trash accumulation in each case. Respondents will be probed on the benefits and problems associated with each trash receptacle, covering the following issues:

  - General appropriateness for specific trash

  - Adequacy of size

  - Weight or thickness of material

  - Durability/breakage

  - Leakage

  - Sanitation

  - Closure

  - Odor containment

  - Aesthetic concerns

  - Disposability

  - Cost

  - Other problem issues

- Respondents will then be asked to empty trash containers as they normally do: into larger bags, garbage cans, or whatever the final disposal container may be. If respondents usually sort trash in any way, they will be asked to do so at this point.

- The interviewer will probe respondents on these issues as they relate to final disposal containers.

- In addition, the interviewer will probe respondents concerning trash collection or disposal at the local dump. What are the local requirements concerning the receptacles respondents use? Does what they use meet these requirements adequately? If not, why not?

- Interviews will last approximately one hour.

- Taped interviews will be sent by express mail each night to House Calls for logging and analysis and further direction, if required. The client will be informed continuously as the assignment proceeds about significant attitudes and practices.

Deliverables for Clients

- House Calls will provide a summary of the findings. A transcript of the interview excerpts and videotape narration will be prepared. The summary tape will contain a detailed examination of habits, practices, and attitudes concerning trash disposal. The tape will note problem areas, product gaps, potential positioning opportunities for current products, and areas for new product development.
- Raw footage of all interviews will be available as well.

5.2.4 Developing the Ethnographic Research Team and Ground Rules

An ideal ethnographic research team is cross-functional, which means that the team consists of people from different backgrounds, coming from different departments. The team members should commit to work together throughout the project. Demographics, job functions, and experience in ethnography should be balanced among team members. A professional ethnographic researcher should play a leading role—he or she should act as research designer, coach, trainer, and facilitator. The team should also include chief product design engineers, and people from the marketing department.

Because the knowledge gained through the ethnographic process is exceptionally rich, and only a fraction of it can be communicated to people who have not been in the field, the actual product development team should perform the field research and participate in the analysis whenever possible. This process helps a development team break away from incremental learning and gain a new perspective on the product by refocusing on the problem as experienced by the customer. The lifeline of innovation is "dirty" field knowledge. First-hand field knowledge of customers' needs and desires brings with it insight into how to develop new products, improve product designs, or fix problems.

After the research team has been selected, it should be broken down into smaller field teams consisting of no more than three people who will conduct the individual site visits. Teams of more than three people tend to overwhelm customers and make them uncomfortable.

It is imperative that those who participate in ethnographic research be coached and instructed by a trained ethnographer on what to expect in the field. An experiential learning format is the best way to train a team to conduct ethnographic research. Their training should cover the following:

- Basic theories and techniques in ethnographic research methods
- How to observe and listen
- How to develop open-ended, story-laden, and nonleading questions and probes
- How to take notes, use other methods of data collection, build a field record, and debrief
- How to sustain disciplined subjectivity to understand and manage personal and corporate biases

After the ethnographic research team is established and team members are trained, the ethnographic fieldwork guides must be set up. These are the guides for interviews, observations, and data collection. The following rules of thumb should be kept in mind when putting together those guides:

- Allow time at the beginning of the conversation to set expectations for the site visit. Introduce the field team members and their roles; review the purpose of the research, how long the site visit will last, and what the customer will be asked to do. Begin to build rapport with the customer.
- Start with a tour of the entire site, home, office, or factory.
- Start the visit by asking about broad topics and then move to more specific topics. Save the sensitive and personal topics until closer to the end of the conversation.
- Ask open-ended questions that allow customers to give as much detail as they want and that allow the researcher the opportunity to ask more specific or probing follow-up questions.
- Do not ask questions that lead or direct the respondent toward a particular answer. Leading questions tend to start with words such as *do, are, can, could,* and *would.*
- Allow time at the end of the site visit for the respondent to ask questions of the field research team. At this time, the research team can physically stop taking notes but mentally needs to be extremely aware. This is a cue to the respondent that the research is over, and often the customer will open up.
- At the close of the time together, thank the customers, tell them what is planned for the information collected, and compensate them for their time.
- Have sample and photo notes in your hand, so they are not forgotten.

After the guide is created, review it for any biases. Ask yourself what hypotheses and filters have already been created in your mind about the research outcomes. Then rework the guide to make it more objective.

5.3   Ethnographic Project Execution

After the ethnographic project planning is complete, the research team can enter the field and execute the project. As was mentioned in previous sections, the popular ethnographic methods include interviews, participant observations, and ethnographic data collection. In this section, we will look at how those methods are used in real ethnographic projects.

5.3.1   Ethnographic Interviews and Documentation

As in regular VOC data collection, interviews with selected customers, called informants, are an important tool for getting valuable information. However, there are several distinct differences between ethnographic interviews and regular customer interviews:

- Ethnographic interviews usually happen in the "field," for example, the customer's home, car, office, or shopping places.

- Ethnographic interviews do not have a fixed list of interview questions. The ethnographic researcher is trying to let informants to talk as much as they can and guide the talk to all possible aspects about the products or issues to be investigated. The researchers want to get all the possible information from the informants.

- Ethnographic interviews will go side by side with participant observations. Ethnographic researchers will work and play with informants using the products and see what happens in real time.

- Ethnographic researchers will look for and find problems while customers are using the products.

- Ethnographic researchers will generate reality-based product ideas, modifications, and improvements with their "field work."

5.3.1.1   Guidelines for Conducting Ethnographic Interviews   While interviewing the informants, ethnographic researchers must act carefully, nonintrusively, and positively in order to let informants say what's on their minds. At the same time, the researchers must very carefully watch the informants' behavior and the circumstances, as well as take notes.

The following guidelines are helpful in structuring the interview process:

- Do not make the respondent feel self-conscious by calling attention to a behavior or practice; for example, "You just skipped that part of the program." Try not to probe at all until a behavioral routine is reaching completion.

- Researchers should be quiet, laid-back, and unobtrusive. They should be good observers and listeners, actively interested in everything that the informant does or says, but being careful not to impose. The last thing you want is a researcher behaving like a marketing person, taking control and directing the interview.

- Minimize asking why a respondent is doing or not doing something in a particular way; for example, "Why didn't you clean behind the TV set?" In general, avoid making the respondent react defensively. With patience and keen observation, you will eventually discover whether that area was avoided because the respondent believed that it was dangerous, because no current product was adequate, because the respondent didn't care about the out-of-sight area, or any other reasons. The immediate challenge is likely to provoke a defensive reaction, regardless of the true reason for the behavior.

- Keep the same interviewer throughout the whole project if possible. There is a cumulative learning process that takes place from interview to interview that works very much to the advantage of the project.

- Avoid asking respondents directly to explain or describe what they are doing step-by-step, let them to do things in their own way and watch them carefully instead.

- Probe behavior gently and indirectly by asking respondents to describe their goals or intentions; for example, "Please describe what you are trying to accomplish" or "Please tell me how you expect this to go?" Another way to probe is to ask respondents: "Describe what is going through your mind."

- If equipment is taken to the interview, it should be as compact as possible. No attempt should be made to hide a camera or tape recorder when a research team is in a consumer's home. The person handling the equipment should be up-front about showing it to the informants, so they can quickly get used to its presence.

- Respect the rights, property, and privacy of the participants. Avoid interfering with the respondent's other responsibilities and relationships. For example, if a friend normally comes by to visit while dinner is being prepared, it is to your benefit to preserve that pattern of interaction. Be patient and tolerant if a nonresearch responsibility (for example, a telephone call from a family member) suddenly breaks the respondent's attention.

- Participants must he thoroughly briefed about how the research will be implemented. Enough background information should be provided without informing respondents to such a degree that they lose spontaneity during the visits. Try to avoid surprises that are likely to confuse or disorient respondents.

- Try to minimize your own impact on the environment being studied so that you can observe it as naturally as possible. Do not help participants with tasks or provide advice or otherwise influence the way that the respondent naturally behaves. For example, if the respondent has been asked to prepare a meal, it is incorrect to accept and eat that meal—no matter how tasty it may appear—because the researcher's satisfaction, rather than the respondent's behavior, becomes the focus of the interaction. Similarly, if the respondent must carry unwieldy supplies from one location to another during the course of product usage, it is inappropriate for any site visitor to help.

- Be patient and tolerant about the course of activities during a site visit. Let the respondent set the pace; the ethnographer should not rush activities or try to stop something already in motion.

- Because a high level of intimacy normally develops between the ethnographic researcher and respondent, matters that are entirely extraneous to the subject under investigation will come up for discussion: the news of the day, family composition, social activities, and personal interests. It is only natural for visiting researchers to share details about their own children or a recent vacation destination. Innocuous small talk supports rapport building especially when people find they share some commonality. Nevertheless, researchers should be careful about discussing their own opinions about their profession, personal matters, or the category under investigation, because these would also be out of role.

- Even though the observation guide will be used as a map or a blueprint to the process or behaviors you are exploring, it is not a strict questionnaire or program of events. Try to follow the guide as much as possible; but be ready to stray from the guide as needed to follow, engage, or understand a respondent.

- There should be no more than two or three interviewers or observers so that they don't become intrusive in the household.

- Respondents often incorrectly view ethnography visitors as authorities and specialists on the category being studied. It is only natural for respondents to ask interviewers to evaluate their behavior or make suggestions for solving vexing problems they have experienced.

Decline graciously, because playing the authority will diminish the respondent's sense of autonomy and aptitude. A good way to deflect a request for an evaluation is to make a noncommittal remark about the respondent's effort without commenting on results. For example:

Respondent: Doesn't this product put a great shine on my stovetop?
Researcher: You like the stovetop shining, don't you?

If the respondent is given space to feel like the host and expert and the ethnographic researcher remains modest and behaves like a privileged guest, the ethnographic encounter will prove to be enriching and satisfying for both parties.

5.3.1.2 Observing and Documenting Ethnographic Interviews

In ethnographic interviews, most of the talk and all of the actions should come from the respondent. The interviewer should play a secondary role. The questions should be brief and should serve as triggers to start the flow of actions and words. During the interview, observing informants' facial expressions, body language, and actions can be very revealing, as shown in Example 5.6.

Example 5.6: Baking Soda Brand   When a baking soda-based underarm deodorant was first introduced to the market, an observational study was made of potential respondents. Many were women who were using baking soda as a deodorant for their refrigerators, but it often took the respondents 5 or 10 minutes to make the connection between odor protection for food and odor protection under the arms. Their uncomprehending facial expressions made the manufacturer aware that they could not assume consumers would instantly see the value of a baking soda underarm deodorant. The personal deodorizing power of baking soda had to be played up both on the package and in the advertising.

During the interview, anything related to the product under consideration should be studied and recorded in the interview, which includes how the product is stored, how the product is related to the customer's lifestyle, and any inconvenience related to the use of the product. These detailed observations can lead to substantial new product ideas and improvements, as shown in Example 5.7.

Example 5.7 Pocket-Sized Camera   A manufacturer of single-use cameras made a research visit to a popular zoo and handed out single-use cameras to people who hadn't used this particular compact, pocket model before. When watching respondents take pictures of their families, researchers observed that they moved their fingers all around the camera, sometimes obscuring the lens, before they found the shutter button. In addition, many respondents with larger hands were unsure about how and where to grip the camera. Their fingers seemed to overwhelm the equipment.

A brief interview with each respondent after the pictures had been taken confirmed the observation. They didn't see where the shutter was at first. A pervasive concern was that, with such a small camera, their fingers might get in the way of the lens.

Clearly, the pocket-sized model needed to be more user-friendly. The result was a redesign of the camera, making the model a little less compact (but still pocket-sized) and the shutter more obviously accessible.

Interviewing while observing can also reveal subtle aspects of key customer requirements. In one ethnographic research study, researchers watched people cleaning the bathroom, and they heard many use the word "shiny" or a variation on it. Informants spoke of "shining up" the tub or toilet, or pointed to a shiny sink with pride. Probing what they meant, researchers found that shine was a synonym for cleanliness. If a bathroom surface was shiny, it was assumed to be clean. Furthermore, in many informants' minds, a shiny surface also meant a disinfected surface free of germs. This special meaning of shiny can provide a key input in the functional design of bathroom detergent, as well as in a marketing campaign.

In another study for a household pesticide, the principal scientist was very proud of the effectiveness of a newly developed chemical. It could kill many kinds of bugs, such as ants, effectively. However, after putting this product on the market, the sales figures were very disappointing. When a marketing person asked customers what was wrong with the product, the customers said: "It does not work!" The marketing person was surprised because it had been proved that the pesticide did work remarkably well. When the marketing person prodded further, he found that the customers sprayed the product on a bunch of ants and pointed at them saying: "See, these ants are still moving; they are not dead, so this product doesn't work." The marketing person explained that the ants would die in a couple of days, but the customer replied: "The bugs come and go. After I use this product, I still see bugs, I can't tell whether they are the old bugs that I sprayed or new bugs." This example shows that in customers' minds, the effectiveness of the pesticide is measured by its ability to kill bugs instantly.

Knowledge of lifestyles, especially lifestyle changes, is very important in designing breakthrough products. One of the trends in lifestyles is that people are getting busier. More and more women continue to join the workforce, so the amount of time people have to spend doing household chores is reduced. Two very successful products, Arm & Hammer's Clean Shower and WD-40's 2000 Flushes (Figure 5.2), are examples of products that are designed to fit these busier lifestyles.

Figure 5.2 Clean Shower and 2000 Flushes

Clean Shower is a spray that prevents soap scum and mildew from forming, reducing the need for cleaning. 2000 Flushes is a very durable clip-on toilet flush cleaner that can last for four months. In the market, both products are very successful because they relieve customers from tedious, time-consuming housecleaning work.

5.3.2 Ethnographic Observations in Shops

Another "field" for ethnographic research is where customers shop, such as in supermarkets, specialty shops, and fast-food restaurants. There are several ways to conduct ethnographic research in stores. One is to recruit shoppers beforehand and follow them around stores while they select items they want, with the researcher interviewing them throughout. Another way is to watch shoppers as they shop, and record their shopping processes. You can select shoppers randomly for interviews, with incentives.

This kind of ethnographic observation can show you the primary, secondary, and other motivations that prompt customers to select certain products at the point of sale. You can also observe the effectiveness of the packaging, promotion, and advertisement of the products. Based on this valuable information, you can improve your product design, as well as the sales and promotion. Ethnographic observation can also reveal which store traffic patterns and displays attract customers to specific areas and merchandise, so both the retailer and manufacturer can benefit.

In ethnographic in-store studies for general merchandise stores, such as supermarkets, the primary tools are note-taking, videotaping, and interviewing selected customers. Such a study could have many purposes. One common objective is to identify the primary, secondary, and other motivations when customers do their shopping: What is more important, the brand? The quality? The price? The next two examples illustrate these objectives.

Example 5.8 Difficult-to-Access Goods   In an in-store study, shoppers were observed hunting high and low for a well-known brand of single-use cameras in a giant West Coast drugstore. When they finally found what they were looking for, it was locked up in a glass case, and they had to ask the clerk to open the case for them.

Example 5.9 Disconnection Between Advertisement and Packaging of the Goods   Shopper after shopper in a supermarket passed by a new, well-advertised category of concentrated beverages. Interviews at the shelf revealed that shoppers remembered the advertising but were not connecting it to the display on the shelf.

Example 5.10 shows how ethnographic research revealed the key factor in juice shopping.

Example 5.10 Juice Beverage Shoppers   The following excerpts are from actual interviews taken in an ethnographic research project concerning juice shoppers in large supermarkets at various locations. The material was recorded right at the shelf. First, the shoppers were filmed with a hidden camera and the same shoppers were then interviewed in front of an open camera. These excerpts show the kind of information you can expect to obtain from random in-store interviews. The numbers represent codes differentiating the interview portions.

Climbs Up Shelf to Get Capri Sun

MIS   5: 23:00

Shows 8-pack boxes of Capri Sun Red Berry and Wild Berry

MIS   5:23:15

My three kids drink them. Ages 15, 11, and 7. They ask for flavors and make me write them down.

MIT   6:30:40

I know my daughter likes the punch flavor.

Daughter shows boxes.

MIS   1:12:13

The Juicy Juice boxes

MIS   1:12:28

Plan to buy this specific one? Not this specific one.

DISS TO MIS   1:12:38

What made you select this particular one? She grabbed it. Grabs it all the time.

Father and Son

NJS   2:17:10

I let him decide what juice, because I'll buy juice and he won't drink it. I know he'll drink apple—his favorite.

He likes Juicy Juice and Mott's.

Mother and Daughter Look Through Boxes for Flavor

MIT   6:04:00

Daughter: *I look for flavor.*

MIT   6:07:00

Mother: *We look for juices. Fun juices. Whatever appeals to them.*

MIT   5:25:32.

To Daughter: *I saw you're the one who actually selected it. How did you select it?*

Daughter: *I don't know, it looked good.*

T   5:25:44

I had Mott's before, and I decided to try a different flavor.

T   5: 26: 54

To Mother: *How do you decide?*

Mother: *Generally, whatever she likes.*

In the above notes, MIS, MIT, NJS, and so on are initials of interviewers or note takers. By observing the above notes we can see clearly the strong influence that kids have in the selection of juice beverages. As many in-store studies do, this one uncovered proprietary information that has helped the beverage manufacturer, who sponsored it, enhance marketing and packaging practices.

Shopping ethnographic research can tell you how and why a customer selects a brand or a product on a particular day. It will not tell you how and why the customer chooses that brand or product on an ongoing basis. In order to get a complete picture of purchase motivations and usage patterns, you also need to observe customers in their homes, offices, and wherever the products are used.

5.3.3   Ethnographic Observations in Product Usage Processes

The most important ethnographic observations usually happen where the products are used. By observing customers using the products, especially paying attention to where and how customers struggle in using products, you can find the possible weak points in the current product design, and thus improve on the design.

This kind of ethnographic research is also called usability research. It is especially effective for understanding the user interface; that is, the interaction between people and technology. Usability research originated in the military as *human factors research*, and it seeks to assess how easily users manage their prospective technological tools, and how much satisfaction that engagement produces (Fox and Fisher 2002; Nielsen 2000). It can be applied to such technologies as web sites, computer software, ATM dialogs, automobile dashboard technologies, and cell phone functionality.

5.3.3.1 Key Issues in Usability Research   In usability research, there are several key issues: whose equipment is being used, what factors will be analyzed, and what mix of objective and subjective measures need to be acquired.

Whose Equipment Is Being Used?   It is only natural to start by considering the location of the study: whether within special usability laboratories sponsored by technology companies or research facilities, or in the respondent's own home or at work. Researchers gain the advantage of consistency and uniformity when they invite respondents to their own labs.

What Factors Will Be Analyzed?   Specific user-interface metrics that may be observed include:

- **Ease of learning**   How quickly can users acquire skills for negotiating the technology?
- **Efficiency of use**   Are the steps associated with achieving a user-defined objective, such as sending and receiving instant messaging, economical, and logical?
- **Memorability**   Can users recall how to complete tasks after some time has elapsed, or, like most of us, do they have to relearn functions?
- **Error frequency and severity**   When and why are mistakes made? Do the errors represent some kind of mistaken assumptions about how people naturally use the product?
- **Subjective satisfaction**   Do users like the product? Once factors associated with successful product implementation have been learned, they become part of the standard knowledge about a category and may be engineered into successive versions and alternative technology designs.

For example, if you were studying the usability of the user interface of a web site, the following factors should be studied:

- **Navigation**   Does the site produce the results that users seek? Do they know where they are at every moment? Can they find what they want?

Does the search mechanism produce the desired results? Is there a high level of association between web site terminology and users' terminology so that confusing jargon does not misdirect users?

- **Structure of the web site**   Is it organized in a logical and meaningful hierarchy?

- **Layout**   Is the page pleasing to the eye? Is there sufficient white space so that each page can be managed without strain? Do repeating elements have a clear relationship with changing elements? Are graphics and images aesthetically pleasing, and do they avoid downloading complications?

- **Error messages**   Are users empowered to take action following error messages, or do the messages produce paralysis?

What Mix of Objective and Subjective Measures Need to Be Acquired?   When completing a usability test, respondents must generally complete tasks as behavioral challenges. Objective measures may include the time needed to complete tasks and the error rates. Subjective analysis may involve user comments and behavioral observations. Sometimes, the "think-aloud" technique is applied, in which respondents are encouraged to divulge what is going on in their minds continuously while completing a task.

5.3.3.2 Conducting Usability Research   To get a clear picture of how ethnographic observations are conducted in product usage research, we'll look at some typical ethnographic observation scenarios concerning the use of cars, computers, and workplaces.

Cars   In recent years, car manufacturers have learned a lot about improving car design by putting a video camera in the front seat and recording the driver on an entire commuting or shopping trip. How do drivers handle the controls? Are the controls well placed? Are they easy or difficult to use? Researchers have learned from the expressions on drivers' faces as they manipulate the car through traffic and park. What frustrates them? What annoys them? Are there blind spots that might cause safety problems or that make parking harder than it should be? Do drivers have to take their eyes off the road for any length of time to read the gauges on the dashboard or use the radio? Do they have to crane their necks to see if they're going to fit into that tight space at the curb?

Loading the car, either with people or packages, is a natural subject for observational research. Is there enough room for the family—in the front and the back? Is the vehicle easy to get in and out of? Does the trunk or rear portion of a minivan provide problem-free access for grocery bags and other supplies?

Just the basic knowledge of where a driver takes his or her car each day—the kind of traffic, the length of the trips, the frequency of loading and unloading—can be highly useful to car makers designing and marketing cars for specific consumer segments. It's a good idea either to have an interviewer riding along with the driver, or to take the time afterward to go over the video recording with the driver. The interviewer should probe in detail why the driver does what he or she does. At the same time, the interviewer can investigate the relationship between car and driver.

The observational researcher can tap into many kinds of meaning—especially in a one-on-one conversation in the intimate, private space of a respondent's automobile. Questioning both indirectly and directly, a researcher should be able to explore issues of status, personal identity, power, freedom, and more as they touch on car ownership. Are these motivations as much of a driving force for a leased car as they are for a car that's owned? This is an interesting question for both a car manufacturer and a dealer's association.

There is also a wealth of meaning in the way a car is displayed—or not displayed—when it's not being used. Is it parked in the driveway, close to the road so that the neighbors and passersby can see it? Or is it kept within the protection (or secrecy) of a garage? How often is it washed? How often is it tuned up? Is the inside clean and relatively free of junk? More than a simple matter of pride, the display and condition of the car reveals something of the way the owner feels about himself or herself and about his or her readiness for the next automobile purchase.

It is also very valuable to observe what goes on in the dealership during the car-buying process. We can record the whole process of the negotiation in an auto salesman's cubicle, watching and listening with one camera trained on the buyer and another on the salesperson —this would point out effective and ineffective selling and negotiating techniques and provide the basis for a training video for dealerships, dealer associations, and car manufacturers.

Ethnographic researchers can also study car sales on the Internet. It would be interesting to know how Internet shopping affects the car-buying process. How much of the buyer's decision is made in front of the computer? How much is made when he or she visits the dealership? Following an online shopper through the entire search and buying process would point out ways the marketing process—particularly on a local level—can be made more efficient.

Computers   There are a number of computer-related consumer experiences in which observational research can be useful: setting up a new computer (known as the *out-of-the-box experience)*, installing and learning new software, using a web site, and buying online. For most people, setting up a computer can be a full day's worth of frustration, with at least three frantic calls to the service number for help. Observational research may not be able to find a way completely out of this technical morass, but it could help.

By observing and recording—moment by moment—the points of frustration, the inadequate digestion of dubious instructions, the mismatching of cable to port, the actions and words (even if unspeakable) that result from a printer's refusal to print, the whole process of dialing for help and being asked to stand by while other customers are being served, and having to explain the problem to one of those voices with an ever-so-subtle implication that you're just another non-tech moron, a company can go a long way toward correcting the out-of-the box experience and making it user-friendly.

Today's software programs are often even more formidable than the hardware, and the experience of installing and learning to use them can take hours and weeks of trial and error. There is an ever-increasing opportunity for observational researchers to help make the software world a kinder, gentler place.

Observation in this case may be conducted over a period of weeks, with the observer present at program installation and then returning at intervals to record progress (or lack of it) as the respondent begins to use the new software. The objectives of the research needn't only be oriented toward picking up on and solving problems. The researcher could enlist the user to figure out ways the program could be made better, easier to install, faster to learn, with fewer steps and keystrokes.

Frustration seems built into the process of getting from one web page to another, exploring links, and finding exactly what you want. Watching people bumble from one icon to another while following inadequate directions, revealing their feelings through facial expressions and grunts, and the obverse—watching them use an intelligently designed, easy-to-follow web site—will offer clear insights into the ways your web site can be improved.

The researcher sits with the respondent as he or she attempts to find and navigate the web site under consideration. Every time the researcher notices a hesitation or an error, the respondent is probed. Why the pause? What went through the respondent's mind when he or she clicked on the wrong icon? Where did the respondent really want to go? When all is said and done, was the web site worth the trouble? Would respondents return to it in the future? If not, why not? What would make it more accessible? There is nothing like direct observation of the experience to find reality-based answers to these questions.

Broader studies of computer usage have revealed that, for some families, the computer has replaced the family hearth. When the computer is not used for homework, E-commerce or business, friends or family members often gather around it for games, looking up family-related information such as potential vacation spots, and surfing the net.

Workplaces   Observation should not be confined to the home computer user. Office networks—from the e-mail experience, to the transfer of files, to the control of security and more—can benefit from close study of flow and process as employees communicate with their coworkers and with distant offices.

Researchers working with systems engineers *before* a new network or new hardware is installed can help save organizations months of frustration and expense. How are employees using the current system? What are the problem points? What needs to be changed? What parts should remain unchanged? Day-to-day observation of individual users throughout the organization will provide valuable background for the open-minded engineer assigned to upgrade the network.

Web site design and architecture have become more and more in demand as companies use the Internet to sell, to communicate with customers, and to communicate between departments. Here, too, close observation of the stumbling blocks can point the way to improvements.

The computer is not the only potential target of observational research in the office. Imagine a hidden camera facing the copier. Is the machine providing angst-free output, or is the office staff frustrated with its creaky service? What would it take to make the copier more useful for the organization? Watch a time-coded video of just one day's work, and you will probably learn the answers to these questions.

Other busy places and activities such as the cafeteria, the mail room and delivery process, the phone system, and the everyday traffic that flows from one office to another are worthy of observation and study in the interests of efficiency and organizational morale. It is also useful for an outsider to observe and characterize what is commonly referred to as the *culture* of the company. It's hard for those inside the organization to step back and look at the company objectively, just as it's difficult for people to appraise and express their own salient characteristics. A trained ethnographic researcher familiar with the various components and values of an office culture can help a company understand and present itself to the outside world.

5.3.4   Ethnographic Studies of Customer Cultures

Ethnography represents a combination of several social science disciplines, and ethnographic researchers should be good at understanding human behavior. In product and service development, what will be successful depends on whether these products or services will satisfy people's needs in many aspects, such as function, social value, and self image. By observing customers in real life, we can study the lifestyle and cultural aspects of customers. This valuable information can help to uncover customers' unstated needs and generate clues for developing new products.

Culture represents the baseline of our experience as human beings living in society. It is a broad concept encompassing the individual worldviews, social rules, and interpersonal dynamics characterizing a group of people in a particular time and place. Culture works through religious beliefs, language, social institutions, and other group dynamics to create a pattern, a taken-for-granted set of ideas and instincts that delimit and define people in their social settings. Although cultures are highly mutable, they have a conservative tendency and can sometimes be changed only by circumstances of military conquest, or rapid technological change. People tend to defend their cultures and view alternatives as different or even wrong.

Culture operates on both the material and immaterial levels of human experience. It serves as the foundation for the *behaviors, meanings,* and *tools* of all human collectivities. By cultural tools, we mean all of the physical components of a group's way of life, its technology and materials, as well as the fundamental rules, codes, and techniques for accomplishing daily affairs. Cultural behaviors are the totality of activities associated with membership in a group, whether or not they are practical, goal-oriented, sensate, or mystical. Cultural meanings refer to the process of making sense: how we intellectually or emotionally understand the purposes, implications, and associations that underlie all of our behaviors and the tools we use in everyday life.

Some additional explanation will help clarify the definitions of these terms and others that are useful in analyzing culture.

5.3.4.1 Cultural Behaviors   The concept of cultural behavior can include any human action outside of biologically imposed conditions (such as sleep and digestion). Behaviors can be conscious and purposeful or subconscious and non-goal-directed. Here are several ways of classifying behaviors.

Rituals   Rituals are patterned behaviors, usually performed without thought, that are repeated by force of habit or belief. Examples of rituals we may observe in the course of ethnographic practice may include the order in which dishes are served in a meal or one's own particular pattern of personal care and grooming in the morning.

Roles   Through role behaviors, we enact a relationship between ourselves and another person or a group. Being a boss and being a father are examples of roles. Status refers to behavior that dramatizes differential power or prestige in a setting. Thus, being a father implies more than just particular role-related responsibilities, such as nurturance of children.

Particular brands are often purchased to reinforce the roles played by consumers or their status in a group. A luxury brand icon on an automobile or premium cigarette brands may be selected to demonstrate a certain level of authority and power.

Ethnographic findings can often demonstrate opportunities for role-related new products. A major manufacturer of paper goods discovered that paper towels are mostly used by women in kitchens, but there are similar needs in auto repair shops, machine shops, and so on. A new kind of paper towel is developed with higher weight and strength, colored in a blue denim, and it becomes a successful product. This is an example of linking culture roles to product development.

Practical Activities and Goal-Oriented Behaviors These include work, shopping, cooking, cleaning, or anything else that is directed toward accomplishing actions necessary in everyday life. Carrying out these daily tasks involves a range of habits, routines, skills, and styles that have been learned or otherwise channeled through social influence. Understanding the details of these practical activities and goal-oriented behaviors are very important in developing and promoting relevant products. For example, ethnographic researchers made detailed observations on how people are doing paperwork on paper, where people's eyes are moving, and where it is easy to get stuck. This research work helped a great deal in developing all-electronic paperwork software.

Performances Performances are patterned behaviors that are staged for the benefit of an observer. Baking cookies for the family, for example, may be a response to the homemaker's desire to perform the role of "good parent."

Play, Games, and Diversions Although the activities of play, games, and diversions are believed to induce personal relaxation or bonding between individuals, they may also serve functions with greater meaning. They can also provide the basis for various social relationships and either reinforce or negate status differences between individuals. For example, when the boss invites a junior employee to join in a golf outing, he or she is encouraging team commitment and inviting the employee to join an inner circle.

5.3.4.2 Cultural Meanings Meanings are the ideas, emotions, or beliefs that we attach to an object, a behavior, or another idea. Researchers make sense of behaviors by trying to understand the meanings behind them. The purpose or function of a behavior is not always self-evident. Ethnographers typically have to probe the respondent and deeply understand the context if they wish to gain insights into the underlying meanings attached to observed behaviors. The language of cultural patterns that hold or convey meaning includes these concepts:

Symbols   Symbols are things that stand for something else, including simple shapes or marks, such as the cross, which stands for Christianity, the crucifixion, Christ's suffering, salvation, or other religious ideas. Alternatively, symbols can be much less ethereal, such as a corporate logo. Symbols connect an idea or thing to some underlying system of meaning. Thus, for example, a Coca-Cola logo can stand for youth, energy, refreshment, or tradition.

Signs   Signs or markings point to something in the environment. A highway sign, for example, directs you to your exit. The McDonald's arches point consumers to a place where they can enjoy a consistent and predictable family meal and share each other's company.

Language, Jargon, and Slang   Words are tools of communication that have meanings beyond their basic communication value. They serve as code words to distinguish "in" groups from "out" groups; words provide a cipher or code to communicate meanings within closed groups such as business organizations and religious or ethnic subcultures.

Marketing ethnographies provide opportunities to study language in its natural context by listening in on what consumers say to each other, in addition to what they say in response to a question posed by a researcher. We can learn a great deal about beer brands or mobile phones by studying the ways that young consumers talk about these subjects during a night out with friends.

Beliefs and Values   Beliefs and values are meaning filters or standards of truth for everyday life. They provide legitimacy to personal behavior and help people tell right from wrong. Broad patterns of values—for example, the refusal to eat meat or a concern about the environment—can be the basis for consumer behavior across a range of categories.

Attitudes and Opinions   Attitudes and opinions are expressions of a point of view toward people, things, and events, and can range from positive to neutral to negative. They can be intensely upheld or maintained with minimal salience. Attitudes and opinions are occasionally rooted in deeply held principles and values, but, more likely, they are situational and less character-based.

Interpretation   Through interpretation, we make sense of things that are communicated to us. Two people may see the same advertisement or read the same novel but interpret them differently. Interpretation is an essential component of the marketing communication process; it's what consumers *hear* when marketers talk.

Emotions and Feelings   Inner conscious and unconscious emotions and feelings normally occur reflexively; sometimes, we actively attach them to people, behaviors, or ideas. In a marketing context, emotions are consumers' inner response to the external world.

Relationships   Socially constructed ties between things or individuals—for example, being part of a family, nation, or club—have a tremendous impact on our daily life, our work habits, and our beliefs. Many brands are used to tie people into larger communities. Harley-Davidson owners, for example, think of themselves as part of a larger community of somewhat rebellious, adventure-seeking individuals.

5.3.4.3 Cultural Tools   Tools are culturally produced devices—both objects and ideas—that expand human powers. Tools can help us live life more comfortably or securely; they can help us perform our work, defend ourselves, and organize our social ties. To a considerable degree, the practice of marketing ethnography is directed toward evaluating the marketplace tools that are available to consumers and, potentially, creating new tools that expand consumer satisfaction, stimulate productive efficiencies, and boost client profitability. In a broad ethnographic context, tools may take a variety of forms.

Physical Space   The environment, including our homes, workplaces, shopping locations, and cities, is the most basic tool.

Technology—Low to High   The conventional meaning of tool is anything that expands human powers, from paper clips to supercomputers.

Rules   Culture rules provide a systematic framework for the operation of social organizations and processes: nation-states, professions, commercial enterprises, and families. They guarantee that everyday life proceeds with some degree of predictability and security. Rules can be both formal—written, codified, and elaborated—or informal, a set of understandings that guide everyday behavior such as rules for queuing for the commuter bus. A simple company policy statement, the rules of baseball, and the U.S. Constitution are examples of rules that help us carry on our lives.

An important component of ethnographic practice is decoding the rules that are operating in the settings under study. In many cases, particularly where informal rules are active, consumers may not be aware that they are following any set of patterned behaviors until the basis of their actions is questioned. In studying home barbecue grilling, for example, we found that knowledgeable consumers generally followed a highly prescriptive set of rules for the steps in tasks such as how meats should be prepared and how charcoal fires should be started and maintained. Less successful chefs were less aware of the rules and, consequently, were susceptible to switching to gas grills, a competing technology, because of the larger tolerance for error that gas grilling offered.

Techniques   Ways of getting things done are the "how-tos" of our daily lives. Techniques work together with technologies and rules to advance human potential. The techniques that people use to complete the tasks of daily living in a culture may serve as clues to how well people are satisfied with the technologies available to them. Many consumers clean floors on their hands and knees in the belief that nothing else produces their desired state of cleanliness. The ways in which we complete everyday tasks may be creative and innovative; alternatively, they may follow conventional rules quite ritualistically. Ethnographic study proves its value to marketers when it can use findings about techniques to structure innovations and incremental improvements in product formulations and delivery.

A Case Study on Ethnographic Research on Customer Culture and Product Development   This case study is about Lipton's effort in bringing its brand and product into China. Lipton is the world's biggest tea brand, selling both hot and iced tea around the world. It started as a British tea company, founded by Sir Thomas Lipton. The brand name is now owned by Unilever. Lipton Tea represents around 10 percent of the world market for tea.

China is one of the top tea producing and consuming countries of the world. After the late 1970s, the Chinese market gradually opened to the outside world. As a top tea company, Lipton wanted a part of this huge Chinese tea market. Lipton is famous for its tea bags and black tea. In many Chinese people's minds, black tea is foreign. Because of Lipton's long history and brand image; it did not take long for Lipton to take 80% of the tea bag market.

However, black tea in bags accounts for less than 2% of the overall Chinese tea market. The predominate portion of the tea market in China is taken by green tea and jasmine tea. In these two market segments, most people in China think green tea and jasmine tea are mostly Chinese; Lipton has nothing to do with it. Secondly, Lipton is well known for its tea bags, and most Chinese people who consume green tea or jasmine tea serve their tea in the form of "loose leaves," as illustrated in Figure 5.3.

In the West, most hot tea is made with tea bags (Figure 5.3). Tea bags do have several clear advantages over loose leaves:

■ Tea bags are easier to dispose of.

Figure 5.3 Loose tea leaves and tea bags

- When tea is served in a cup with loose tea leaves, some tea leaves tend to float on top, so it is easy for people to end up swallowing tea leaves, an uncomfortable experience. Tea bags prevent these problems.
- Tea cups are easier to clean with tea bags than with loose tea leaves.

However, in spite of these apparent advantages, Chinese people have a hard time accepting green tea or jasmine tea in tea bags. One important reason is that China has a unique and rich tea culture.

In China, tea is not only a popular drink but also serves many cultural rituals:

- **As a sign of respect** In Chinese society, the younger generation always shows its respect to the older generation by offering a cup of tea. Inviting and paying for their elders to go to restaurants for tea is a traditional activity on holidays. In the past, people of lower rank served tea to higher-ranking people. Today, as Chinese society becomes more liberal, sometimes at home parents may pour a cup of tea for their children, or a boss may even pour tea for subordinates at restaurants. The lower-ranking person should not expect the higher-ranking person to serve him or her tea on formal occasions, however.
- **For a family gathering** When sons and daughters leave home to work and get married, they may seldom visit their parents. As a result, parents may seldom see their grandchildren. Going to restaurants and drinking tea, therefore, becomes an important activity for family gatherings. Every Sunday, Chinese restaurants are crowded, especially when people celebrate festivals. This phenomenon reflects Chinese family values.
- **To apologize** In Chinese culture, people make serious apologies to others by pouring them tea. That is a sign of regret and submission.

The way Chinese tea is brewed, served, and consumed is another sophisticated ritual. There are many different ways of brewing Chinese tea depending on variables like the formality of the occasion, the means of the people preparing it, and the kind of tea being brewed.

The tea can be brewed in teapots or in cups. The formal way is to brew tea in pots, and in one form is considered to be a kind of art (Figure 5.4):

1. Wash the teapot.
2. Put tea leaves into teapot.
3. Put hot water into the teapot, and wait about 1 to 2 minutes.
4. Pour out the tea in cups; this is the first infusion.
5. Put hot water in teapot again, and wait about 2 minutes.
6. Pour out the tea into cups; this is the second infusion.
7. Repeat step 3 and keep going for about two more infusions.
8. In Chinese tea culture, the first and second infusions are the best "shots"; it is believed that the first infusion is the best in terms of aroma, and the second infusion is the best in terms of flavor and color.

This process of brewing and drinking tea is an essential part of Chinese tea culture. In many Chinese people's minds, drinking tea is not just for quenching thirst; it is a process for experiencing different tea tastes, and for sharing the experience with others.

A less informal way of brewing tea is in tea cups. The procedure is similar to that of teapots, and people enjoy watching the changes in shapes of the tea leaves. In the beginning, the tea leaves are dry, but after a couple of infusions, the tea leaves are in full blossom and float to the top, and finally the tea leaves fall down to the bottom. "Reading tea leaves" is a part of the tea-drinking process.

After studying the Chinese tea-drinking culture, Lipton researchers realized that the tea bags' advantages would not be enough to capture the green tea and jasmine tea market, for the following reasons:

- In the West, the tea brewing is very simple. First, you put a tea bag in the cup; second, you pour hot water in; third, you wait a couple of minutes, and then you drink. After one infusion, you throw away the

Figure 5.4 Teapots and the tea brewing process

tea bag. In this case, it is desirable to make the tea in such a way that most of its contents, aroma, and flavor should be getting into the hot water in one shot. Clearly, this design doesn't match the Chinese tea culture.

- In the West, the tea bag is brewed only once, in China, it will be brewed many times, as the part of tea culture. So the tea bags used in Chinese green tea or jasmine tea should be more sturdy and durable.

- Tea bags are good for disposal and cleaning, but the tea bag takes away all the cultural meanings of "reading the tea leaves."

Based on this analysis, Lipton came up with three major design improvements:

- Design the teabag in such a way that the tea contents, aroma, and flavor will be gradually dissolved into hot water, not in one shot.

- Make the tea bag of a more sturdy and durable material, so it can withstand multiple brewings and infusions.

- Make the shape of the tea bag like a pyramid, so it is a tea bag, but you can still "read the tea leaves" (Figure 5.5).

It is reported that this new design achieved some degree of success.

Figure 5.5   Pyramid tea bags and tea leaves reading

6

VOC Data Processing

Chapters 4 and 5 discussed many effective methods of capturing VOC information, and in a comprehensive VOC study, we might use several different approaches, such as ethnographic observation, notes, taping, customer surveys, and lead-user interviews. Regardless of the techniques, though, we will have a lot of data of different kinds. In this chapter, we will look at how to analyze various types of VOC data and derive meaningful results.

6.1 Types of VOC Data

The kinds of data you get from VOC-capturing activities will depend on the methods you use. In general, the ethnographical research methods described in Chapter 5 will produce the following types of data:

- Notes of open-ended interviews
- Field observation notes
- Video and audio recordings

This kind of data is generally not quantitative in nature.

If you use predesigned surveys (Chapter 4), you will get customers' answers to the survey questions. Most surveys use multiple-choice questions, with some of the choices involving attribute data (Yes/No answers) where there is no numerical value associated with the answer, and some choices being numerical, such as ratings from 1 to 5. If the survey results are intended to derive product functional requirements (as discussed in Chapter 3), much more detailed information will need to be collected, such as "What jobs do you want this product to do?" and "What outcomes do you want the product to achieve?" In this case, we will get some very descriptive, detailed, technical data. In general, there will be three kinds of data:

- Words and notes
- Attributes and numerical data
- Data that can lead to design specifications

6.2 Analyzing VOC Data

The usual purposes of analyzing VOC data (as discussed in Chapters 1–3) are as follows:

- **Getting sufficient and accurate input for the product development process**   In this case, the purpose is to determine design requirements for the product based on the VOC data. In Six Sigma practice, a common outcome is critical-to-quality characteristics (CTQ), or critical-to-satisfaction characteristics (CTS). CTQ and CTS are sets of quantitative and actionable key product-quality measures that can be readily transformed into design specifications. Sometimes only limited improvements are made to an existing product, so the scope of the task is smaller than for new product design, but it is still necessary to derive some CTQs. (The details of deriving CTQs from raw VOC data will be discussed at the end of this chapter.)

- **Evaluating a product's competitive position, customer value rating, and so on**   The output of this kind of VOC data analysis is usually a traditional statistical analysis report. In this case, we need to analyze survey data— both attributes and quantitative data.

6.2.1 Methods of Analyzing VOC Data

For data in the form of "words and notes," the objective is to discover meanings and patterns in the data. The affinity diagram or KJ method is very good for this. Affinity diagrams will be discussed in the next section.

For "attributes and numerical data," the objective is to quantify the data's statistical measures and possibly conduct statistical inferences. We will look at statistical analysis tools for this kind of data in Section 6.3.

The procedures for deriving and quantifying CTQs from data that can lead to design specifications will be covered in Section 6.4.

6.2.2 Affinity Diagram—KJ Method

The affinity diagram is also called the KJ Method, named after the Japanese anthropologist, Jiro Kawakita, who developed a method of establishing an orderly system from chaotic information (Kawakita

1977, 1991). The KJ method is a four-step process for organizing and summarizing a large amount of data (ideas, issues, solutions, problems) into logical categories so that it is possible to understand the essence of a problem or solution.

When using the KJ method, we write all relevant facts and information on individual cards, which we collate, shuffle, spread out, and read carefully. We shuffle the cards because these cards could come from several sources. We then review, classify, and sort the cards based on the similarity, affinity, and characteristics of the ideas. The four steps of the KJ method are illustrated in Figure 6.1.

6.2.2.1 Step 1: Collecting Data and Preparing Notes First, prepare some index cards or sticky notes for use. Then write each idea from the "words and notes" raw data on one of the cards or notes—one idea for each card.

6.2.2.2 Step 2: Sorting Ideas into Related Groups Sort the cards into groups using the following process:

1. Put all the cards in one pile.
2. Draw one card at a time from this pile, and examine its content. If the idea from this card is related to the idea from the card that you drew before, put them together; these two cards belong to a group.

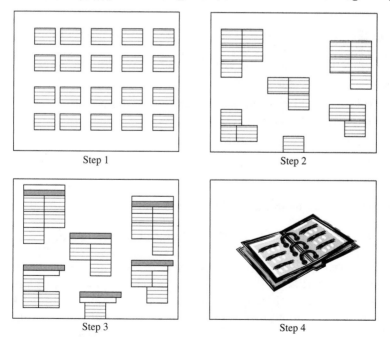

Figure 6.1 Four steps for building affinity diagrams (KJ method)

3. Keep drawing the cards one at a time. Examine the card to see if the idea from the card is similar to any existing group; if the answer is yes, put this card into that group. If the answer is no, this card can start a new group. Keep doing this step until all the cards are drawn and all of the ideas are sorted into groups.

4. After all cards are drawn, re-examine all the groups. You can move some cards around so that each group's ideas look more coherent. If an idea seems equally applicable to two groups, create a duplicate of that card and place one in each group.

5. It is possible for one card to stand alone and form a group.

6. The ideal grouping result should have the following features:

   a. The ideas within a group should be closely related.

   b. There should be significant differences between groups.

6.2.2.3 Step 3: Creating Header Cards for Each Group   Create header cards for each group. A header is a title that captures the essential link among the ideas contained in a group of cards.

a. The header should be the best word or phrase that describes the meaning of each group. The meaning of the header should stand alone and be clear to outside readers without reading the contents of the cards in the group.

b. During the process of creating headers, it is possible to regroup the ideas so that the headers will have clear and better meanings. It is also possible that a big group will be subdivided into several small groups under different headers.

c. It may take several iterations to finalize the header process in order to best capture the meaning of each group.

d. Clarify and finalize headers through consensus.

e. It is possible that hierarchical groups or multilevel groups will be adopted.

Figure 6.2 gives an example of idea groups and headers.

6.2.2.4 Step 4: Writing Reports and Doing Further Analysis   Affinity diagrams can help to reveal hidden groups and structures in a sea of fragmented notes and words, so you can see how everything fits together. Once you have done this initial analysis, you will need to write a report outlining the meaning of these groups, and this information can be used in the product development process. One way to figure out the meaning of the idea groups is to try to connect different groups with arrows or lines that indicate relationships among the groups. Figure 6.3 shows

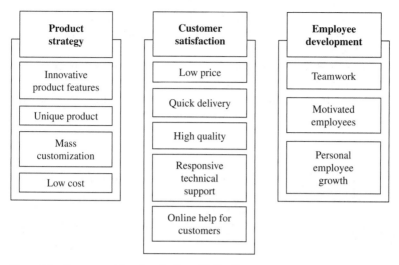

Figure 6.2   Groups and headers in the affinity diagram

such an example; it connects different idea groups related to a hand-held telecom product.

The structure of affinity diagrams can even be used to analyze whether you have captured enough VOC information in your study. If you were

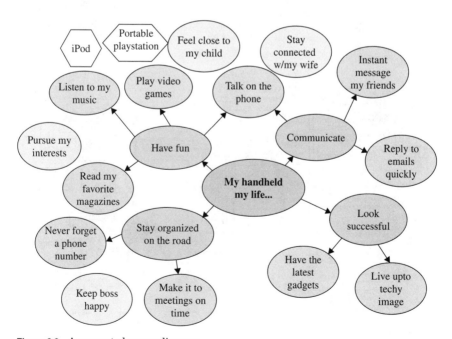

Figure 6.3   A connected arrow diagram

expecting some idea groups to show up in the results, but they are absent, this is an indication that you haven't captured enough VOC information. Figure 6.4 illustrates how this works.

In the real world, we may stick Post-It notes to a whiteboard to construct an affinity diagram, as shown in Figure 6.5.

Example 6.1 (from Ramaswamy 1996) shows an affinity diagram being used to study VOC data for a restaurant chain.

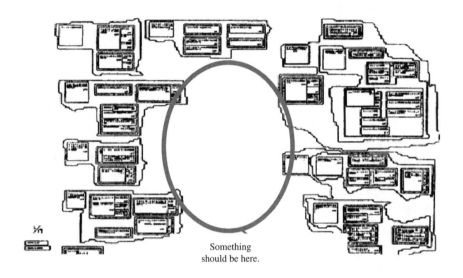

Something
should be here.

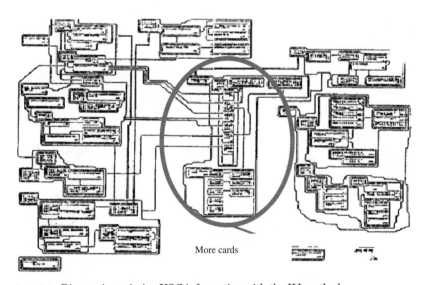

More cards

Figure 6.4 Discovering missing VOC information with the KJ method

Voice of the customer	Interpretation	Critical to Quality (CTQ)
"Wait in the lobby for too long"	Short wait	Waiting time
"Sit in triage for a long time"	Quick response from nurses and doctor in emergency care	Response speed in emergency care unit
"Too many errors in my bill"	Billing accuracy	Percentage of erroneous invoices
"It takes me several visits to get it right"	Accuracy in diagnosis	Percentage of erroneous diagnosis
"Nurses don't care about the patients"	Patient Care	Customer satisfaction rating
"It takes so long to see the expert"	Slow Process	Expert utilization rate

Figure 6.5 A finished affinity diagram

Example 6.1 Customer Needs for Restaurant Service   VOC data can be used to derive key CTS (Critical to Customer Satisfaction) metrics in the restaurant business. The VOC data is often disorganized, nonspecific, and nonquantitative, as shown in Table 6.1, which lists customer needs for restaurant service.

TABLE 6.1   List of Customer Needs for Restaurant Service (Ramaswamy 1996)

1. Food tastes good	16. Don't want noisy atmosphere
2. Unusual items on menu	17. Want smoke-free atmosphere
3. Hot soup, cold ice cream	18. Wide choice of food
4. Feel full after the meal	19. Enough time to read menu
5. Don't feel overfull after meal	20. Can order quickly
6. Food looks appetizing	21. Know how long the wait for a table will be
7. Food courses arrive on table at right time	22. Food is healthy
8. Don't feel hungry one hour after meal	23. Menu items easy to understand
9. Clean restrooms	24. Prompt delivery after ordering
10. Clean tables	25. Get what was ordered
11. Clean plates and silverware	26. Get the correct bill
12. Clean, well-dressed employees	27. Billed as soon as meal is over
13. Light not too bright	28. Shouldn't feel rushed out of restaurant
14. Light not too dim	29. Make me feel at home
15. Shouldn't feel too crowded in space	30. Order additional items quickly

TABLE 6.1   List of Customer Needs for Restaurant Service
              (Ramaswamy 1996) (*Continued*)

31. Errors and problems quickly resolved	35. Waiter should be patient while ordering
32. Errors and problems satisfactorily resolved	36. Fill water glass promptly without asking
33. Staff willing to answer questions	37. Polite, friendly staff
34. Greeted immediately on being seated	38. Short wait for table

The affinity diagram or KJ method (Shigeru 1988) can be used to analyze and organize the VOC data into a CTS tree. A CTS tree is a refined multilevel table of attributes that identify critical characteristics for customer satisfaction. By using an affinity diagram, the two levels of affinity groups described in Table 6.2 can be discovered.

TABLE 6.2   Two Levels of Affinity Groups for Restaurant Service (Ramaswamy 1996)

First Level	Second Level	Original Notes
Satisfying food	Tasty food	Food tastes good
		Balance of flavors
		Hot soup, cold ice cream
		Food looks appetizing
		Food is healthy
	Enough food	Feel full after meal
		Don't feel overfull after meal
		Don't feel hungry one hour after meal
	A lot of variety	Wide choice of food
		Unusual items on menu
Clean and attractive surroundings	Clean facility	Clean restrooms
		Clean tables
		Clean plates and silverware
		Clean, well-dressed employees
	Comfortable atmosphere	Light not too bright
		Light not too dim
		Shouldn't feel crowded in space
		Don't want noisy atmosphere
		Smoke-free atmosphere

TABLE 6.2 **Two Levels of Affinity Groups for Restaurant Service (Ramaswamy 1996)** (*Continued*)

First Level	Second Level	Original Notes
Good service	Friendly and knowledgeable staff	Make me feel at home
		Staff willing to answer questions
		Polite, friendly staff
		Waiter should be patient while ordering
		Menu items easy to understand
		Shouldn't feel rushed out of restaurant
		Fill water glass promptly without asking
		Enough time to read menu
	Quick and correct service	Short wait for table
		Know how long the wait for a table will be
		Can order quickly
		Greeted immediately on being seated
		Prompt delivery after ordering
		Get what was ordered
		Order additional items quickly
		Food courses arrive on table at right time
	Accurate billing	Get the correct bill
		Billed as soon as meal is over
	Problems and complaints addressed effectively	Problems quickly resolved
		Problems satisfactorily resolved

6.3 Quantitative VOC Data Analysis

Quantitative VOC data can come from customer surveys, competitive benchmarking (collecting competitors' product specifications and performance data), lead-user interviews, and so on. There are two types of quantitative VOC data: attribute data and variable data.

Attribute data are either categorical or discrete. Examples of categorical data include gender, color, social class, and so on. Discrete data means that the data values can only be integers (that is, no fractions or decimal points), such as the number of defective units, paint chips per unit, number of scratches, and so on.

Variable data, or continuous data values, can be any real number, such as -1.238, 78.45, 0.02875, and so on. Examples of continuous data include length, volume, time, weight, and so on.

Another important characteristic of data is its measurement scale. Table 6.3 describes four measurement scales with the latter ones being more useful for statistical analysis.

For data with different measurement scales, the two important statistical metrics—the central tendency and dispersion—are different. Table 6.4 shows the statistical measures for data with different measurement scales. (Commonly used statistical basics, measures, and analysis methods are discussed in Chapter 11.)

TABLE 6.3   Four Measurement Scale Levels

Scale	Description	Example
Nominal	Data consists of names or categories only. No ordering scheme is possible.	A parking lot has cars of the following colors: Red 5 White 4 Blue 7 Black 6
Ordinal (Ranking)	Data is arranged in some order but differences between values cannot be determined or are meaningless.	A survey question: Ice cream is good for breakfast: 1. Strongly disagree 2. Disagree 3. Neither agree nor disagree 4. Agree 5. Strongly agree The difference between Strongly agree (5) and Agree (4) does not have the same meaning as the difference between Disagree (2) and Strongly disagree (1).
Interval	Data is arranged in order and differences can be found. However, there is no inherent starting point and ratios are meaningless.	The temperature of three heated metal pieces is 300°C, 600°C, and 900°C respectively. Three times 300°C is not the same as 900°C in temperature measurement.
Ratio	An extension of the interval scale that includes an inherent zero starting point. Both the difference between values and the ratios are meaningful.	Product A costs \$400; product B costs \$200.

TABLE 6.4   Statistical Measures for Data with Different Measurement Scales

Measurement Scale	Central Tendency	Dispersion
Nominal	Mode	Bar chart
Ordinal	Median	Percentage distribution
Interval	Arithmetic mean	Standard deviation or range
Ratio	Geometric or harmonic mean	Coefficient of variation

Various statistical analyses can be performed on quantitative VOC data, and it is important to take the measurement scale of the data into consideration to make sure you are doing the right analysis.

6.4 Critical-to-Quality Characteristics (CTQ)

The ultimate goal of product development is to design a product that customers really want. That is why it is important to spend so much effort capturing the VOC information. However, we cannot design a good product simply by using the raw VOC information. For example, suppose you want to design a household power saw for yard work, and when you summarize the findings from the VOC data, you have a list like this: "We want a saw that cuts wood quickly and easily"; "We want a saw that does not get stuck easily"; "We want a saw that is easy to carry." When you give these findings to design engineers, they will likely ask: "What do you mean by cutting wood quickly?"; "How quickly is quick enough?"; "What kind of wood you are talking about?" and so on. Clearly, this raw VOC information won't give design engineers enough information to set technical specifications.

The raw VOC data needs to be developed into clear, specific, quantitative requirements in order to be really helpful in product development. These kind of requirements are called critical-to-quality characteristics (CTQs) in Six Sigma practice. CTQs are the product or service characteristics that the customer considers important, and they are measurable characteristics whose performance standards or specification limits must be met to satisfy customer requirements.

For example, "light in weight" may be one of the customers' key requirements for a power saw, but the statement "light in weight" is not a CTQ, because it does not give either a performance standard or specification limit. The statement "the weight of power saw should be no more than 3 kg" is a CTQ because weight is a key performance factor that is important to customers, and this statement gives a very specific performance specification.

A typical CTQ is illustrated in Figure 6.6. It usually has four components: characteristic, measure, target, and specification limits.

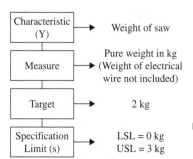

Figure 6.6 Components of a CTQ and example

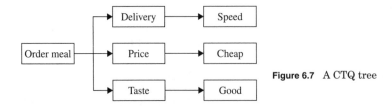

Figure 6.7 A CTQ tree

For a product or service item, there are usually several CTQs. For example, a fast-food restaurant meal order process may have three CTQs: delivery, price, and taste. For customer satisfaction, the delivery needs to be fast, the cost needs to be low, and the taste needs to be good, as illustrated in Figure 6.7. This is also called a CTQ tree. Of course, for a complete CTQ tree, quantitative target and specification limits are required. The delivery time has a numerical target (in minutes and seconds), the cost target should be in dollars and cents, and the taste can be quantified by using customer-rating scores.

The next question is how to derive CTQs from raw VOC data. We can follow these steps:

1. Listen to statements and comments made by clients.

2. Determine what factors really drive a client to turn on or turn off. What product features, service features, attributes, dimensions, and characteristics (reliability, availability, and so on) cause a client to be excited or angry?

3. Determine the best way to measure performance relative to these factors and specifications.

4. Determine what targets we would need in order to meet the client's factors and specifications.

5. Assess current performance relative to these factors and specifications. Table 6.5 shows the process of translating some raw VOC data to CTQs. Each row of Table 6.5 shows a three-step transformation from a raw VOC item to a CTQ. The VOC items in Table 6.5 are all related to patients' expectations when they are in the hospital.

The last question in CTQ development is how to derive target values and specifications for CTQs. There are several ways to do this:

1. **Customers' inputs** Though customers are usually not experts in the product or service they are purchasing, you can still find their "specification limits." For the example illustrated in Figure 6.7, when asked what speed and price they consider good, fast-food restaurant customers may answer: "Faster than McDonald's, cheaper than Wendy's." In that case, you can determine the appropriate serving

TABLE 6.5 Translating VOC to CTQ for Hospital Patients

Voice of the Customer	Interpretation	Critical to Quality (CTQ)
"Wait in the lobby for too long"	Short Wait	Waiting Time
"Sit in Triage for a long time"	Quick Response from Nurses and Doctor in Emergency Care	Response Speed in Emergency Care Unit
" Too many errors in my bill"	Billing accuracy	Percentage of Erroneous Invoices
"It takes me several visits to get it right"	Accuracy in Diagnosis	Percentage of Erroneous Diagnoses
"Nurses don't care about the patients"	Patient Care	Customer satisfaction rating
"It takes so long to see the expert"	Slow Process	Expert Utilization Rate

speed and price by finding out what is being done in McDonald's and Wendy's. If customers give very clear outcome statements, you can derive specifications from this, for example, if customers who are buying weed killer say that "the weed should die within a day after application."

2. **Benchmarking** Benchmarking is comparing your approach to existing approaches or competitive designs. One easy way to do this is to check your competitor's specifications. You can also look for existing ways of doing your function or for similar designs. Check the Internet, the library, and patents.

3. **System constraints** Government regulations, safety concerns, and physical limitations will all set limits on specifications.

Example 6.2 shows how to derive CTQs (adapted from Ramaswamy 1996)—it is a continuation of Example 6.1.

Example 6.2 Deriving Quantitative CTQs for the Restaurant In Example 6.1, we derived three groups of customer requirements from the KJ method:

- Satisfying food
- Clean and attractive surroundings
- Good service

The second-level groups explained the first-level groups and they are the aggregated categories of original customer statements.

To derive CTQs, however, we still need quantitative measures. For example, "short wait for a table" is a specific customer requirement, but how short is short? Five minutes or ten minutes? Similarly, just what does "food tastes good" mean? Does the food taste good enough now? Also, we need to know the relative importance of each CTS item. For example, which is more important to the average customer: the taste of the food, or the nutritional quality of the food? All this needed information can be found by various means, such as using specially designed customer surveys, benchmarking competitors, and using the "mystery customer" approach.

The mystery customer approach involves hiring either employees or temporary helpers as "mystery customers" who are paid to visit competitors' facilities, as well as your own facility, and to fill out a number of specially designed questionnaires after each visit. Some stopwatch activities may also be included to record such measures as the service waiting time, time to deliver the meal, and so on. With the help of "mystery customers," clear, quantitative measures can be developed. For example, if the waiting time of the best competitor is no longer than five minutes, and if waiting time is really important to customers, you need to set a goal of waiting time being less than five minutes. If "mystery customers" found that your competitor does offer better-tasting food, you need to work on making your food taste better.

After processing the original affinity diagram, the following CTQ measures are determined:

- Degree of waiter patience
- Degree of waiter responsiveness
- Degree of waiter knowledge
- Degree of waiter friendliness
- Time between seating and menu delivery
- Time between menu delivery and order taking
- Time between ordering and meal delivery
- Percentage of bills produced without errors

Mystery customer studies and competitive bench marking on several competitors yielded the results in Table 6.6.

This example illustrates how you can develop two different kinds of measurable performance metrics. One is an evaluation score type, such as "degree of waiter patience," and the other is a measurable performance metric, such as the time between ordering and meal delivery. Benchmarking competitors can help in designing performance specifications, which are listed as "our desirable performance." The performance gaps on the performance metrics can be used to guide redesign practice.

TABLE 6.6 Benchmarking Results for Restaurants

	Degree of Waiter Patience	Degree of Responsiveness	Degree of Knowledge	Degree of Friendliness	Time between Seating and Menu Delivery	Time between Menu Delivery and Order Taking	Time between Ordering and Meal Delivery	Percent of Bill produced Without Errors
Performance Gap	One grade short	One grade short	Two grades short	One grade short	Achieved	3-minute gap	2-minute gap	5% gap
Our Desired Performance	Exceptional	Exceptional	Exceptional	Exceptional	< 5 minutes	< 5 minutes	10 minutes	95%
Our Restaurant	Excellent	Excellent	Good	Excellent	< 5 minutes	< 8 minutes	12 minutes	90%
Vive la France	Good	Exceptional	Excellent	Good	< 10 minutes	< 10 minutes	20 minutes	>90%
Downtown Steakhouse	Excellent	Excellent	Excellent	Exceptional	< 5 minutes	< 8 minutes	15 minutes	>90%
Sarah's Seafood House	Good	Good	Good	Good	< 5 minutes	<5 minutes	10.5 minutes	>92%

Quality Function Deployment (QFD)

Quality Function Deployment (QFD) is a planning tool that can be used to translate customer needs and expectations into appropriate design actions. QFD stresses problem prevention and places its emphasis on achieving results in customer satisfaction, reducing design cycle time, optimizing the allocation of resources, and ensuring that minimum changes are required. Together with other quality tools and concepts, QFD makes it possible to release products at Six Sigma level. Since the customer defines quality, QFD develops customer and technical measures to identify areas for improvement. It translates customer needs and expectations into design requirements by incorporating the voice of the customer into all phases of the product development process, through production and into the marketplace.

In the context of product development, the real value of QFD is its ability to direct the application of other quality tools to those design tasks that will have the greatest impact on the team's ability to design a product, service, or process that satisfies the needs and expectations of the customers, both internal and external. QFD is best viewed as a planning tool that relates a list of VOC parameters to functional design requirements. Customers define the product using their own expressions, which usually do not carry any significant technical terminology. With the application of QFD, possible relationships between quality characteristics, as expressed by customers, and *substitute quality requirements,* expressed in engineering terms, can be explored (Cohen 1988, 1995; Clausing 1988). In the context of product development, these requirements are called *Critical-To* characteristics, which include subsets like *Critical-To-Quality (CTQs), Critical-To-Delivery (CTDs),* and others.

The VOC items will be developed into a list of needs used later as input to a relationship diagram, which is called QFD's *House of Quality*.

Correct market predictions are of little value if the requirements cannot be incorporated into the product design at the right time. Wresting market share away from a viable competitor is more difficult than capturing market share by being the first producer into a market. One major advantage of QFD is that it shortens the development cycle by deploying customer expectations into product design in the early stage, avoiding major design changes or redesigns.

The other significant advantage of QFD is increased customer satisfaction. The team will employ marketing and product planning inputs to incorporate customer expectations in the design process, production planning, and in all functional departments. This will ensure that issues are resolved, and the design is kept lean and focuses on those innovations that are important to the customer.

Figure 7.1 shows that a company that is using QFD places more emphasis on responding to problems early in the design cycle; it will consume more resources in the early design stage to make sure the design concept is sound. A company that does not use QFD will consume fewer resources in the early design stage, but it will incur more resources later in the design stage to fund design changes, and it usually will spend a lot of money to fix problems after design release.

This chapter will explain how Quality Function Deployment works. Specifically, I will start with QFD notations and the QFD matrix, and then I will show how to fill a QFD matrix. Finally I will discuss the Kano model and four phases of QFD.

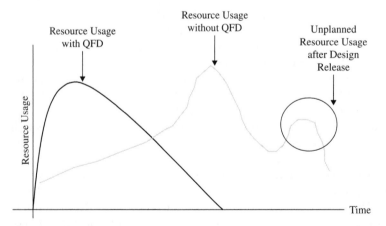

Figure 7.1   Effect of QFD on project resources

7.1   History of QFD

QFD was created by Mitsubishi Heavy Industry at Kobe Shipyards in the early 1970s. Stringent government requirements for military vessels coupled with the large capital outlay per ship forced Kobe Shipyard's management to commit to upstream quality assurance. The Kobe engineers drafted a matrix, which related all the government regulations, critical design requirements, and customer requirements to technical characteristics the company could control for and achieve. In addition, the matrix also depicted the relative importance of each entry, making it possible for important items to be identified and prioritized so they would receive a greater share of the available resources.

Winning is contagious. Other companies adopted QFD in the mid-1970s. For example, the Japanese automotive industry first applied QFD to the rust problem. Since then, QFD usage has become a well-rooted methodology in many American businesses. It has become familiar because of its adopted commandment: "Design it right the first time."

7.2   QFD Benefits, Requirements, and Practicalities

The major benefit of QFD is customer satisfaction. QFD gives customers what they want: Development cycles are shorter, failures and redesign peaks during pre-launch (shown in Figure 7.1) can be avoided, and customer demand knowledge is preserved and transferred to subsequent design teams.

Certain things must be done before QFD can be implemented. They include forming a multidisciplinary product development team, which includes people from the marketing department, engineering department, and others; spending time beforehand understanding customer needs and expectations; and defining the product or service in detail.

Many practical concerns must be addressed in order to implement QFD successfully. For example, departments represented in the team often don't tend to talk to one another. Also, problem prevention is not traditionally rewarded as well as problem solving or "fire fighting," so convincing the management to allocate manpower and resource on a problem-prevention methodology such as QFD is a challenging task. Usually, it is easier to use QFD on an incremental design than a brand new creative design, because both marketing and engineering people would have a better idea of the voice of the customer for an existing product. So it is better to develop your first QFD application as an incremental design task.

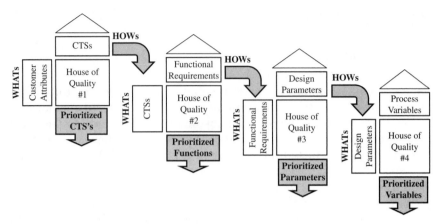

Figure 7.2   Four phases of QFD in the product development process

7.3   QFD Methodology Overview

Quality Function Deployment is accomplished by multidisciplinary product development teams using a series of charts to deploy critical customer attributes throughout the phases of design development. QFD is usually deployed in multiple phases. Figure 7.2 shows the typical four-phase deployment in a product development application.

These four phases focus on planning CTS characteristics, functional requirements, design parameters, and process variables.

In a typical service industry setting, QFD can be deployed in two phases, focusing on planning CTS characteristics and operation variables, as illustrated in Figure 7.3.

QFD incorporates many techniques to make it easier to handle the large numbers of functional requirements that might be encountered. Applications involving 130 functions multiplied by 100 customer features have been recorded (Hauser and Clausing 1988). One typical

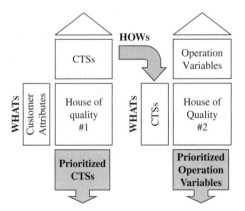

Figure 7.3   Multiple phases of QFD in a service application

grouping technique that may initially be used in a QFD study is the *affinity diagram*—a hierarchical grouping technique that consolidates multiple unstructured ideas generated by the voice of the customer (affinity diagrams are explained in Chapter 6). Affinity diagrams are based on intuitive similarities from low-level stand-alone ideas to arrangements of classes of ideas. This bundling of the customer features is critical. It requires a cross-functional team that can brainstorm, evaluate, and reconsider existing ideas in pursuit of identifying logical (not necessarily optimum) groupings in order to organize the overall list of needs into manageable classes.

Another technique is the tree diagram, which goes a step beyond the affinity diagram. The *tree diagram* is used mainly to fill the gaps not previously detected in order to achieve a more complete structure, which in turn can lead to more ideas. Such expansion of ideas will allow the structure to grow, but will also provide more insight into the voice of the customer (Cohen 1988).

The *House of Quality* (see Figure 7.4) is the most important template produced in a QFD study. The House of Quality looks like a house; the center box of the house is a matrix diagram, which quantitatively relates rows and columns. The rows on the left of the center usually represent the customer attributes, or what the customer wants (WHATs);

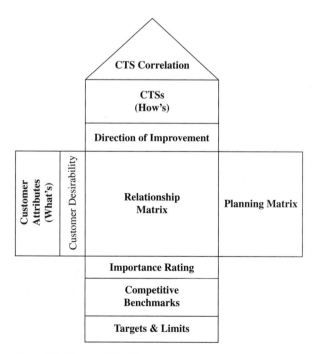

Figure 7.4   House of Quality

the columns on the top of the center box are usually refined product requirements (HOWs), or how to design the product to satisfy the customer attributes. The quantitative relationships between "WHATs" and "HOWs" are the key information produced in a QFD study. The other portions in the House of Quality provide other supporting information in the QFD study. Employing the house will result in improved communication, planning, and design activity. This benefit also extends beyond the QFD team to the whole organization—customer wants defined through QFD can be applied to many similar products and form the basis of a corporate memory on the subject of critical-to-satisfaction requirements (CTSs). As a direct result of the use of QFD, *customer intent*, or what the customer wants, will become the driver of the design process as well as the catalyst for modifying product design solutions.

In Figure 7.4, we have the following components that constitute the *House of Quality* (Cohen 1988):

- Customer attributes (WHATs)
- CTSs (HOWs)
- Relationship matrix
- Importance ratings (Customer Desirability Index and Technical Importance Ratings)
- Planning matrix
- CTS correlation (HOWs correlation)
- Targets and limits (HOW MUCH)
- Competitive benchmarks (Customer Competitive Assessment and Technical Competitive Assessment)
- Other optional QFD chart extensions

7.3.1 Customer Attributes (WHATs)

Customer attributes are obtained from the VOC data collected by surveys, claim data, warranty, and promotion campaigns. Usually customers use fuzzy expressions in characterizing their needs and refer simultaneously to many dimensions that need to be satisfied—affinity and tree diagrams may be used to complete the list of needs. I use "WHATs" here because most QFD literature uses "WHATs" to represent "what customers want," so WHATs are really customer attributes. Most of these WHATs are very general ideas that require more detailed definition. For example, customers often say they want a "stylish" or "cool" look when they purchase a product. "Cool" may be a very desirable feature, but since it has different meanings to different people, it cannot be acted upon directly.

Legal and safety requirements or other internal requirements, such as company policy, or company internal safety requirements, are considered extensions of the WHATs. The WHATs can be characterized using a Kano Model (discussed in Section 7.4).

7.3.2 CTSs (HOWs)

The HOWs are design features derived by the product development team to answer the WHATs. Each of the initial WHATs needs operational definitions—the objective is to determine a set of Critical-To-Satisfaction requirements (CTSs) that can materialize the WHATs. This process translates customer expectations into design criteria such as speed, torque, and time to delivery.

For each WHAT, there should be one or more HOWs that describe a means of attaining customer satisfaction. For example, a "cool car" can be achieved through a new stylish body, improved seat design, more leg room, and requirements for lower noise, harshness, and vibration levels.

At this stage only overall requirements that can be measured and controlled need to be determined. These substitute for the customer needs and expectations and are traditionally known as Substitute Quality Characteristics.

Teams should define the HOWs in a solution-neutral environment and not be restricted by listing specific parts and processes. Just itemize the means (the HOWs) whereby the list of WHATs can be realized. In addition, each HOW will have some direction of goodness or improvement as shown in the following illustration:

Direction of Improvement		
Maximize	▲	1.0
Target	●	0.0
Minimize	▼	−1.0

For example, if one of the WHATs is the fuel efficiency for the automobile, and one of the HOWs is the weight of the car, because less weight in the car will save fuel consumption, then "minimize" or "down arrow" is the direction of improvement. The circle represents the nominal best target case.

7.3.3 Relationship Matrix

The process of relating WHATs to HOWs often becomes complicated by the absence of one-to-one relationships, as some of the HOWs affect more than one WHAT. In many cases, they adversely affect one another. HOWs that could have an adverse effect on another customer want are important. For example, "cool" and "stylish" are two of the WHATs that a customer would want in a vehicle. The HOWs that support "cool" are lower noise, increased roominess, and seat design requirements, among others. These HOWs will also have some effect on "stylish" as well.

In the relationship matrix, the HOWs (in columns) and the WHATs (in rows) form a grid. The relationship in every (WHAT, HOW) cell can be displayed by placing a symbol representing the cause-and-effect relationship strength in that cell. When employees at the Kobe Shipyards developed this matrix in 1972, they used the local horse racing symbols as relationship matrix symbols—solid circles mean a strong relationship, one circle means medium strength, and a triangle indicates a weak relationship. Symbols are used instead of numbers because they can be identified and interpreted easily and quickly. Different symbol notations have been used, but this one is more common than others:

Standard 9-3-1		
Strong	●	9.0
Moderate	○	3.0
Weak	▽	1.0

After determining the strength of each (WHAT, HOW) relationship and marking it in the cells, the product development team should take the time to review the relationship matrix. For example, blank rows or columns indicate gaps in either team's understanding or a deficiency in fulfilling customer attributes. A blank row shows a need to develop a HOW for the WHAT in that row, indicating a potentially unsatisfied customer attribute. When a blank column exists, then one of the HOWs does not impact any of the WHATs. Delivering that HOW may require a new WHAT that has not been identified, or it might be a waste. The relationship matrix gives the product development team the opportunity to revisit their work, leading to better planning and therefore better results.

What is needed is a way of determining to what extent the CTS in a column contributes to meeting the customer attribute in the row. This requires a subjective weighing of the possible cause-effect relationships. To rank the CTS and customer features in order, multiply the numerical value of the symbol representing the relationship by the *Customer Desirability Index*. The Customer Desirability Index is the numerical rating of the relative importance of a given customer attribute. For example, if a WHAT is about safety, customers usually think it is very important, so this will have a very high rating. This product, when summed over all the customer features in the WHATs array, provides a measure to the relative importance of such CTSs to the product development team and is used as a planning index to allocate resources and efforts, comparing the strength, importance, and interactions of these various relationships. This importance rating is called the Technical Importance Rating.

7.3.4 Importance Ratings

Importance ratings are a relative measure indicating the importance of each WHAT or HOW to the design. In QFD, there are two importance ratings:

- **Customer Desirability Index** This is obtained from VOC activities such as surveys, clinics, and so on, and is usually rated on a scale from 1 (not important) to 5 (extremely important) as given in the following illustration.

Importance		
Extremely Important	●	5.0
Very Important	◑	4.0
Somewhat Important	◑	3.0
A little Important	◐	2.0
Not Important	○	1.0

- **Technical Importance Ratings** These are calculated as follows:

  1. By convention, each symbol in the relationships matrix receives a value representing the strength in the (WHAT, HOW) cell.
  2. These values are multiplied by the Customer Desirability Index, resulting in a numerical value for each symbol in the matrix.
  3. The Technical Importance Rating for each HOW can then be found by adding together the values of all the relationship symbols in each column.

The technical importance ratings have no physical interpretation and their value lies in their ranking relative to one another. They are utilized to determine which HOWs are priorities and should receive the most resources allocation.

7.3.5 Planning Matrix

The planning matrix is the area in the House of Quality (see Figure 7.4) that is located on the right side of the relationship matrix. The planning matrix is used to make comparisons of competitive performance and identification of a benchmark in the context of ability to meet specific customer needs. It is also used as a tool to set goals for improvement using a ratio of performance (goal rating/current rating).

Hauser and Clausing (1988) view this matrix as a perceptual map in trying to answer the following question: How can we change the existing product or develop a new one to reflect customer intent, given that the customer is more biased toward certain features? The product of *customer value*, the *targeted improvement ratio* for the raw feature, and the *sales point*, which is a measure of how the raw feature affects sales, will provide a weighted measure of the relative importance of this customer feature to be considered by the team.

7.3.6   CTS Correlation (HOWs Correlation)

Each cell in the roof is a measure of the possible correlation of two different HOWs. The use of this information improves the team's ability to develop a systems perspective for the various HOWs under consideration. The correlation matrix is one of the more commonly used optional extensions over the original QFD developed by Kobe engineers.

Traditionally, the major task of the correlation matrix is to make trade-off decisions by identifying the qualitative correlations between the various HOWs. This is a very important function in the QFD because HOWs are most often correlated. For example, a matrix contains "quality" and "cost." The design engineer is looking to decrease cost, but any improvement in this aspect will have a negative effect on the quality. This is called a negative correlation and it must be identified so that a trade-off can be addressed.

Trade-offs are usually accomplished by revising the target values (HOW MUCHs). These revisions are called realistic objectives. Using the negative correlation example discussed previously, in order to resolve the conflict between cost and quality, the cost objective would be changed to a realistic objective.

In the correlation matrix, once again, symbols are used for ease of reference to indicate the different levels of correlation with the following scale:

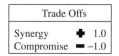

If one HOW directly supports another HOW, a positive correlation is produced.

7.3.7   Targets and Limits (HOW MUCH)

For every HOW shown on the relationship matrix, a HOW MUCH should be determined. The goal here is to quantify the customers' needs and expectations and create a target for the design team. The HOW MUCHs also create a basis for assessing success, so HOWs should be measurable. It is necessary to review the HOWs and develop a means of quantification.

A target orientation to provide a visual indication of target type is usually optional. In addition, the tolerance around targets needs to be identified based on the company marketing strategy and contrasting it with the best-in-class competitor.

7.3.8   Competitive Benchmarks

Competitive assessments are used to compare the competition's design with the team design. There are two types of competitive assessments:

- The Customer Competitive Assessment is found to the right of the relationships matrix in the planning matrix. VOC activities (for example, surveys) are used to rate the WHATs of the various designs in a particular segment of the market.

- The Technical Competitive Assessment is located below the relationship matrix. It rates HOWs from several competitors from a technical perspective.

7.4   Kano Model of Quality

In QFD, the voice-of-the-customer activities, such as market research, provide the array of WHATs that represent the customer attributes. These WHATs are "spoken" by the customer and are called "performance quality." For example, gas mileage of a car is a performance quality; it is also "the more, the better." However, more WHATs have to be addressed than just those directly spoken by the customer. As Figure 7.5 shows, there are also "unspoken" WHATs.

Unspoken WHATs are the basic quality features that the customer automatically assumes will be in the design. Such WHATs are implied

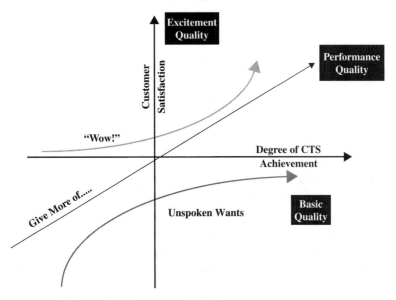

Figure 7.5   Kano model of customer attributes

in the functional requirements of the design or assumed from historical experience. For example, customers automatically expect their lawnmower to cut grass to the specified level, but they wouldn't discuss it on a survey unless they had trouble with one in the past. Unspoken wants have a "weird" property—they don't increase customer satisfaction, but if they are not delivered, they have a strong negative effect on customer satisfaction.

Another group of "unspoken" WHATs can be categorized as excitement quality—innovations or delighters. These pleasant surprises increase customer satisfaction in a nonlinear fashion. For example, in the automotive industry, van owners were delighted by the second van side door and by baby-seat anchor bolts.

Design features may change position on the Kano model over time. In the 1990s, the second side door in a van was a pleasant surprise for customers. Now, on most models, the second door is standard and is expected to be installed without a specific request. The ideal product development plan would include all three types of quality features: excitement quality (unspoken latent requirements), performance quality (spoken and one-dimensional requirements), and basic quality (unspoken or assumed requirements).

7.5   QFD Analysis

The completion of the first QFD House of Quality may give the product development team a false impression that their job is completed. In reality, they have simply created a tool that will guide future efforts toward deploying the VOC information into the design. QFD matrix analysis in every phase will identify design weaknesses, which must be dealt with by a continuous improvement effort.

Here is a relatively simple procedure for analyzing the House of Quality:

- **Blank or weak columns**   These indicate HOWs that don't strongly relate to any customer attribute.

- **Blank or weak rows**   These are customer attributes that are not being strongly addressed by a HOW.

- **Conflicts**   These are technical competitive assessments that are in conflict with customer competitive assessments.

- **Significant points**   These are HOWs that relate to many customer attributes, safety/regulatory, and internal company requirements.

- **"Eye Opener" opportunities**   These are areas where the team's company and competitors are doing poorly. The QFD project team should seize the opportunity to deliver on these sales points, which may initially be treated as delighters in the Kano model.

- **Benchmarking** This points out opportunities to incorporate the competitor's highly rated HOWs. The team can modify and incorporate these using benchmarking and not resort to creation.

7.6 Example 7.1 Information System Design

This example (from Yoruk 2003) deals with a QFD guided information system design for a trucking company, Migros Inc. The key function of this information system is to create management reports that capture all important aspects of trucking company operation.

7.6.1 Ranking Customer Input

The first step in this QFD project was to survey the important aspects of trucking company operation relevant to this information system. The QFD project leader found three groups of people in the trucking company relevant to this information system: partners (the trucking company's customers), site managers, and data workers. A mini-QFD project was done to determine the relative importance of these three groups of people in giving customer input, and the findings are shown in Figure 7.6.

The precise QFD scores calculation is shown in Table 7.1, where X_1, $X_2,...,X_m$ are the items in the QFD rows; $x_1, x_2,..., x_m$ are the corresponding scores for $X_1, X_2,..., X_m$; $Y_1, Y_2,..., Y_n$ are the items in the QFD rows; and $y_1, y_2,..., y_m$ are the corresponding scores for $Y_1, Y_2,..., Y_n$.

	WHATs / HOWs	Weight	Partners	Site Manager	Data Workers
			Customers		
CRITERIA	Intensity of Use	5	△	◉	○
	It Knowledge	1	○	○	
	Influence on Use	5	◉	◉	△
	Number of People	3	△	○	◉
	Influence on Project	5	◉	◉	
	ABSOLUTE IMPORTANCE		101	147	47
	RELATIVE IMPORTANCE		34.	49.	15.
	RANK		2	1	3

MATRIX WEIGHTS
Strong ◉ 9
Medium ○ 3
Weak △ 1

Figure 7.6   QFD table for weighting key customers

TABLE 7.1   QFD Scores Calculation

Items		Y_1	Y_2	....	Y_j	....	Y_n
	Scores	y_1	y_2	....	y_j	....	y_n
X_1	x_1	p_{11}	p_{12}		p_{1j}		P_{1n}
X_2	x_2	p_{21}	p_{22}		p_{2j}		P_{2n}
.							
X_i	x_i	p_{i1}	p_{i2}		p_{ij}		p_{in}
.							
X_m	x_m	p_{m1}	p_{m2}		p_{mj}		p_{mn}

The relationships between y_j's and x_i's are given in the following equation (7-2):

$$y_j = \sum_{i=1}^{m} p_{ij} x_i \qquad (7\text{-}2)$$

In Figure 7.6, the column "Weight" gives the scores for the relative importance of customer selection criteria: 1 = lowest importance, 5= highest importance. The items in "weight" are X_is, and the scores in the weight column are corresponding x_is. The three key customers, (partners, site managers, and data workers) correspond to Y_is. The row "absolute importance" corresponds to the scores of key customers, that is, y_is, and they are calculated based on equation (7.2), as follows:

Partners' absolute importance:

$$y_1 = \sum_{i=1}^{5} p_{i1} x_i = 1 \times 5 + 3 \times 1 + 9 \times 5 + 1 \times 3 + 9 \times 5 = 101$$

Site managers' absolute importance:

$$y_2 = \sum_{i=1}^{5} p_{i2} x_i = 9 \times 5 + 3 \times 1 + 9 \times 5 + 3 \times 3 + 9 \times 5 = 147$$

Data workers' absolute importance:

$$y_3 = \sum_{i=1}^{5} p_{i3} x_i = 3 \times 5 + 1 \times 5 + 9 \times 3 + 9 \times 5 = 47$$

The row "relative importance" is the normalized importance scores as follows:

$$\text{Partners' relative importance} = \frac{101}{101 + 147 + 47} = 0.34 = 34\%$$

$$\text{Site managers' relative importance} = \frac{147}{101 + 147 + 47} = 0.50 = 50\%$$

$$\text{Data workers' relative importance} = \frac{47}{101 + 147 + 47} = 0.16 = 16\%$$

Finally, based on the importance scores, the site manager is ranked number 1, the partner is ranked number 2, and the data worker is ranked number 3.

7.6.2   Ranking the Functional Requirements

In the next step, the QFD team rated the relative importance scores of the following functional requirements by weighting the inputs from these three key customers, that is, partners, site managers, and data workers. The results are summarized in Table 7.2.

In Table 7.2, the relative importance score calculation is still based on equation (7-2), for example:

The relative importance score of "Fair truck run distribution" =

$0.34 \times 3 + 0.5 \times 9 + 0.16 \times 0 = 5.5$

The relative importance score of "Prevent fuel theft" =

$0.34 \times 9 + 0.5 \times 9 + 0.16 \times 0 = 7.6$

and so on.

TABLE 7.2   Relative Importance Scores for Functional Requirements

	Partner	Site Manager	Data Workers	Weighted Importance
Functional Requirements	34%	50%	16%	
Fair truck run distribution	3	9		5.5
Prevent fuel theft	9	9		7.6
Prepare and print waybills	1	3	9	3.3
Minimize truck idle time	9	9		7.6
Calculate payment to drivers	1	3	3	2.3
Eliminate unreliable trucks	3	9		5.5
Prepare invoice data for Migros	9	3	1	4.7
Prepare confirmation report to Migros	9	3		4.6
Minimize fuel consumption	9	3		4.6
Follows runs to the market	3	9	1	5.7
Follow actual costs of runs	9	3		4.6

At the next step, the QFD team developed a QFD House of Quality that linked the functional requirements of the information system with the management report. This QFD is illustrated in Figure 7.7.

Again, in Figure 7.7, the importance score calculation is also based on equation (7-2).

7.7 QFD Case Study 1: Global Commercial Process Design

Project Objectives:
Design a global commercial process with high performance.

Project Problem Statement:
- Sales cycle time (lead generation to full customer setup) exceeds 182 business days. Internal and external customer specifications range from 1 to 72 business days.

- Only 54% of customer service requests are closed by the commitment date. The customers expect 100% of their service requests to be completed on time.

- The current commercial processes are not standardized, and none of the current processes are capable of standardization.

FUNCTIONAL REQUIREMENTS	WHATs	IMPORTANCE	Failures	Truck Usage	Market Visits	Expenditures	Hour Statistics	Payments	Fuel Consumption	Run Distribution	Distance Traveled
TRUCK	Fair truck run distribution	5.5	△	●							●
	Prepare & print waybills	7.6			●						
	Minimize truck idle time	3.3	○				●				
	Calculate payments to drivers	7.6			●			●			●
	Eliminate unreliable trucks	2.3	●								
	Follow actual cost of runs	5.5				●		○	○		△
FUEL	Prevent fuel theft	4.7				△		△	●		
	Minimize fuel consumption	4.6							●		
CUSTOMER	Prepare invoice data	4.6		○	●	●				●	○
	Aid in confirmation with Migros	5.7		○	●	●				○	
	Follow destinations of runs	4.6		●	○					○	
ABSOLUTE IMPORTANCE			36	121	243	146	29	89	100	72	137
RELATIVE IMPORTANCE			3.6%	12.4%	24.9%	15.0%	3.0%	9.1%	10.2%	7.3%	14.0%
IMPORTANCE RANK			8	4	1	2	9	6	5	7	3

(HOWs — MANAGEMENT REPORTS)

Figure 7.7 QFD house that links functional requirements to management report

Business Case:

- There is no consistent, global process for selling to, setting up, and servicing accounts.

- Current sales and customer service information management systems do not enable measurement of accuracy and timeliness on a global basis.

- Enterprise-wide customer care is a "must be" requirement—failure to improve the process threatens growth and retention of the portfolio.

Project Goals:

- Reduce prospecting cycle time from 16 to 5 business days.

- Reduce discovery cycle time from 34 to 10 business days.

- Reduce close-the-deal cycle time from 81 to 45 business days (all sales metrics net of customer wait time).

- Reduce setup cycle time from 51 to 12 business days.

- Increase the percentage of service requests closed by commitment date from 54% to 99.97%.

7.7.1 QFD Steps

The following are the QFD steps:

1. Identify the WHATs & HOWs and their relationship
2. Identify the HOWs and relationship matrix

7.7.1.1 Identify the WHATs & HOWs and Their Relationship The QFD project team identifies customers and establishes customer wants, needs, delights, and usage profiles. In addition, corporate, regulatory, and social requirements should be identified also. The value of this step is to greatly improve the understanding and appreciation that team members have for customer, corporate, regulatory, and social requirements.

The QFD project team, at this stage, should be expanded to include market research. A market research professional might help the black belt assume leadership during startup activities and perhaps later remain an active participant as the team gains knowledge about customer engagement methods. The team leader should put plans in place to collaborate with identified organizations and/or employee relations to define tasks and plans in support of the project, and to train team members in customer processes, that is, forward-thinking methods such as brainstorming, visioning, and conceptualizing.

The QFD project team should focus on the key customers in order to optimize decisions around them, and try to include as many additional customers as possible. The team should establish customer environmental conditions, customer usage and operating conditions; study customer demographics and profiles; conduct customer performance evaluations; and understand the performance of the competition. In addition, the team should

- Establish a rough definition of an ideal service.
- Listen to the customer and capture wants and needs through interviews, focus groups, customer councils, field trials, field observations, surveys, and so on.
- Analyze customer complaints and assign satisfaction performance ratings to attributes.
- Acquire and rank these ratings with the Quality Function Deployment (QFD) process.
- Study all available information about the service including marketing plans.
- Create innovative ideas/delights and new wants by investigating improved functions and cost of ownership, and by benchmarking the competition to improve weak areas. Create new delights by matching service functions with needs, experience, and customer beliefs.

The following WHATs are used:

Direction of Improvement
Available Products
Professional Staff
Flexible Processes
Knowledgeable Staff
Easy to Use Products
Speedy Processes
Cost-Effective Products
Accuracy

7.7.1.2 Identify the HOWs and Relationship Matrix   The purpose of this step is to define a "good" product/process in terms of customer expectations; to benchmark projections, institutional knowledge, and interface requirements; and to translate this information into CTSs. These will then be used to plan an effective and efficient QFD project.

One of the major reasons for customer dissatisfaction and warranty costs is that the design specifications do not adequately reflect customer use of the product or process. Too many times the specification is written after the design is completed, or it is simply a reflection of an old specification that was also inadequate. In addition, poorly planned designs commonly do not allocate activities/resources in areas of importance to customers and waste engineering resources by spending too much time in activities that provide marginal value. Because missed customer requirements are not targeted or checked in the design process, procedures for handling field complaints for these items are likely to be incomplete. Spending time overdesigning and overtesting items not important to customers is a waste. Similarly, not spending development time on areas important to customers is not only a missed opportunity, but significant warranty costs are sure to follow.

In good design practice, time is spent up front understanding customer wants, needs, and delights together with corporate and regulatory requirements. This understanding is then translated into CTSs, which then drive product and process design. The CTSs (HOWs) are given in the following illustration, as well as, the relationship matrix to the WHATs, shown in Figure 7.8.

| Importance to the Customer |
| Meet Time Expectations |
| Know My Business & Offers |
| Save Money / Enhance Productivity |
| Do It Right the First Time |
| Consultative |
| Know Our Products & Processes |
| Talk to One Person |
| Answer Questions |
| Courteous |
| Adequate Follow-Up |

A mapping begins by considering the high-level requirements for the product or process. These are the true CTSs, which define what the customer would like if the product or process were ideal. This consideration of a product or process from a customer perspective must address the requirements of higher-level systems, internal customers (such as manufacturing, assembly, service, packaging, and safety), external customers, and regulatory legislation. This diagram, which relates true quality characteristics to substitute quality characteristics, is called a relationship matrix.

The mapping of customer characteristics to CTS characteristics is extremely valuable when done by the QFD team. A team typically begins differing in opinion and sharing stories/experiences. When a QFD study is completed, the entire team understands how product and process characteristics that are detailed on drawings relate to functions that are important to customers.

The full Phase I and II QFDs are given in Figures 7.8–7.10. An analysis of Phase I follows in Section 7.7.3, and you are encouraged to perform such an analysis on the other phases as well.

7.7.2 The HOWs Importance Calculation

Importance ratings are a relative comparison of the importance of each WHAT or HOW to the quality of the design. The weight scales of 9, 3, and 1 are used in the relationship matrix. Theses values are multiplied

Direction of Improvement

Maximize ↑	1.0
Target ●	0.0
Minimize ↓	−1.0

	#	Importance to the Customer	Meet Time Expectations	Know My Business & Offers	Save Money / Enhance Productivity	Do It Right the First Time	Consultative	Know Our Products & Processes	Talk to One Person	Answer Questions	Courteous	Adequate Follow-Up
		1	1	2	3	4	5	6	7	8	9	10
Direction of Improvement	1		↑	↑	↑	↑	↑	↑	↑	↑	↑	↑
Available Products	1	2.0	○		○		▽		○			
Professional Staff	2	3.0		▽		▽	○	●	▽		●	
Flexible Processes	3	4.0	○					●				
Knowledgeable Staff	4	4.0	○	●	○	●	●	●	●	●		▽
Easy to Use Products	5	4.0	○		○	○		▽	○			▽
Speedy Processes	6	5.0	●		●	○		○	○	○		▽
Cost-Effective Products	7	5.0	○	●	●	○	●	○				
Accuracy	8	5.0		●		●						

Figure 7.8 The WHATs, the HOWs, and relationship matrix

by the customer importance rating obtained from customer engagement activities (like surveys) resulting in a numerical value. The HOWs importance rating is summed by adding all values of all relationships.

For example, in Figure 7.8, the first HOW importance rating is calculated as: $2.0 \times 3.0 + 4.0 \times 3.0 + 4.0 \times 3.0 + 4.0 \times 3.0 + 5.0 \times 9.0 + 5.0 \times 3.0 = 102$. Other HOWs importance ratings can be calculated similarly.

7.7.3 Phase I QFD Diagnostics

From the Phase I QFD diagram illustrated in Figure 7.9, the following problems are identified:

- **Weak WHATs** The black belt needs to identify WHATs with only weak or no relationships. Such situations represent a failure to address a customer attribute. When this occurs, the company should try to develop CTS(s) to address this WHAT, although the team may sometimes discover that present technology can't satisfy the WHAT. The QFD project team should resort to customer survey and assessment for review and further understanding.

  No such WHAT exists in our example. The closest to this situation is "Available Products" in row #1 and "Easy to Use Products" in Row #5. Row 1 and Row 5 were highlighted as the weakest WHATs, but not weak enough to warrant the analysis in Figure 7.9. However, the team is encouraged to strengthen this situation by a CTS with strong relationship.

- **Weak HOWs** The team needs to look for blank or weak HOWs (all entries are inverted deltas). This situation occurs when CTSs are included that don't really reflect the customer attributes being addressed by the QFD. The QFD project team leader and his team may consider eliminating the CTSs from further deployment if they do not relate to basic quality or performance attributes in the Kano model.

  In our example, The CTS "Adequate Follow-Up" is weak (rated 13 on importance rating). However, the WHAT "Easy to Use Products" has no strong relationship with any CTSs, and eliminating "Adequate Follow-Up" may weaken the delivery of this WHAT even further.

- **Conflicts** The DFSS team needs to look for cases where technical benchmarking rates their product or service high but the customer assessment is low. Misconception of customer attributes is the major root cause. The team together with marketing can remedy the situation.

  In our example, the "Cost-Effective Products," a WHAT, is addressed by many CTSs, including "Save Money/Enhance Productivity." The customer rates our design as Weak (rating 2), while the technical assessment is rated the highest (rating is 4). Who is right? Conflicts may be

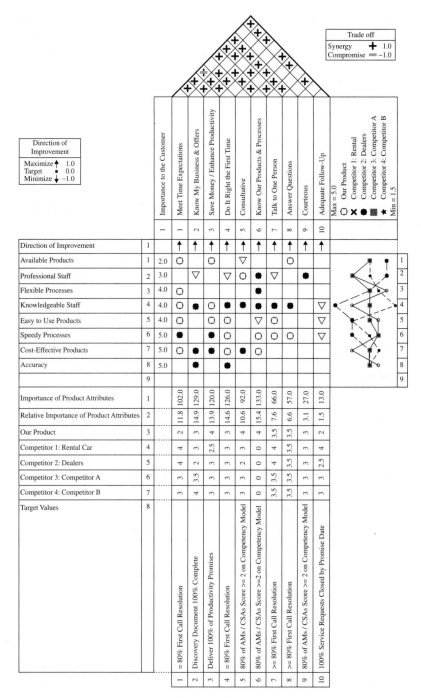

Figure 7.9 Phase I QFD

a result of a failure to understand the customer and must be resolved prior to further progress.

- **Strengths** By identifying the CTSs that contain the most "9" ratings, the DFSS team can pinpoint which CTSs have a significant impact on the total design. Changes in these characteristics will greatly affect the design, and such effects propagate via the correlation matrix to other CTSs, causing positive and negative implications.

The following CTSs are significant as implied by their importance ratings and number of "9" ratings in their relationships to WHATs: "Meet the Expectations," "Know My Business & Offers," "Save Money/Enhance Productivity," "Do It Right the First Time," and "Know Our Products & Processes." By examining the Correlation Matrix in Figure 7.10, it is clear that we have positive correlations throughtout except in the "Do It Right the First Time" and "Meet Time Expectations" cells.

- **Eye Openers** The DFSS team should look at customer attributes where

1. Their design as well as their competitors' design is performing poorly.

2. The WHATs are performing poorly compared with their competitors based on competitive benchmarking.

3. CTSs need further development in Phase II.

We can pencil "Flexible Processes" in the first category and "Accuracy" and "Easy To Use Products" in the second category. The CTSs that deliver these WHATs should receive the greatest attention as they represent potential pay-offs. For the WHATs where the competitors are rated significantly higher than in our own design, it is recommended that we should learn our competitors' design and adopt a similar design strategy. This saves design and research time.

The highest CTSs with the largest importance ratings are the most important. For example, "Know Our Products & Processes" has the highest rating at 133. This rating is so high because it has three strong relationships to the WHATs. The degree of difficulty is medium (rating equal to 3) in the technical benchmarking. In addition, any CTS that has negative or strong relationships with this CTS in the correlation matrix should proceed to Phase II.

7.8 QFD Case Study 2: Yaesu Book Center

The Yaesu Book Center is a bookstore in Japan. When it first opened, the store had few employees experienced in bookselling. Most of the business was conducted by employees who had recently graduated from school.

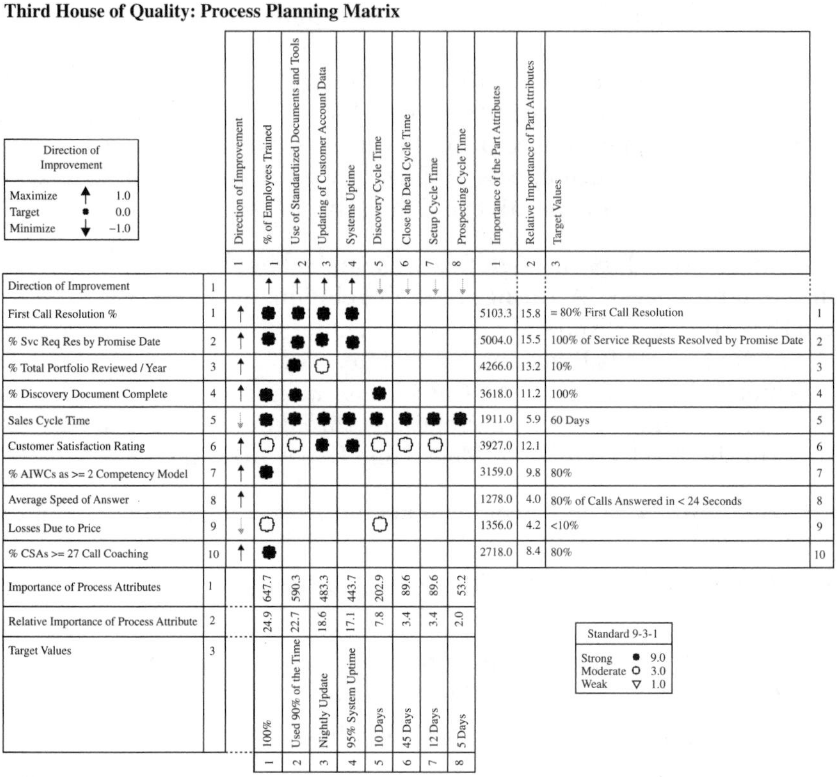

Third House of Quality: Process Planning Matrix

		Direction of Improvement	% of Employees Trained	Use of Standardized Documents and Tools	Updating of Customer Account Data	Systems Uptime	Discovery Cycle Time	Close the Deal Cycle Time	Setup Cycle Time	Prospecting Cycle Time	Importance of the Part Attributes	Relative Importance of Part Attributes	Target Values	
Direction of Improvement	1		↑	↑	↑	↑	↓	↓	↓	↓				
First Call Resolution %	1	↑	●	●	●	●					5103.3	15.8	= 80% First Call Resolution	1
% Svc Req Res by Promise Date	2	↑	●	●	●	●					5004.0	15.5	100% of Service Requests Resolved by Promise Date	2
% Total Portfolio Reviewed / Year	3	↑		●	○						4266.0	13.2	10%	3
% Discovery Document Complete	4	↑	●	●			●				3618.0	11.2	100%	4
Sales Cycle Time	5	↓	●	●	●	●	●	●	●	●	1911.0	5.9	60 Days	5
Customer Satisfaction Rating	6	↑	○	○	●		●	○	○	○	3927.0	12.1		6
% AIWCs as >= 2 Competency Model	7	↑	●								3159.0	9.8	80%	7
Average Speed of Answer	8	↑									1278.0	4.0	80% of Calls Answered in < 24 Seconds	8
Losses Due to Price	9	↓	○			○					1356.0	4.2	<10%	9
% CSAs >= 27 Call Coaching	10	↑	●								2718.0	8.4	80%	10
Importance of Process Attributes	1		647.7	590.3	483.3	443.7	202.9	89.6	89.6	53.2				
Relative Importance of Process Attribute	2		24.9	22.7	18.6	17.1	7.8	3.4	3.4	2.0				
Target Values	3		100%	Used 90% of the Time	Nightly Update	95% System Uptime	10 Days	45 Days	12 Days	5 Days				

Direction of Improvement:
Maximize ↑ 1.0
Target ● 0.0
Minimize ↓ −1.0

Standard 9-3-1
Strong ● 9.0
Moderate ○ 3.0
Weak ▽ 1.0

Figure 7.10 Phase II QFD

In spite of that, Yaesu Book Center attracted a great deal of attention and was highly regarded by book lovers.

Yaesu Book Center has its own quality control (QC) circle. In the QC circle, the area managers are also group leaders. The QC circle determined that the following three things are essential to satisfy customers' needs:

- To have enough books available
- To have enough product information
- To provide enough service

The QC circle also found that they did not have enough information to figure out how to accomplish these objectives. They did not know the following:

- Specific customer demands were not clear.
- There were no specific quantitative measurements for customer demands.

Product Trade-offs

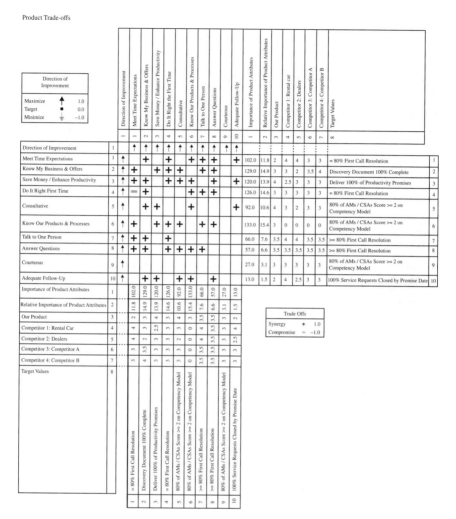

Figure 7.11   Correlation

- The relationship between the customers' demands and Yaesu Book Center's service product was not clear.

To solve these problems, VOC data was collected and a two-phase QFD was conducted by the Yaesu Book Center to improve the bookstore operation. This QFD study was conducted by performing the following steps:

1. Determine customer attributes (WHATs)

2. Determine quality characteristics (HOWs)

3. Assign degree of importance to customer attributes

4. Determine operations items

5. Two-phase QFD analysis for Yaesu Book Center

7.8.1 Determine Customer Attributes (WHATs)

First, a lot of information about customers' demands was collected by customer surveys and interviews. The raw customers' demands were in their own words. In a brainstorming session, this raw customer information was translated into a set of better-defined customer attributes. The following procedures were used in this translation process:

- Vague comments were changed into precise expressions.

- Comments expressed in negative conditions were changed into positive comments.

- Comments were grouped into subcategories and similar subcategories were combined.

- Customer attributes were fitted into tree diagrams.

Figure 7.12 shows a partial tree diagram that organizes the customer attributes in this case.

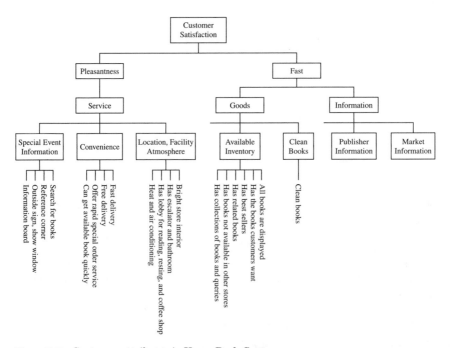

Figure 7.12   Customer attributes in Yaesu Book Center

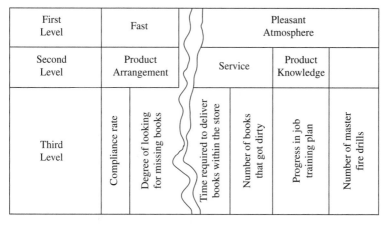

Figure 7.13 Quality characteristics (HOWs)

7.8.2 Determine Quality Characteristics (HOWs)

Quality characteristics are extremely important, so the Yaesu Book Center QC circle made an effort to identify them. A partial list of quality characteristics is shown in Figure 7.13. A three-level hierarchy of quality characteristics was used.

7.8.3 Assign Degree of Importance to Customer Attributes

The survey questionnaire in Table 7.3 was used to collect customer importance ratings for each customer attribute, and it served as a basis for determining importance ratings in the QFD study.

7.8.4 Determine Operations Items

In this case, operations items are what bookstore management and employees are actually doing in their work. The QFD study should provide guidelines for the best ways to do this work, and indicate what they didn't do enough before, and how much they should do it now. These operation items are organized in a tree diagram and illustrated in Figure 7.14.

7.8.5 2 Two-Phase QFD Analysis for Yaesu Book Center

Based on the work in the first four steps, two QFD House of Quality charts were developed for the Yaesu Book Center. The first relates customer attributes to quality characteristics, and the second relates quality characteristics to operation items. The partial listings of these two houses of quality are illustrated in Figure 7.15 and Figure 7.16.

TABLE 7.3 Survey Questionnaire

How important are the following items in a bookstore? Please rank from 1 to 5.

Questions	1 Not important at all	2 Not important	3 Neither	4 Somewhat important	5 Important
1. Has a good variety of best sellers	1	2	3	4	5
2. Scheduled date for availability of out-of-stock books is clear	1	2	3	4	5
3. Has a good variety of art books	1	2	3	4	5
4. Has a variety of books on sociology, literature, science, and history	1	2	3	4	5
....					
13. The store clerks look hard for books for the customers	1	2	3	4	5
14. Can easily find books you want	1	2	3	4	5
15. Book classifications are easy to understand	1	2	3	4	5
16. Attractive, easy-to-find book displays	1	2	3	4	5
17. Books are always clean	1	2	3	4	5

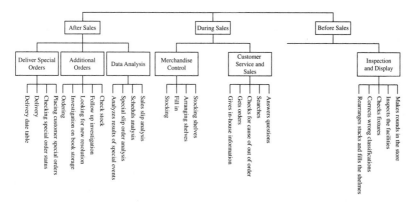

Figure 7.14 Operation items

Customer Attributes	Importance Rating	Clear classification	Immediate response	Ratio of availability	Degree of satisfaction	Easy to find	Service satisfaction	Waiting time	Product knowledge	Customer reception	Order lead time	Delivery sale	Procurement time	Cleanliness	Lighting	Damage rate
Can tell the book is in stock	2.9	●	●		●											
Can tell why not in stock	6.0	●		●												
Can tell if book is available		○														
Can give date of availability	0.1	○						○								
Can tell detailed book description									●	○						
Can find related book	1.8								●	○						
Offer information on the contents									●	○						
Information list of book available							○		○							
Has large variety and volume of books	4.0		●	○	●											
Has book not available in other stores	1.3		●	○	●								●			
Has newly published books	0.2		●													
Has many speciality books			●													
Easy to find books	4.3					●										
Classification is easy to understand	5.8						●									
Signs are easy to see	5.0					●	●									
Display is easy to see	2.5						●									
His clean books	2.5															●
Has product knowledge	3.4							○	●	○						
Kind and polite	3.8							●		○						
	0.1								●	○						
Assistance in looking for books	3.4								●	○						

Figure 7.15 House of Quality Phase 1

		Operation Items																			
		Before sales						During sales				After sales									
		Make rounds in ?	Inspects the ?	Checks fixtures	Correcting wrong classifications	Rearrange stacks/fill shelves	?	Searches	?	Get orders	Gives in-house information	Sales slip analysis	Schedule analysis	Special order slip analysis	Analyzes result of special orders	?	?	?	?	?	?
Quality Characteristics	Clear classification											○									
	Immediate response											○	●		○						
	Rate of availability											○									
	Degree of satisfaction	●		●										○	○						
	Easy to find	●				●					○										
	Service satisfaction							●													
	Waiting time							●	●												
	Product knowledge							●	●												
	Customer reception																				
	Order lead time											○	○	○		○					
	Delivery rate																				○
	Procurement time									●				●				●	●		
	Cleanliness																				
	Lighting																				
	Damage rate	●	●																		

Figure 7.16 House of Quality Phase 2

After constructing the two QFD charts, the QC circle found that there was one major problem in the book center: The customer attribute "book classification is easy to understand" was rated very high in customer survey results, but by QFD analysis, the rating for this was not high with current operation items. The following corrections were made:

1. If a book could fall into more than one category, the book would be displayed in all these categories.

2. Point-of-purchase clerks were placed in the boundary areas between book sections.

8

Customer Value Creation by Brand Development

Famous brand names make a big difference in the marketplace. T-shirts can be made of exactly the same fabric, in the same style, and with the same workmanship, but because they have different brand names, the retail prices of these T-shirts are vastly different.

A good brand name brings extra value to the product and to the company that makes the product. McDonald's, Coca-Cola, Disney, Kodak, and Sony are among the most globally recognized names in the world (Kochan et al. 1997). The name recognition of these brands brings tremendous marketplace success and high profitability. Brand development has become one of the key sources of competitive advantage for companies worldwide. Brands are regarded as among the most valuable assets owned by a company (Batra 1993, Davis 2000). Some brands are valued so highly that companies have paid huge amounts to acquire the rights to them. For example, in 1988 Philip Morris bought Kraft, the maker of cheese products, for $12.9 billion, a sum that was four times the value of the assets of the company (Murphy 1989). Companies that have good brand names can sometimes defend their market positions for a long time (Arnold 1992), as illustrated by Table 8.1.

What is a brand? Why do brand names have such power? What is the importance of a brand name in developing a service product? These are some of the questions that we'll look at in this chapter.

Merriam-Webster's dictionary defines a brand as a mark made to "attest manufacture or quality or to designate ownership" or "a characteristic or distinctive kind." Philip Koltler (1991) defines a brand as "a name, term, symbol, or design, or a combination of them, which is intended to signify the goods or services of one seller or group of sellers and to differentiate them from those of their competitors." Davis (2000) defines a brand as "an intangible but critical component of what a company stands for," and, as "a set of promises, it implies trust, consistency, and a defined set of expectations. The strongest brands in the world own a place in the consumer's mind, and when they are mentioned almost everyone thinks of the same things." Mercedes-Benz stands for prestige and the ultimate driving experience; Ralph Lauren stands for classic looks, high status, and pride.

The strongest brands usually stand for superior functions, benefits, and quality; without these, a brand will not be among the strongest in the world. However, superior functions, benefits, and quality alone will not make a strong brand. Does McDonald's offer much better food than Burger King? Does Starbucks offer much better coffee than Caribou Coffee? Probably not—the difference between the top brand and second-tier brands is mostly psychological. Research in psychology has shown that name recognition alone can result in more positive feelings toward nearly everything, whether it is music, people, words, or brands. In a study, respondents were asked to taste each of three samples of peanut butter. One of these samples contained an unnamed superior peanut butter (preferred in blind tests 70 percent of the time). Another contained an inferior peanut butter (not preferred in taste tests) that was labeled with a brand name known to the respondents but neither purchased nor used by them before. Surprisingly, 73 percent of respondents selected the inferior brand name option as being the best-tasting peanut butter. These results clearly show the power

TABLE 8.1   Leading U.S. Brands from 1933 to 1990

Brand	Market
Eastman Kodak	Cameras/film
Del Monte	Canned fruit
Wrigley	Chewing gum
Nabisco	Baked goods
Gillette	Razors
Coca-Cola	Soft drinks
Campbell's	Soup
Ivory	Soap
Goodyear	Tires

of brand name recognition. Mere name recognition can make people think that an inferior peanut butter tastes better than a better-tasting peanut butter. Consumers' psychology plays a very important role in brand name strength, and Davis (2000) calls this psychological reaction to brand names *PATH*, which is an acronym for promise, acceptance, trust, and hope. A strong brand makes promise, acceptance, trust, and hope tangible.

The benefits from strong brands are numerous; Davis (2000) lists the following benefits:

- Seventy-two percent of customers say that they will pay a 20 percent premium for their brand of choice, relative to the closest competitive brand. Fifty percent of customers will pay a 25 percent premium. Forty percent of customers will pay up to 30 percent premium.

- Twenty-five percent of customers state that price does not matter if they are buying a brand that owns their loyalty.

- Over 70 percent of customers want to use a brand to guide their purchase decision and over 50 percent of purchases are actually brand-driven.

- Peer recommendations influence almost 30 percent of all purchases made today, so a good experience by one customer with your brand may influence another's purchase decision.

- More than 50 percent of consumers believe a strong brand allows for more successful new product introductions, and they are more willing to try a new product from a preferred brand because of the implied endorsement.

These benefits clearly indicate that strong brands do create tremendous value for the companies that own the brands, so creating strong brands should be an integral part of the product development strategy. The making of a strong brand, usually called brand development, is a very elaborate process; it involves the coordinated efforts of product development, marketing, promotion, customer service, and corporate leadership. Since the strength of the brand is directly related to the value of service products, the people who work on product development must understand the basics of brand development.

This chapter will cover the important aspects of brand development. Section 8.1 will dive deep into the question of "what is a brand?" Section 8.2 will discuss the brand development process. Section 8.4 will discuss the role of brand development in product development practice.

8.1 The Anatomy of Brands

Strong brands have an almost magic power to add value to products and inspire customer loyalty. It is very important to understand how strong brands influence consumers' opinions and what the essential components of a strong brand are. In this section we will look at all important aspects of brands.

8.1.1 People's Buying Behavior and Brands

According to Arnold (1992), the power of brands can be explained by how people make buying decisions:

- Most customers, especially consumers in mass markets, do not understand a product or service as well as the company selling it. Most customers have only superficial knowledge of the product or service, and many are not even interested in the details.

- Customers will perceive a product or service in their own terms. Since customers usually have imperfect knowledge of a product or service, they will base their opinions on the attributes most obvious to them. For example, airline customers may rate the airline based on what they see. If they see stains on the flip-down table, or washrooms that aren't perfectly clean, they may doubt the operation of the entire airline, including airplane engine maintenance. Customers may judge a detergent by its smell, not by how well it cleans. Different customers will notice different attributes; every customer has a personal view.

- Customers' perception often focuses on the benefits of the product or service—what the product or service can do for the customer. Customers will see different benefits. Some may see some functional benefits, and some may be more interested in emotional benefits. For example, some kids will want to buy cereal because there is a sports star on the box (emotional benefit), and not because of the taste of cereal itself (functional benefit).

- Customer perceptions are not always conscious. If you ask a customer why they chose a particular product or service, you will sometimes get a rational answer, and sometimes not. Even if there is a rational answer, it may not tell the whole story. Feelings about a product or service may not be easily articulated, because these feelings are complex and hard to explain, and they are sometimes subconscious and irrational.

Because the relationship between customers and the things they buy is complex, brand names become a shortcut for customers when they are choosing products or services. When customers gradually develop positive perceptions about a product or service, the thing they remember is the brand name. These feelings and perceptions are often contagious, and customers will spread them to friends, family members, and other people whom they can easily influence, and this will create a snowball effect. Watkins (1986) used the diagram in Figure 8.1 to illustrate the model of customer choice.

Customers' buying decisions involve a complex set of perceptions and demands. Therefore, a brand should also address many elements of customer perception and demand in order to be successful. The following criteria for a successful brand are adapted from Arnold (1992):

- The products or services associated with a brand must deliver the functional benefits to meet market needs at least as well as the competition. No product or service will survive in the long run if it does not perform. A brand is not merely the creation of advertising and packaging.

- A first-of-its-kind product or service is a strong basis on which to build a brand. However, the brand will not be successful in the long run if it cannot satisfy customers. When competing in a crowded market with similar products from existing brands, a newcomer has to provide significant advantages in some area, whether functional, emotional, or price, in order to compete effectively.

- Besides the functional benefits, a brand has to offer intangible benefits, such as emotional, belonging, prestige, style, and so on, in order to shine. People will pay a high price for a top brand T-shirt for the sake of pride, belonging, and prestige.

- The benefits offered by a brand should be consistent with each other and present a unified character or personality. If the benefits offered from a brand are too confusing, or change from time to time, it will drive the customer away because customers will often come to quick and superficial conclusions when purchasing a product. Customers form stereotypes about brands quickly, and if they like a brand, they will stick with it. For example, both McDonald's and Chinese restaurants provide food and both have loyal customers based on customers' perception of the food. If a McDonald's restaurant started offering

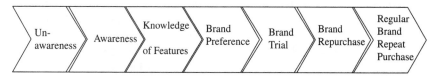

Figure 8.1 Model of customer choice

good Chinese food, it would send a very confusing signal to customers, and many customers might turn away. To maintain a brand, a company must actively manage the personality of the brand to keep it clear and consistent over time.

■ The benefits offered by a brand must be wanted by the customer. No brand image, however clear and consistent, is of any use unless it meets customer wants. If people's wants change, the benefits offered by the brand will have to change.

Brands have a magic power, and brand building is an important element in customer value creation. To build a strong brand, we first need to know what the essential elements of a brand are, and how these elements are related to each other. In the next two sections, we will look at two essential elements of a brand: brand identity and brand equity.

8.1.2 Brand Identity

Customers' perception about brands is much like people's perceptions of other people. A person's name is simply a symbol, and we form opinions about a person based on our perceptions of what the person is good at, what his or her personality is like, what the person looks like, what he or she stands for, what his or her core values are, and so on. These characteristics form a person's identity. Brand identity, according to Aaker (1996), "provides direction, purpose and meaning for the brand":

> Brand identity is a unique set of brand associations that the brand strategist aspires to create or maintain. These associations represent what the brand stands for and imply a promise to customers from the organization members. ... Brand identity should help establish a relationship between the brand and the customer by generating a value proposition involving functional, emotional, or self-expressive benefits.

There are several models that describe what brand identity is. Aaker (1996) proposed a brand identity model based on four perspectives: brand as product, brand as organization, brand as person, and brand as symbol. Davis (2000) created the brand image model, which involves two components: brand association and brand persona.

8.1.2.1 Aaker's Brand Identity Model
The Brand Identity section in Figure 8.2 illustrates the framework of Aaker's brand identity model. This model describes brands from four perspectives: brand as product, brand as organization, brand as person, and brand as symbol. A brand need not actually employ all these four perspectives, however. For brands that relate to larger corporations and its products, it is very likely that all four perspectives will be employed.

The perception of brand identity in customers' minds is the result of the customer's total experience with the products and services of that brand. What perception customers have depends on the products/services themselves, how customers are treated by the company, and advertisements and promotions; ultimately, it depends on the company's business strategy and business operation. To build a strong brand, the company should carefully design a good brand identity for their products/services and make this ideal brand identity a reality.

The first step in creating a brand is to perform a strategic brand analysis (as shown in Figure 8.3). A strategic brand analysis consists of three components: customer analysis, competitor analysis, and self-analysis. The purpose of this analysis is to provide a basis on which to design an appropriate brand identity for the product/service offered by the company. (The details of the strategic brand analysis will be discussed in Section 8.2.3.)

To design a good brand identity, we need to consider all four perspectives, though we may not deploy all of them. The following sections describe these four perspectives in detail.

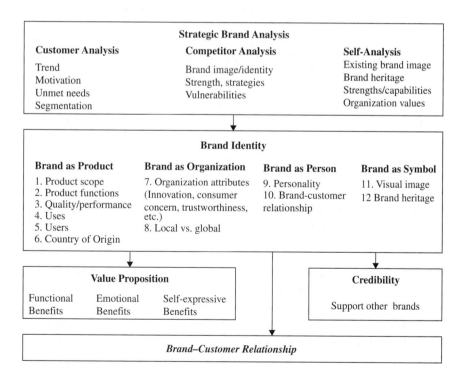

Figure 8.2   Brand identity model (adapted from Aaker 1996)

The Brand as a Product: Product-Related Associations   Product-related associations are always an important part of brand identity, because customers are buying the product. Product-related associations have the following aspects:

- **The product scope**   This aspect deals with the product class the brand is associated with. For example, McDonald's is associated with the fast-food product class; VISA is associated with credit cards; Hertz is associated with rental cars. A brand is unsuccessful if, when the brand name is mentioned, most people don't know what product class it is related to. When a nondominant brand name is mentioned, people know what product class it is related to. In the case of a dominant brand, when the product class is mentioned, most people will recall the name of the brand. For example, when "soft drink" is mentioned, the name Coca-Cola will at least be thought about. On the other hand, when Faygo, a nondominant brand, is mentioned, some people will recognize that it is a soft drink brand; however, when the soft drink product class is mentioned, many people will not think of Faygo.

- **Product functions**   This aspect deals with what functional benefits (and some emotional benefits) the product or product class can provide to customers. How well this aspect of the product performs depends on how well the customers' needs (identified and unidentified) are met. For example, McDonald's functional benefits include all their breakfast and ordinary meal items, hamburgers, fries, soft drinks; fast purchase cycle time (time from ordering to getting the food); happy meals, toys, and playgrounds; unrivaled worldwide product consistency; clean washrooms; and so on. McDonald's emotional benefits include friendly service, being child-friendly, and so on.

- **Quality/performance**   This aspect deals with how well and how consistently the functional and emotional benefits are provided. For example, fast purchase cycle time is a key functional benefit for McDonald's—the speed and consistency of this cycle time is a matter of quality and performance. McDonald's is famous for its unrivaled worldwide product consistency, which is also a matter of quality. For products with similar functional benefits, the performance depends on how well these functional benefits are delivered. For example, Mercedes, Buick, and Kia are all cars, but their performance levels are quite different.

- **Uses**   This aspect deals with the particular use or application associated with a brand. For example, Gatorade specializes in providing a drink for athletes who want to maintain a high level of performance.

- **Users**   This aspect deals with the type of users targeted by a particular brand. For example, Motherhood Maternity targets pregnant women and its products are for pregnant women.

- **Country of Origin** Associating a brand with a country of origin will add credibility if the country that the brand relates to has a good reputation in this product area. For example, French fashion products are better regarded than French electronic goods.

Product-related association is an important part of brand identity. After all, people are buying products and the benefits are related to product functions. However, if the brand identity is only associated with product attributes, the brand will have serious limitations. Here are some specific examples (Aaker 1996):

- **Failure in brand differentiation** A product attribute can be extremely important to customers, but if all brands are perceived to be adequate on this attribute, it does not differentiate the brand. For example, in the hotel business, cleanliness is always rated as one of the most important attributes to customers. Thus it would be appropriate for cleanliness to be a part of Hilton Hotels' brand identity. However, because all hotels are expected to be clean, cleanliness will not be a brand differentiator. Without brand differentiation, the brand name will not stand out in customers' minds when a product is needed.

- **Easy to copy** Product functional benefits are easy to copy. A brand that relies on superior performance in functional attributes will eventually be beaten, because functional attributes are transparent, fixed targets. If the brand name does not have a psychological dimension, a low-cost competitor could easily nudge another brand out of the marketplace.

- **Limitation on brand extension** An overly strong association with particular product attributes may limit the ability of a brand name to extend to other fields. Both GM and Ford have strong financial arms and they make a profit from them. However, they are mostly related to automobile financial operations, due to the overwhelming brand association with the auto industry. It is really hard for these companies to expand into non-auto-related financial operations in a big way.

- **Limitation on business strategy change** An overly strong association with particular product attributes will also limit a brand's ability to respond to changing markets. The Atkins brand is closely associated with Atkins diet theory, and it is doing fine today. However, if the Atkins diet theory goes out of favor, this brand will have a big problem.

Therefore, product-related association alone will not be adequate for the constantly changing marketplace—it is important for a brand to address other perspectives of brand identity.

The Brand as Organization   The brand as an organization consists of the brand attributes that feature the organization that manufactures the brand as opposed to the products or services related to the brand. Specifically, the brand-as-organization attributes should portray the desirable and appropriate organization images that organization leaders want the general public to see, such as corporate visions, beliefs, core values, innovation, and so on.

Some organization-related attributes can also be related to the product. For example, innovation and quality could also be related to product design. However, when these attributes are related to an organization, they usually mean different things, such as the culture and values that create the innovation and quality. For example, Toyota's lean manufacturing principles are easy to copy from a procedure point of view; however, it is Toyota's culture that is really difficult to imitate and copy, which is why so many companies want to implement lean manufacturing but fail to reach its full benefit (Liker 2004).

Organizational attributes are more enduring and more resistant to competitive claims than are product attributes. First, it is much easier to copy a product than to duplicate an organization with unique people, value, and culture. Second, organizational attributes usually apply to a set of product classes, and a competitor with a single product class is difficult to match. For example, a low-end cloth manufacturer can imitate a Ralph Lauren T-shirt easily; however, it is very difficult to actually create a brand with equivalent value because Ralph Lauren has provided so many kinds of products, as well as sponsorships for sporting events over many years. Third, organizational attributes, such as a culture of innovation and quality, are difficult to measure and communicate, so it is difficult for competitors to convince consumers that they have closed the perceived gap.

The Brand as Person   The brand as a person is the brand attribute that adds personality components, such as uplifting, youthful, energetic, upscale, trustworthy, casual, and so on. This brand attribute is more lively and personal than those related to products and organization.

The most important concept here is the brand personality, which can be defined as the set of human characteristics associated with a given brand. Thus it includes such characteristics as gender, age, and socioeconomic class, as well as such classic human personality traits as warmth, concern, and sentimentality.

A brand personality can create a strong brand in several ways. First, it can help customers to express their own personality. For example, a rich person may want to drive a Mercedes-Benz to show his affluence and pride. Second, brand personality can be the basis of a relationship between the customer and brand. For example, Harley-Davidson has a personality of rugged, free-spirited, outdoor-oriented men, which the buyer can use as an identifier for his own personality. Third, a brand personality may help communicate product attributes.

The brand personality perceived by customers is created by many factors, some of which are related to the product, and others that are non-product-related. Table 8.2 summarizes these factors.

Product-related characteristics could be the primary drivers of a brand personality. These characteristics include

- **Product class** The product class can affect the personality. For example, a bank or insurance company tends to assume a "banker" personality (competent, serious, male, older, upper class).

- **Package** The Huggies package always features healthy, happy children.

- **Price** Price is a complex factor in affecting brand personality—if the price is low, it may attract low-end buyers and thus increase sales, but it also gives the brand a "cheap" image. If the products related to a brand have high performance and top quality, a higher price may actually give an image of a prestigious brand.

- **Attributes** Product attributes also affect brand personality. For example, the strong flavor in Marlboro brand cigarettes gives them a rugged male personality.

Non-product-related characteristics can also affect brand personality.

- **User image** User image refers to either the profile of typical users (the people who use the brand) or idealized users (as portrayed in advertising and elsewhere). User image can be a powerful driver for

TABLE 8.2   Brand Personality Drivers (Aaker 1996)

Product-Related Characteristics	Non-Product-Related Characteristics
Product class	User image
Package	Sponsorship
Price	Symbol
Attributes	Age
	Ad style
	Country of origin
	Company image
	CEO
	Celebrity endorsers

brand image; for example, the Marlboro Man is the defining image of Marlboro's brand personality, a free-spirited, rugged man.

- **Sponsorship** Sponsorship of particular events will influence a brand's personality. For example, Nautica's sponsorship of Olympic swimming events gives its brand and swimwear a personality of "world-class swimmer."

- **Symbol** Brand symbols can have a powerful influence on brand personality because people see them every time they see an advertisement or see the product. The Intel Inside symbol created a very strong psychological impact for buyers—without the Intel CPU, it wouldn't be perceived as a good computer. Marlboro Country and the Maytag repairman are among the most successful brand symbols that give their brands desirable stereotypes in the customers' minds.

- **Age** How long a brand has been in the market (age) can affect its personality. An old brand tends to have a "traditional" or "reliable but old-fashioned" brand personality. A newcomer tends to have a younger brand personality.

- **Ad style** Advertisement style will affect brand personality as well. The selection of actors/actresses and the selection of themes should reflect the brand personality.

- **Country of origin** Country of origin is also a very powerful opinion-shaping factor; for example, a German brand might capture some perceived characteristics of the German people (precise, serious, hardworking, and so on).

- **Company image** Company image will affect brand personality. The brand personality should be consistent with company image.

- **CEO** The personality of the CEO, such as Bill Gates of Microsoft, influences people's perception of the company and its products.

- **Celebrity endorsement** Celebrity endorsements, such as Michael Jordan's endorsement of Gatorade, give the brand a similar personality—a strong, thirsty athlete, in this case.

Value Proposition The purpose of products or product classes under a brand name is to provide customers with benefits, and there are many kinds of benefits. In Aaker's brand identity model, illustrated in Figure 8.3, three kinds of benefits are listed: functional benefits, emotional benefits, and self-expressive benefits. The benefits that are offered by each brand will be different, and Aaker (1996) call this the *value proposition*. Specifically, a brand value proposition is a statement of the functional, emotional, and self-expressive benefits offered by the brand. An attractive brand value proposition should lead to market share increase.

The concepts of functional, emotional, and self-expressive benefits are explained as follows:

- **Functional benefits**   Functional benefits are the aggregated product functions that a product provides to customers. The functional benefits of a car include moving from point A to B, changing direction and speed, providing a nice driving environment, providing styling, and so on. Besides some "must-have" functions, a brand often provides some functional benefits that are special features. For example, Volvo provides safety and durability; 7-Eleven provides convenience; Nordstrom provides customer service.

  Functional benefits are important—if a brand can dominate a key functional benefit that customers really care about, it can dominate its product class. The challenge is to select functional benefits that will ring a bell with customers. However, just delivering this functional benefit is not enough, because customers buy products based on *perceived* quality and *perceived* functional superiority. Convincing customers that you are truly the leader in a key functional area might be more challenging than simply delivering these key functional benefits.

- **Emotional benefits**   When purchasing or using a particular brand gives the customer a positive feeling, that brand is providing an emotional benefit to customers. For example, you feel safe when you drive a Volvo; you feel important when you shop at Nordstrom. A strong brand value proposition often includes emotional benefits, on top of functional benefits.

  If a brand only has functional benefits, it is vulnerable because a low-cost producer could duplicate the same functional benefits with a lower price. Emotional benefits are more complex and much more difficult to copy. Emotional benefits are intertwined with functional benefits, and it is important to study the relationship between functional attributes and emotional benefit.

- **Self-expressive benefits**   Customers can also use a brand to show off. We call this the self-expressive benefit. For examples, some youngsters want to buy fashions from the Gap to show off; successful business people drive Lincoln, Lexus, or Mercedes-Benz.

Example 8.1 McDonald's Brand Identity   McDonald's is one of the most successful global brands. Its brand identity can be summarized as follows:

Brand as product

Product scope: Fast food, children's entertainment, eating spaces

Product functions:

Variety of fast-food items: hamburgers, Big Mac, Happy Meals, Egg McMuffin, and so on

Service: Fast, accurate, friendly, and hassle-free

Cleanliness: Spotless in eating spaces, restrooms, and counter

Low prices

Quality / Performance: Consistent in temperature, taste, portion, layout, and decoration; cleanliness in all its worldwide locations

User: Family and children are the focus, but the restaurant serves a wide clientele.

Country of Origin: USA

Brand as Organization

Convenience: McDonald's is the most convenient quick-service restaurant; it is located close to where people live, work, and travel; it features efficient, time-saving service; and it serves food that is easy to eat.

Brand Personality

Family-oriented, all-American, genuine, wholesome, cheerful, fun

Brand as Symbol

Logo: Golden Arches
Characters: Ronald McDonald; McDonald's doll and toys

Value Proposition

Functional benefits: Good-tasting burgers, fries, and drinks; extras such as playgrounds, prizes, and games

Emotional benefits: Children—fun via excitement of birthday parties; joy from toys and playgrounds; the feeling of special family times

8.1.2.2 Davis's Brand Image Model Davis (2000) also developed a brand image model. The brand image has two components: the brand associations and the brand persona. Brand associations describe the kinds of benefits the brand delivers to customers and the role it plays in their lives. Brand persona is a description of the brand in terms of human characteristics. The brand image model is a concise model for the brand identity.

Brand Associations Brand associations relate to the product, service, and organization aspects of the brand. Brand associations describe a hierarchy of benefits that a brand provides to its customers.

Brand associations are best described by the brand value pyramid, illustrated in Figure 8.3.

The Features and Attributes layer is at the bottom of this pyramid. Here the features and attributes are the most essential product functions, features, and quality levels that must be delivered to customers in order to survive in the marketplace. The Benefits layer is in the middle and includes the additional functional and/or emotional benefits that the brand provides to its customers, beyond the features and attributes covered in the bottom layer. The Beliefs and Values layer is at the top of the pyramid, and this layer represents the emotional, spiritual, and cultural values that are addressed by the brand.

Many brands will not be able to fill all the layers of the brand value pyramid. If a brand cannot fill the bottom layer, it cannot even deliver the most basic benefits to its customers for this kind of product, and this brand will fail in the long run. If a brand can only fill the bottom layer, it is a very marginal brand—it is essentially a commodity, such as raw cotton, or raw sugar. It is simply the equivalent of the no-brand T-shirt in Figure 8.1, and its market survival is mostly dependent upon low price. If a brand fills or somewhat fills the middle layer, it becomes a surviving brand. It is better than just a commodity. The most powerful brands should fill all the layers in the brand value pyramid.

Figure 8.4 shows the Ralph Lauren brand value pyramid. Ralph Lauren has achieved the strongest brand status in its product class. Many brands can deliver the features and attributes illustrated in

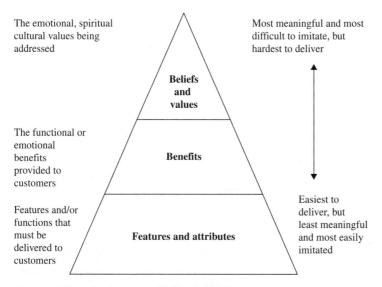

Figure 8.3   Brand value pyramid (Davis 2000)

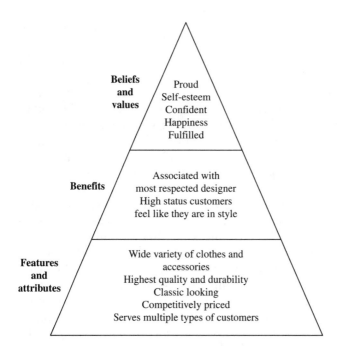

Figure 8.4 Ralph Lauren's brand value pyramid (Davis 2000)

Figure 8.4, that is, offering high-quality, durable, and classic-looking clothes. But few of them can say their clothes allow their customers to make a statement. Wearing Ralph Lauren clothes is like driving a Mercedes-Benz in its appeal to social status. The psychological benefits of this kind of brand name usually take years to evolve; they are difficult to explain, and they are even more difficult to duplicate.

Brand Persona   According to Davis (2000), brand persona is the set of human characteristics that consumers associate with the brand, such as personality, appearance, values, likes and dislikes, gender, size, shape, ethnicity, intelligence, socioeconomic class, and education. The brand persona brings the brand to life and customers subconsciously decide if they want to be associated with this brand, just as they do about whether they want to be associated with other people. If a brand persona is unpopular or unattractive, that will affect the sale.

The brand persona here is very similar to the brand personality in Aaker's brand identity model.

Example 8.2 Brand Personas for Mail Services   Table 8.3 lists the brand personas of three major mail carriers: Federal Express (FedEx), the U.S. Postal Service, and United Parcel Service (UPS).

TABLE 8.3   Brand Personas for Three Mail Carriers

FedEx	U.S. Postal Service	UPS
Male or Female	Male	Male
Young	Old	Middle-aged
Friendly	Grumpy	Friendly
Dependable	Unreliable	Inconsistent
High technology	Low technology	Unionized
Athletic	Sluggish	Evolving
Prompt	Slow	Brown uniforms
Energetic	Unsophisticated	Mediocre service
Professional	Rigid	Professional
Motivated	Complacent	International
Problem solvers	Problem makers	Problem solvers

8.1.3   Brand Equity

Brand equity is the set of assets (and liabilities) that is linked to a brand name and symbol. The brand equity adds (or subtracts) the value provided by a product or service to a firm and/or that firm's customers (Aaker 1996).

During the 1980s, a lot of research was done to define and estimate the true value of brands to the competitive position of enterprises (Keller 1993, Aaker 1991, Farquhar 1989, Tauber 1988). There were two reasons for this: The first reason was an accounting one, and it was to estimate the value of brands more precisely for the balance sheet, especially in cases of mergers, acquisitions, and divestitures. The second reason was a strategy-based motivation to improve marketing productivity (Keller 1993).

Brand equity provides a mechanism for capturing the marketing effects that are uniquely attributable to the brand (Keller 1993). Aaker's brand equity model (Aaker 1991: 269) is one of the best-known models of brand equity. It is a tool for understanding the linkage between the brand and the value it provides to the firm and its customers beyond what is inherent in the functional attributes of the products and services. The brand equity model defines five dimensions of value that the brand provides the firm: brand loyalty, brand awareness, perceived quality, brand associations, and other proprietary brand assets (see Figure 8.5). Each of these dimensions is important in influencing the decision of the customer during their purchasing process, and thus is a contributor towards the viability of the enterprise. The strategic role of each of these brand equity dimensions is described in the following sections.

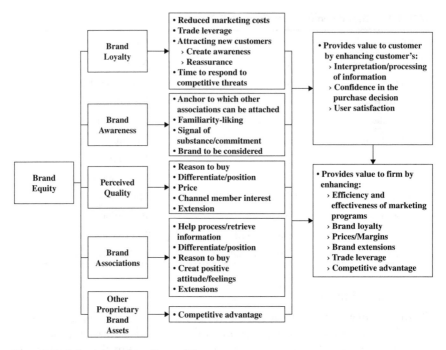

Figure 8.5 Aaker's brand equity model

8.1.3.1 Brand Loyalty In a competitive environment the ability of a company to retain its existing customers is of key importance. Brand loyalty is first a key factor influencing the repeat-buying behavior of customers (Arnold 1986, Keller 1993), and it reduces vulnerability to competitive actions in the marketplace. Secondly, it reduces the company's cost of doing business because it is more expensive for a business to acquire new customers than to retain existing ones, especially when the existing ones are satisfied with the brand (Aaker 1991). Thirdly, brand loyalty can be powerful leverage for negotiating more favorable terms in the distribution channels (Aaker 1991).

8.1.3.2 Brand Name Awareness Brand name awareness relates to the likelihood that a brand name will come to mind, and the ease with which it does so. Brand name awareness consists of two dimensions: brand recall and brand recognition.

■ Brand recognition reflects a familiarity gained from past experience with the brand (Aaker 1996). Studies have shown that people will often buy a brand because they are familiar with it (Aaker 1991).

■ Brand recall refers to how strongly the brand comes to mind when the consumer thinks about that product category, or the needs fulfilled by that product category (Keller 1993).

Brand name awareness plays an important role in consumer decision making because it allows the brand to be included in the consideration set, which is a prerequisite for its eventual choice.

8.1.3.3 Perceived Quality   Perceived quality is part of the human experience, and it is developed entirely from the perspective of the consumer based on those product attributes that are important to them. As a result, perceived quality may be different from the actual quality of the product. Studies have shown that the customer's perception of quality has one of the greatest impacts on the financial performance of a company (Buzzell et. al. 1987: 7, Jacobson 1987, Anderson et. al. 1994, Aaker 1994). A study of 33 publicly traded companies over a four-year period demonstrated that perceived quality had an impact on stock return (Aaker 1996). Perceived quality is a key strategic variable for many companies (Aaker 1996), and it is a key positioning dimension for corporate brands.

8.1.3.4 Brand Associations   Brand associations can be anything that connects the customer to the brand (Aaker 2000). These associations help determine the brand image with the customer and marketplace. Brand associations can be hard attributes, related to specific perceptions of tangible functional attributes such as, speed, user-friendliness, taste, and price. Brand associations can also be soft emotional attributes like excitement, fun, trustworthiness, and ingenuity (Biel 1993). Apple is an example of a brand with values that have resonated with those of its target audience. Emphasizing values such as fun, excitement, innovation, and humor (Kochan 1997), the company has succeeded in carving out a niche for itself in the highly competitive personal computer marketplace (Levine 2003).

8.1.3.5 Other Proprietary Brand Assets   Beyond their use as a tool in achieving a competitive advantage, brands are also a financial asset to a company. Successful brands can be traded, or can be used to increase the valuation of a company during a corporate acquisition.

8.2 Brand Development

Strong brands can create tremendous value for the companies who own them. Developing strong brands is one of the most important goals for companies who strive to excel in the marketplace. The objective of a brand development process is to create a brand that achieves and maintains the intended position in the minds of customers within the targeted market group. In other words, the brand development process must create a brand image that reflects the brand identity defined by the company, in the mind of customers.

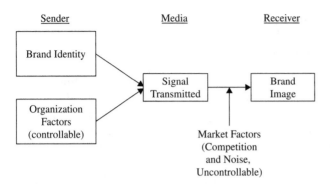

Figure 8.6 Brand identity transmission process

Creating this brand position in the mind of the customer involves creating the brand identity, transmitting the brand image to customers, and having customers receive and accept this image, as illustrated by Figure 8.6.

This process is influenced by a variety of factors, which include the nature of the brand identity, organization factors, the communications media, and market forces. In this section, we will first look at the factors that influence brand development, and then at the several key steps in the brand development process.

8.2.1 Key Factors in Brand Development

There are many factors that affect the brand development process. These factors can be categorized into two classes: controllable factors and uncontrollable factors. The controllable factors are those over which the company that owns the brand has some degree of control. The uncontrollable factors are ones over which the company has little or no control. These controllable and uncontrollable factors are also illustrated in Figure 8.7.

8.2.1.1 Controllable Factors   There are three classes of controllable factors:

- Brand identity
- Marketing mix factors, or the four Ps: product, price, promotion, and place
- Time to market

In brand development, these factors can be used to shape a desirable brand image and transmit this image to customers.

Brand Identity   Brand identity has been thoroughly discussed in Section 8.1.2. If we use Aaker's brand identity model, there are four perspectives: brand as product, brand as organization, brand as person-ality, and brand as symbol. Clearly, the company that owns the brand has full control of the brand symbol, a good degree of control in product development, and relatively good control over organization behavior. Brand personality takes years to form; it is more difficult to change.

In brand development, the company should design a desirable brand identity based on thorough analysis of the marketplace, competitors, and the relative strengths and weaknesses of the company itself, in order to achieve the best brand image possible.

Product When customers are making their purchasing decisions, they are not buying the brand symbol; they are buying the product. Initially, customers may be influenced by advertisements or friends' advice to try a product from a certain brand. However, the product has to perform to customers' expectations. If the product performs equal to or better than customers' expectations, the perceived brand image will be confirmed by their experience. This will trigger word-of-mouth recommendations, sales will grow, and the positive brand image will spread among more and more customers.

In order to accomplish this, it is very important that the actual prod-uct characteristics, such as functions, performance, and quality levels, be consistent with the brand identity. Therefore, product development has to go hand-in-hand with brand identity design.

Price The role of price in brand development is quite interesting. The brand price is related to the benefits that the brand provides, as illustrated by Figure 8.7.

If a price is too high relative to the benefits that the brand provides, the perceived value in customers' minds will be low. Customers will think the brand is overpriced. However, if the benefits that the brand provides are high, but the price is low relative to the benefits, the cus-tomers' reaction can be quite complex. Theoretically, customers will be happy to get more and spend less, but the perception that it is a "cheapo" product may creep in, and this may undercut the brand image.

Usually, benefits are the main focus for brand identity creation. If you have more benefits and customers are happy with them, it is always

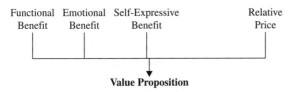

Figure 8.7 The relationship between price and benefits of a brand (Aaker 1996)

easy to raise the price to match the benefits. If the benefits are really low in comparison with other brands, the price will have nowhere to go but lower.

Promotion (Communication)   The effectiveness of the communications campaign is a critical factor in creating the desired brand position and image (Aaker 1996). The purpose of marketing programs, such as advertising, is to transmit the brand image in order to increase brand awareness and create a strong, favorable, and unique brand identity in the customer's mind. The strength of a brand image is greatly influenced by communications in the marketing programs, particularly by the effectiveness with which the brand identity is integrated into the marketing programs (Keller 1993). Given the many choices of media available, selecting the correct media mix to reach the targeted audience and ensuring that the message is integrated is a key factor in creating a distinct brand image.

Place (Distribution)   The distribution channels used to deliver the products to customers are also very important factors for brand development. It is important that the distribution channels deliver the goods to the right customers. The key variables in the selection of distribution channels include the type of channels, the number of outlets, the locations of outlets and stock levels, and so on. The distribution process should be synchronized with the promotion activities so that the desired customers will know what the products are and where they are available. An efficient distribution process will give customers the service benefits and reduce the hassles in obtaining the products, thus enhancing the brand image.

Time to Market   Brands that are able to position themselves first in the minds of their customers have the best chance of achieving the highest brand awareness (Reis 1981). Being first in the marketplace will affect the company's ability to create a strong brand. However, this factor is not totally controllable, because it also depends on competitors' activities.

8.2.1.2 Uncontrollable Factors   The uncontrollable factors in brand development are mostly marketing factors, as illustrated in Figure 8.7.

The demographics of the target population for the brand is an uncontrollable factor; it includes age group, income level, sex, marital status, area of residence, location, social group membership, stage in life, and so on. The variation in demographics makes the targeted customers very nonhomogeneous, so the brand image that the company tries to communicate to customers will be perceived in different ways. For example, a particular brand image might be very attractive for one age group, but it might be unattractive for another age group.

Cultural factors are also important uncontrollable factors. People from different cultural backgrounds, such as business clients from different corporate cultures, will see things in different ways. These cultural factors may affect how customers perceive the brand image.

Competitive activity is clearly a very important uncontrollable factor. The number of competitors and their relative strength will have a tremendous impact on brand image. Therefore, it is critical to conduct a competitor's analysis at an early stage of the brand development process.

The customer is another uncontrollable factor, but the company can try to understand the customers' needs. The ability of a product or service to meet the needs of its customers is a critical factor in creating the brand loyalty that will determine the success or failure of the brand.

8.2.2 The Brand Development Process

Different processes have been proposed for brand development, but they are fundamentally based on two paradigms. The first views the brand development process as being closely associated with the development and marketing of new products (Watkins 1986). In this paradigm, the brand development process follows the classical brand management model (Aaker 2000), and it consists of the following steps:

1. Market exploration
2. Preliminary financial analysis and screening
3. Formal business analysis and planning
4. Product and brand development
5. Product testing
6. Product launch

More recently, the second paradigm has elevated brand management to a more strategic position within the organization, as defined by the brand leadership model (Aaker 2000) and brand asset management process (Davis 2000). The differences between the two brand development paradigms are shown in Table 8.4.

The new paradigm for brand development and management is more focused on managing the brand as a strategic asset of the company (Arnold 1992, Davis 2000, Aaker 1996). The brand development process defined by this paradigm generally consists of the following phases (shown in Figure 8.8):

1. Brand strategy analysis phase
2. Brand strategy development phase

TABLE 8.4 Paradigms of the Brand Development Process (Adapted from Aaker 2000)

Features	Classical Brand Management Model	Brand Leadership Model
Perspective	Tactical and reactive	Strategic and visionary
Brand Manager Status	Less experienced, short time horizon	Higher in the organization, longer time horizon
Conceptual Model	Brand image	Brand equity
Focus	Short-term financials	Brand equity measures
Product–Market Scope	Single products and markets	Multiple products and markets
Brand Structures	Simple	Complex
Number of Brands	Focus on single brands	Category focus—multiple brands
Country Scope	Single country	Global perspectives
Brand Manager's Communication Role	Coordinator of limited options	Team leader of multiple communication options
Communication Focus	External/customer	External and internal
Driver of Strategy	Sales and share	Brand identity

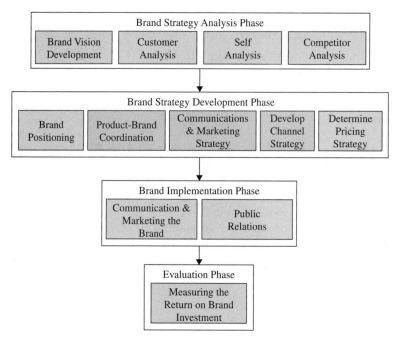

Figure 8.8 Brand development process

3. Brand implementation phase

4. Brand evaluation phase

In the following sections, we will look at these four phases in detail.

8.2.3   Strategic Brand Analysis

The strategy analysis phase of brand development focuses on understanding the brand's competitive position in the marketplace. It also looks at the ability of the company to influence this positioning through its capabilities and the attributes of its products and services. The steps of the strategy analysis phase are brand vision development, customer analysis, self-analysis, and competitor analysis.

8.2.3.1   Brand Vision Development

The brand vision is a short, succinct statement of what the brand intends to become and to achieve at some point in the future, often stated in competitive terms. Brand vision refers to intentions that are broad, all-inclusive, and forward-thinking. It is the image that a business must have of its goals before it sets out to reach them. It describes aspirations for the future, without specifying the means that will be used to achieve those desired ends.

This is IBM's brand vision:

> At IBM, we strive to lead in the creation, development, and manufacturing of the industry's most advanced information technologies, including computer systems, software, networking systems, storage devices, and microelectronics. We translate these advanced technologies into value for our customers through professional solutions and service businesses throughout the world.

According to Davis (Davis 2000), a good brand vision should have four components: a statement of the overall goal of the brand, the target market that the brand will pursue, the points of differentiation that the brand will strive for, and the overall financial goals for which the brand will be accountable. Davis gives an example of such a brand vision statement:

> Around the world, our eye care brand will stand for leadership in visual care. Consumers and the professional channel will recognize us as the industry leader in visual care solutions, including the best service, follow-up, expertise, and product innovation. Our brand will help us fill one-third of our stated financial growth gap through price premiums, better relationships with the channel, and close-in brand extensions.

Developing a brand vision links the brand development process to the strategic objectives of the company, which is an important step in ensuring the necessary top management and financial commitment to the brand (Davis 2000). During this step the strategic and financial goals of the brand are defined, and the commitment of senior management to the goals and objectives of the brand are obtained.

8.2.3.2 Customer Analysis The activities in the customer analysis step are focused on understanding the trends, motivation, and unmet needs of the various segments of the customer market (Aaker 1996). This step creates an understanding of how the "customer thinks and acts and why and how they make a purchasing decision" (Davis 2000). The objectives of this step are fourfold (Aaker 1996):

1. To determine the functional, emotional, and self-expressive benefits that the customer seeks when they buy and use the brand.

   Customer surveys are usually needed to determine these benefits as well as the relative importance of these benefits. The following three questions are used to assess the functional benefits:

   - What functional benefits are relevant to customers?
   - What is the relative importance of each functional benefit?
   - Can benefit segments be identified?

   Emotional and self-expressive benefits are subtle and thus more difficult to detect than functional benefits.

2. To understand the different segments of the customer market and their different needs, wants, and behaviors.

   We need to develop a deep understanding of the needs of various market segments. Because different market segments may have different needs in functional, emotional, and self-expressive benefits, they may respond differently to a brand promotion program. There are many possible segmentation schemes, but in the brand development process, the major task is to find which segments are the most attractive targets for the brand and are most relevant to brand identity development. Therefore, the commonly used segmentation schemes include segmentation by benefits sought, segmentation by price sensitivity, and so on.

3. To understand trends occurring in the customer markets, so that the current and future positioning of the brand can be better assessed.

   By trends we mean the dynamics of the market, and how the demand pattern will change. Analysis of market data, such as sales volume trends and profitability prospects of the submarkets, may help to understand market trends. Understanding the market trends provides insight into changing motivations and emerging segments with strategic importance. For example, in the coffee market, from 1962 to 1993, the sales of regular supermarket brands have been declining, but gourmet coffee and coffeehouse sales are increasing. If a company is in the coffee business, this information will certainly help to position its brand in the future market.

4. To identify customer needs that current products do not meet.

Unmet needs are customer needs that are not met by existing products in the market. Unmet needs are strategically important because they can represent opportunities for the company to make beneficial moves in the market.

For example, Black & Decker organized a focus group of 50 power tool owners. The executives of Black & Decker visited focus group members' homes and found several major unmet needs. One of the problems with a cordless drill is that it can run out of battery power before the job is done. Black & Decker responded by offering detachable battery packs that can be recharged quickly. Tool owners can have several battery packs charged and when one battery runs out, they can replace it with another one in no time and recharge the run-out battery while they are working. Several design changes of this kind gave Black & Decker's Quantum brand a core identity and competitive advantage.

8.2.3.3 Self-Analysis   The objective of the self-analysis step is for management to examine the strengths and weaknesses of their brand's current situation, so as to understand how the brand is positioned in the marketplace and what circumstances contributed to its current position (Arnold 1992). The areas analyzed during this step include

- The current brand image, that is, what is the perception of the brand in the marketplace?

   This analysis can be done by using customer surveys (see Chapter 4). The following types of questions should be included in the survey:

   - How is the company's brand perceived?
   - What associations are linked with the brand?
   - Why do customers like the brand? Why do customers not like the brand?
   - How is the company's brand different from competing brands?
   - What benefits do customers get from the brand?
   - Has the company's brand changed over time? If yes, how?
   - For different market segments, does the company's brand image differ? If yes, then how does it differ?

- Does the company's brand have a personality? If yes, what is it?
- What are the intangible attributes and benefits of the brand?

In assessing the current brand image, it is important that the customer research should study not only the product-related attributes but also non-product-related attributes, such as organizational association, brand personality, brand customer relationships, and emotional and self-expressive benefits.

- The fundamental values of the brand, that is, what does the brand stand for? Is it for fun, luxury, or an active lifestyle? What is the heritage of the brand?

Besides studying the current image of the brand, it is important to understand the heritage of the brand. Any surviving brand has some reasons why it survived; it must have done something right. Many brands get into trouble because they deviate from their heritage. Arbitrary changes in brand identity may hurt a brand more than they help. The answers to such questions as "Who were the early pioneers of the brand?"; "How did it originate?"; and "What was the brand image when it first started?" can help in understanding the brand heritage.

- Links and associations to other brands.

Some companies offer several brands of products or services. In this case, a change in one brand position may affect other brands offered by the company. Brand position decisions shouldn't be made in isolation— each brand should have a well-defined role, and all brands offered by the company should work together in a synergistic manner.

- The strengths and weaknesses of product and service offerings, and the capabilities of the organization, that is, what is the organization good at?

For a realistic brand strategy, the desired brand identity should be supported by the organizational strength of the company. It is necessary to find out the company's and the products' strengths and weaknesses. We need to find out what the company is and is not good at, and how and to what degree these weaknesses can be changed. If a company is pursuing a goal that cannot be substantiated due to the company's weakness, the goal will not be achieved.

8.2.3.4 Competitor Analysis   The purpose of the competitor analysis is to understand the current image, positioning, strengths, and weaknesses of competitive brands in the marketplace, as well as the possible future trajectories of these competitive brands (Aaker 1996). This analysis will develop an understanding of the following aspects:

1. The customer's perception of competitive brands.

   This information is a fundamental input for brand identity determination. It indicates how customers perceive competitors' brands, so it provides a basis for "what our brand has to do to differentiate itself from competitors." It is important to find out what the functional benefits to customers are, how these benefits compare with those of the company's brand, the brand-customer relationship, and brand personality.

   There are two sources for this information. One is a customer survey study on competitors' brands. The other is competitors' own information; for example, competitors' advertisements and advertisement plans can provide clues to what kind of brand image they want customers to perceive.

2. Previous changes in competitive brand positioning and the future market positioning (strategic brand objectives) of competitors.

   In the brand development process, it is important to consider not only the current images of competing brands but also past changes and possible future changes in these images. A thorough examination of such changes can provide useful information about the reasons for such changes and the reality of the competitive environment.

3. Strengths and vulnerabilities of the competitors.

   Information on the strengths and vulnerabilities of competitors is valuable input for a company's brand position. It is difficult and costly to compete head to head with the strong points of competitors. It is much easier to compete in areas where competitors are not strong.

8.2.4   Brand Strategy Development

The purpose of the development phase is to develop the brand strategy. During this phase, the positioning of the brand is developed. Also the channel strategy, pricing strategy, and future extensions to the brand are developed or aligned. The following sections describe these steps.

8.2.4.1   Brand Positioning   "Brand position is the part of the brand identity and value proposition that is to be actively communicated to the target audience and that demonstrates an advantage over competing brands" (Aaker 1996). The effect of the positioning is to create the necessary associations the customers will think of when they recall the brand. Some examples of the associations of some well-positioned brands are shown in Table 8.5.

TABLE 8.5   Examples of Top Brands and Their Attributes (Davis 2000)

Brand	Attributes
Disney	Family fun entertainment
Nordstrom	Highest level of retail service
Saturn	Your car company
FedEx	Guaranteed overnight delivery
Wal-Mart	Low prices and good values
Hallmark	Caring shared

A brand's position should be updated every three to five years, or as often as needed to update the company's growth strategy. Senior management has to lead in developing, updating, and implementing the brand position.

Brand positioning defines the following aspects of the brand's position in the marketplace:

- The target market segment
- The business it provides to the market segment
- Key benefits and points of differentiation between the company's products and other brands in the marketplace
- The contract of the brand with the market, which defines the brand's promises to its customers

Target Market Segment   The company that owns the brand needs to know who the intended customers for its brand are. Customer surveys and self-study can be used to determine the target market segment. The following types of questions should be asked:

- Is the target market both identifiable and reachable?
- Would the current customers be part of our target market segment?
- Will the target market be attracted to our distinct brand identity?
- If we have never served this market segment before, why do we want to serve them now?

Business Provided to the Market Segment   The company that owns the brand needs to know what kinds of businesses they will and will not operate in the target market segment. Customer surveys and self-study can be used to determine this. The following types of questions should be asked:

- What is the category, industry, or business that we compete in?
- How has this changed over time?
- Will the marketplace value and believe our participation in this business?

Key Benefits and Points of Differentiation   The company that owns the brand needs to know what differentiates its brand from other brands in the target market segment, as well as what the key benefits of its brand are to customers. Customer surveys and self-study can be used to determine this. The following types of questions should be asked:

- Are our key benefits important to customers?
- Can we deliver these benefits satisfactorily?
- What are our key points of difference from other brands?
- Can we own these key points of difference over time?
- Are we competing at the Features and Attributes layer, at the Benefits layer, or at the Beliefs and Values layer (see Figure 8.4)?

The Contract of the Brand with the Market   The contract of the brand with the market is also called a brand contract (Davis 2000). It is a list of all promises the brand makes to customers, in terms of its products and service quality. Such a contract is executed internally, but it is defined and validated externally by the marketplace.

The brand contract is derived by analyzing several things:

- The current promises of the brand made to the marketplace
- Positive and negative feedback from customers regarding current promises
- The results of the brand position analysis

Example 8.3 Starbucks' Implicit Brand Contract   Starbucks promises to do the following:

1. Provide the highest quality coffee available on the market today
2. Offer customers a wide variety of coffee options as well as complementary food and beverage items
3. Have an atmosphere that is warm, friendly, homelike, and appropriate for having conversation with a good friend or reading a book

4. Recognize that visiting Starbucks is as much about the experience of drinking coffee as it is about coffee itself

5. Have employees who are friendly, courteous, outgoing, helpful, knowledgeable, and quick to fill customer orders

6. Provide customers the same experience at any one of the several thousand Starbucks worldwide

7. Stay current with the times, meet customer needs, and help customers create the Starbucks experience on their own terms

8. Provide customers with an environmentally friendly establishment

9. Educate customers on the different types of coffee offered

8.2.4.2 Five Principles of Effective Brand Positioning   Davis (2000) proposed five principles of effective brand positioning:

- **Value**   The proposed brand position should provide the targeted customers with values that are superior to those of competitors. It should provide functional, emotional, and self-expressive benefits that a wide range of customers will appreciate and be willing to pay premium prices for.

- **Uniqueness**   The proposed brand position should have some unique attributes that no other competitors can deliver. These unique attributes should be important to customers and be appreciated by them. The uniqueness should make the company's brand stand out in the crowd.

- **Credibility**   The proposed brand position should be implementable in a credible manner, and the company's effort should be able to make customers believe that all the promises will be met. The brand position should be in line with customers' perception of the company's ability.

- **Sustainability**   Once a proposed brand position is implemented, it is desirable that this brand position last as long as it can. Changing brand positioning involves a lot of investment and frequent change reduces the credibility of the brand. A good brand position should be difficult for competitors to copy and should meet customers' changing needs for a long time.

- **Fit**   The brand position should fit the company's objectives and culture.

Example 8.4 Brand Positions of Several Bookstores   Borders, Crown Books, Barnes & Noble, and Amazon.com are four dominant booksellers in the United States. The following table lists their unique brand positions.

Company	Target Market Segment	Business Provided	Point of Difference
Borders	Individuals looking for a community meeting place	Books, music, multimedia, and online	Fun place to go
Crown Books	Price-sensitive individuals, strip mall shoppers	Bookstores	Discount pricing
Barnes & Noble	Individuals looking for a quiet gathering place	Books, music, multimedia, and online	Library-like setting
Amazon.com	Individuals who are active on the Internet and shop online	Online books, music, and many other items	Personalized online service, huge variety

8.2.4.3 Product-Brand Coordination   Customers are buying products or services, and a brand is a symbol and shortcut that helps customers select the products they need. The customers' total experience with the products, including purchasing, consuming, and servicing, has to be in tune with the brand position. For example, if a hospital's brand position is to be "the premier hospital that provides you with an attentive team of caring experts working together for the highest level of professional care," every word, such as premier, attentive, team, caring, and so on, should have concrete actions behind it. Table 8.7 provides an example of a well-coordinated brand and product combination.

These are the key issues in ensuring product brand coordination:

- The brand development team should include product development people.

- Key product development professionals should learn the basics in brand development and management.

- Brand positioning and product development should go hand in hand.

8.2.4.4 Communications and Marketing Strategy   Marketing determines what in the brand's positioning will be communicated, and how it will be communicated to the marketplace (Levine 2003). There are numerous vehicles for communicating the brand to the marketplace, and these include

- Advertising
- Internet
- Public relationships
- Trade and sales promotions

TABLE 8.6   Product-Brand Coordination of a Hospital

Brand Position	Product (Patient Treatment) Attributes
Premier hospital	■ Excellence in all performance metrics ■ Staffed with first-class doctors, administrators, nurses ■ Excellent infrastructure, first-class equipment ■ Modern appearance, spotless, well-organized
Attentive team	■ Reduction of patients' waiting time to industry's best ■ Reduction of unneeded paperwork, tests ■ Clearly explained treatment plan, hospital protocols, discharge procedures ■ Quick feedback to patients' requests
Caring	■ Reduction of patients' waiting time to industry's best ■ Caring nurses ■ Prompt response to all patients' care issues ■ Excellent in-patient facility
Expert	■ Competent doctors ■ State-of-the-art medical equipment and first-class technical support
Highest level of professional care	■ Reduction of treatment error/diagnostic error to minimum ■ Reduction of unneeded treatment to minimum

- Consumer promotions
- Direct marketing
- Event marketing
- Product placement
- Internal employee communications

In order for the brand to achieve its intended positioning in the marketplace, it is very important that the brand image is communicated through various vehicles using an integrated marketing communications strategy (Davis 2000). The message delivered through all these vehicles must be consistent and must be related back to the brand image. The communications strategy determines the best mix of vehicles for communicating the brand image.

8.2.4.5 Determine the Channel Strategy   The objective of this step is to determine the appropriate distribution channel strategy to enhance the brand image, and in the case of existing brands, to leverage the strength of the brand. The selection of the appropriate distribution channel is very important because of the association that is created between the image of the channel and that of the brand. Also, because a strong brand can create a draw to a distribution channel, it is necessary during this step to leverage the power of the brand to create the best distribution arrangements and ensure more control over the distribution of products and services.

8.2.4.6   Determine the Pricing Strategy   This step focuses on determining the correct pricing policy for the brand. A brand's price must be related to the benefits the brand provides (Aaker 1996). An overpriced brand will not be rewarded in the marketplace, and an underpriced brand can negate certain associations in the brand's image. Also, the ability to charge premium prices is one of the benefits of developing a strong brand (Davis 2000), so this must be leveraged in determining the pricing strategy.

8.2.5   Brand Implementation

During the brand implementation phase the plans developed in the brand strategy phase are executed.

8.2.5.1   Communicating and Marketing the Brand   This objective in this step is to communicate the brand to the marketplace using the integrated communication and marketing strategy developed in the previous phase.

During this step care is taken to use all the selected communication vehicles effectively to achieve the optimal sales per dollar spent. Also, in order to improve the effectiveness of the communications, it is important to track all marketing expenditures, by product, promotional tool, stage of the life cycle, and observed effects in order to establish a baseline for improving the use of these tools.

8.5.2.2   Public Relations   The purpose of public relations in the development of the brand image is to encourage positive feelings about the brand among the public. However, unlike marketing, which is very visible in the development of the brand, a well-executed public relations campaign is not visible (Levine 2003: 17). The public relations activity achieves its objectives by encouraging third parties to deliver positive messages about the brand. These messages are usually delivered through news organizations, and print journalists, in the form of news and press releases (Levine 2003: 17). Because the company does not have any control over the news outlet, a challenge during this step is ensuring that the message that is delivered is true to the brand image.

8.2.6   Brand Evaluation

This purpose of the brand evaluation step is to measure the performance of the brand in the marketplace. The classical brand management process emphasizes two metrics: recall and awareness of the brand (Davis 2000). However, these measures alone are not suited for measuring brand performance, as determined by the equity value of the brand. In order to provide information for managing the brand as an asset, brand performance measures should do the following (Davis 2000):

- Provide an understanding of how the brand is performing internally and externally
- Provide information about return on investment of marketing and branding strategies
- Assist the organization in its resource allocation decisions
- Provide information for rewards and incentive systems

Some of the additional brand performance measures include

- Acquired customers
- Lost customers
- Customer satisfaction
- Purchase frequency
- Market share
- Return on advertising
- Price premium

Example 8.5 New Marlboro Cigarette Brand Development   The new Marlboro cigarette brand development process is a brand development legend. This example illustrates a successful brand development.

The Marlboro cigarette brand is now the best-selling packaged cigarette in the world. However, as recently as the late 1950s, it was an old, dying tobacco brand in the United States. In 1954, after careful analysis of the trends in the tobacco market, the management of Philip Morris made a number of key decisions on the changes of brand position, as contrasted in the following table:

Old Marlboro	Marlboro
Mild tar blend	Stronger blend
Less flavor	More flavor
No filter	Filter
White pack design	Red and white pack design
Older image	More modern image
Aimed at women	Aimed at men and women
Product-based advertising	Imagery advertising

To match the newly designed brand image, the product was totally redesigned. At that time, 90 percent of U.S. smokers used unfiltered cigarettes, the company realized that the coming trend would be filters, and this could also help to modernize the image of the brand. To change the perception that the Marlboro cigarette is a mild cigarette for women, the flavor of the cigarette was made stronger, and the filter was covered in tobacco-brown paper, indicating strength and flavor. To shape the new Marlboro brand identity, a new advertisement agency, Leo Burnett, was contracted by Phillip Morris to develop a campaign to relaunch the brand using male role models in tough, rugged jobs, in order to project a new Marlboro brand personality. At the beginning, the images of pilot, deep-sea fisherman, cowboy, and engineer were tried. In 1963, market research indicated that Marlboro needed a more clear-cut identity. The Marlboro Man, symbolized by a cowboy, was established.

Campaign guidelines were laid down as follows:

- The cowboy must symbolize the type of man that other men would like to be, and women would like to be with.
- He must be believable.
- Marlboro Country must always be magnificent, never ordinary.
- Every ad in the campaign must be candid and have impact.
- Variety will be achieved by rotation of cowboy portraits, smoking moments, and magnificent country material.

To the present day these guidelines have been maintained through all media. To ensure projecting a consistent brand image, the Marlboro advertisement and campaign style is highly consistent in worldwide.

After all these efforts, the sales of Marlboro brand cigarette steadily increased, and by 1975, Marlboro had grown to U.S. brand leadership, and is now an international brand leader.

9

Value Engineering

Any product's customer value and satisfaction can be improved by increasing customer benefits and reducing cost. Among the customer benefits, functional benefits are of key importance. People pay for functions, not for hardware or paperwork. For example, people go to a fast-food restaurant to buy such functions as "relieving hunger," "getting nutrition," "getting taste," and so on. People go to hospitals, not to buy the "doctor's time," or "surgery" or "hospital beds," but to buy such functions as "curing disease" and "relieving symptoms." Value engineering is a systematic, team-oriented, creative approach that seeks to deliver functions the customer wants with lower cost.

The Society of American Value Engineers (SAVE) defines value engineering as follows: "Value engineering is the systematic application of recognized techniques which identify the functions of a product or service, establish a monetary value for that function, and provide the function at the lowest cost."

However, value engineering is not merely a cost-cutting program. It only cuts unnecessary costs, which are the costs that can be removed without affecting the functional performance of the product or service. The new design coming out of a value-engineering project should have the same or better functional performance than the old design. It has been estimated that for the average product or service, 30 percent of its cost is unnecessary. This unintentional cost is the result of habits, attitudes, and all other human factors.

In this chapter, we will look at the six phases in value-engineering projects, and then at two value-engineering case studies in the service industry. First, though, we'll look at how value engineering works.

9.1   An Overview of Value Engineering

Value engineering originated at General Electric Company in 1947. Harry Erlicher, Vice-President of Purchases, noted that during wartime, it was frequently necessary to make substitutions for critical materials that not only satisfied the required functions but also had better performance and lower cost. He reasoned that if it was possible to do this in wartime, it might be possible to develop a system that could be applied to normal operations to increase the company's efficiency and profit. L. D. Miles was assigned to study the possibility, and the result was a systematic approach to problem solving based on functional performance, which he called value analysis.

Value analysis, value engineering, value management, value assurance, and value control are all the same in that they make use of the same set of techniques developed by Miles in 1947. In many cases, the title tends to describe how the system is being applied. Value analysis is applied to remove cost from a product. Value engineering and value assurance are applied in the development phase to keep cost out of a product. Value management and value control are overall programs that apply value techniques in business operations.

Value engineering was first applied in product development, manufacturing, and the construction industry, and in the 1970s, value engineering began to be applied in the service industry. David Reeve's case study (1974, 1978) on youth service bureaus was among the first successful case studies in a service organization. Since then, successful value-engineering service case studies have been reported in retail, finance, health care, photo shops, and many other areas.

9.1.1   Collecting Information and Creating Design Alternatives

Value engineering achieves results by following a well-organized approach. It identifies unnecessary cost and applies creative problem-solving techniques to remove it. The three basic steps in this planned approach are

1. Identify the functions (what does the product or service do for customers?).
2. Evaluate the functions (what is the lowest cost to create these functions?).
3. Develop alternatives (what else will do the job?).

9.1.1.1 Identifying Functions   Function is the very foundation of value engineering. The concern is not with the part or the act itself but with what it does; what is its function? It may be said that function is the objective of the action being performed by the product or system.

Function is the property that makes something work or sell. We pay for a function, not for paperwork, or hardware. Hardware has no value in itself. Similarly we pay to retrieve information, not for the papers. Focusing on function tends to break down barriers to visualization by concentrating on what must be accomplished rather than how a task is presently being done. Concentrating on functions opens the way to innovative approaches.

Defining functions is not always easy—it takes practice and experience. Functions must be defined in the broadest possible manner so that the greatest number of potential alternatives can be developed to satisfy the function. A function must also be defined in two words, a verb and a noun. If the function has not been defined in two words, the problem has probably not been properly defined. These are some examples of simple functions: create design, evaluate information, determine needs, grow wealth, and enclose space.

There are two types of functions: basic and secondary. The basic function describes the most important action performed. The secondary function supports the basic function and almost always adds cost.

9.1.1.2 Evaluating Functions
After the functions have been defined and identified as basic or secondary, they must be evaluated to determine if they are worth their cost. This step is usually done by comparing them with something that is known to have a best cost. The best cost is the lowest overall cost for reliably providing a function.

9.1.1.3 Developing Alternatives
Function has been defined as the property that makes something work or sell, and the best cost is the lowest overall cost to reliably provide the function. In value-engineering analysis, if we find that the current cost for providing a function is significantly higher than the best cost, we need to ask: what else will do the job? That is, we must try to develop alternative ways to perform this function.

In order to develop alternatives, we need to make maximum use of imagination and creativity. This is where team activities make a major contribution. The basic tool is brainstorming. In brainstorming, we follow a rigid procedure in which alternatives are developed and listed with no attempt to evaluate them. Evaluation comes later. At this stage, the important thing is to develop revolutionary solutions to the problem.

Free use of imagination means freedom from the constraints of past habits and attitudes. One person's seemingly wild idea may trigger the best solution to the problem from someone else. Without a free exchange of ideas, the best solution may never be developed. A skilled leader can produce outstanding results through brainstorming and by providing simple thought stimulation at the proper time.

9.1.2   Evaluating, Planning, Reporting, and Implementing

The creative phase does not usually result in concrete ideas that can be developed into outstanding products. The creative phase is an attempt to develop the maximum number of possible alternatives to satisfy a function. These ideas or concepts must be screened, evaluated, combined, and developed to produce a practical recommendation. It requires flexibility, tenacity, visualization, and frequently the application of special methods designed to aid in the selection process.

The final recommendations must be accepted as part of a design or plan to be successful. In short, they must be sold. They must show the benefits to be gained, how these benefits will be obtained, and finally, proof that the ideas will work. This takes time, persistence, and enthusiasm.

9.1.3   The Job Plan

A step-by-step approach makes value engineering an effective tool. The approach is called the *job plan*, and it consists of six phases:

1. Information phase
2. Creative phase
3. Evaluation phase
4. Planning phase
5. Reporting phase
6. Implementation phase

Each phase is designed to lead to a solution to the problem after all of the factors are considered.

9.2   Information Phase

The first phase of the value-engineering job plan is the information phase. This is the most time-consuming and the most important phase. In the information phase, we will collect all the necessary raw information for the project, including product descriptions, process flowcharts and layouts, and all relevant cost information. Based on the information collected in this phase, we will produce three important documents for the project:

- **Function list** This is a complete list of all functions that are required in order for the product to work properly. In this list each function is defined and classified.

- **Cost-function worksheet** This is a complete cost breakdown calculation for all the product elements (subtasks, items, or components), as well as for all the functions. The worksheet also lists the actual cost and best cost for each function.

- **FAST (Function Analysis System Technique) diagram** This is a very important diagram that provides the exact logical linkage among all functions. The actual cost and the best cost for each function are also recorded in the FAST diagram.

The information phase consists of three separate parts:

1. Information development
   - Collect information
   - Determine cost visibility
   - Set a goal
2. Function determination
   - Define functions
   - Eliminate duplication
3. Function analysis and evaluation
   - Construct FAST diagram
   - Function/cost analysis
   - Function evaluation
   - Identify problem areas
   - Compare potential benefit to the goal

The work done in the information phase is the basis for developing alternative low-cost methods to perform the required functions. If the functions have not been properly defined and evaluated, the analysis will not be correct and the most satisfactory solution is not likely to be developed. Similarly, if the cost figures are incorrect or incomplete, the low-cost solution will not be identified.

9.2.1 Information Development

The first step in the information phase is the collection of all available information concerning the project. This includes drawings, process sheets, flow charts, procedures, and any other available material. It is important to discuss the project with people who are in a position to provide reliable information, and to verify that honest but wrong impressions are not being collected; that is, information that may have been true at one time but is no longer valid.

It is very important that good human relations be used during this information-collecting phase. Get the person originally responsible for

TABLE 9.1   Project Identification Checklist

1. Flow charts, organization charts

2. Detailed transaction data

3. Facility layout

4. Service product profile

5. Cost data (labor, overhead, material)

6. Work instructions

the project or development to help by showing him how he will be able to profit from a successfully completed study.

Table 9.1 shows an example of a project identification checklist, which details all of the information required for a study. It should be filled out as a first step to identify the project. If the information listed in the chart is not on hand, it will be necessary to obtain it.

9.2.1.1 Cost Visibility   The next step is to complete the cost-visibility section of the cost-function worksheet, as illustrated in Table 9.2, where the items in the leftmost column refer to parts of a product or steps of a service process.

This cost-function worksheet is an important document that should be produced in the information phase. The left side of this worksheet is the cost-visibility portion. This is where all costs are listed in a very detailed fashion, allowing no ambiguity or misunderstanding. Cost visibility does

TABLE 9.2   Cost-Function Worksheet

		Cost Visibility			Cost Function Analysis		
Total Cost $_____			Cost Elements				
Item No.	Name	Material $	Labor $	Burden $	F1 Function 1	F2	. . .
1							
2							
3							
.							
.							
.							
				Cost Total			
				Best Cost			

not tell us where unnecessary costs are; it tells us where high costs are. This is important because it identifies a starting point. The right side of the worksheet is about analyzing the functions delivered by each item and it will be discussed later.

The following definitions are commonly used in cost-visibility analysis.

- **Cost**   The amount of money, time, labor, and so on, required to obtain anything. In business, it is the cost of making or producing a product or providing a service.

- **Fixed costs**   Cost elements that do not vary with the level of activity (insurance, taxes, plant, and depreciation).

- **Actual costs**   Costs actually incurred during the performance of a process. They include labor, materials, and costs related to local ground rules.

- **Incremental costs**   Not all variable costs vary in direct proportion to the change in the level of activity. Some costs remain the same over a given number of production units or transactions, but rise sharply to new plateaus at certain increments. The costs thus affected are incremental costs.

- **Materials**   All hardware, raw materials, and purchased items consumed in producing a product.

- **Labor**   Manpower expended in producing a product or performing a service.

- **Burden (overhead)**   All costs incurred by the company that cannot be traced directly to specific products. The accounting department determines burden rates, which are assigned to individual operations on a formula basis. Burden consists of both fixed and variable categories, and separate rates are often established for each.

  The method of assigning burden differs from industry to industry and even from one company to another within an industry. Any quantifiable product factor may serve as a basis for assigning burden as long as consistent use of the factor across the entire product line results in full and equitable burden distribution.

- **Fixed burden**   Includes all continuing costs regardless of the production volume for a given item, such as salaries, building rent, real estate taxes, and insurance.

- **Variable burden**   Includes costs that increase or decrease as the volume rises or falls. Indirect materials, indirect labor, electricity used to operate equipment, water, and certain perishable tooling are also included in this classification.

- **Allowance**   Costs other than material, labor, and burden that must be included in the total cost of a product, such as packaging materials, scrap, inventory losses, inventory costs, and so on.
- **Total cost** includes production costs plus profit and other expenses.

The following expenses are usually added to production costs by sales and/or accounting departments to make up the total cost:

- **Administrative and commercial costs**   Costs incurred in administering the company, and researching and selling the product. They are usually represented as a percentage of production cost.
- **Freight cost**   Shipping and handling costs.
- **Profit**   Amount earned in producing a product or a service. It is usually applied as a percentage of production cost.

9.2.1.2 Sources of Cost Information   The application of cost-visibility techniques begins with an analysis of total cost, and progresses through analyses of cost elements to component or process costs. To perform these analyses, the best available cost information is required. This information will be available from sources such as

- **Accounting**   Current and historical costs (actual costs)
- **Purchasing**   Cost of purchased items
- **Suppliers**   Estimates and/or quotations, costs, process information, and material prices

In the service industry, labor is usually a big portion of cost. In order to figure out the exact labor cost component in each item, traditional motion-time studies have to be performed. For example, in the health care industry, a doctor's time is an important source of cost because it is very expensive. If we conduct a value-engineering study on emergency care, we may have to use a stopwatch to track the doctor's time spent on patient visits. After recording a sufficient number of patient visits, we can calculate the average doctor time and use that as a basis in computing the doctor's cost.

Review this cost data collected so far and make a preliminary judgment of the potential profit improvement. Consider the factors involved and set a goal that will provide a profitable position. The target should indicate a 30–100 percent cost reduction. It may seem improbable that this can be achieved, but it is a target to work toward. A comparison to this target will be made at the completion of the information phase.

Example 9.1: A Cost-Visibility Worksheet of a Youth Assistance Program (Reeve 1974)   Reeve (1974) did a value-engineering study on the youth assistance program for Oakland County, Michigan. This is one of the very first case studies of value engineering for a government/ service organization. The purpose of the youth assistance program is to help troubled teenagers so they will not become problems for society. There are two major activities in the youth assistance program: prevention and rehabilitation. Each activity is to be accomplished through various meetings, contacts, field visits, and office activities.

Tables 9.3 and 9.4 show the cost-visibility sections of the cost-function worksheets for rehabilitation and prevention, respectively.

In these cost-visibility calculations, the labor cost is computed based on labor hours multiplied by the labor rate. The labor hours are based on the historical records of meeting lengths, interview duration, and so on.

Example 9.2 A Cost-Visibility Worksheet for an Automobile Hood Latch   Table 9.5 shows a hardware cost-visibility worksheet for an automobile hood latch.

This is an example of a cost-visibility worksheet for a product development case.

9.2.1.3 Project Scope   Once the cost-visibility worksheet has been completed, it is possible to make a preliminary determination of the project scope. By considering the project as outlined on the project

TABLE 9.3   Cost Visibility of Rehabilitation per Case

		Cost Visibility			
Total Cost = $109.64/case			Cost Elements		
Item No.	Name	Material	Labor	Burden	Total Cost ($)
1	Client Contact		27.19		27.19
2	Organization Contact		11.04		11.04
3	Secretarial Center Office		6.07		6.07
4	Secretarial Field Office		31.12		31.12
5	Case Management		11.59		11.59
6	General Administration		2.21		2.21
7	Grant Administration		1.38		1.38
8	Others		4.42		4.42
9	Travel Time		5.11		5.11
10	Administration Meetings		1.17		1.17
11	Supervisory Meetings		4.69		4.69
12	Training Meetings		2.35		2.35
13	Statistical Meetings		0.97		0.97
14	Evaluation Meetings		0.34		0.34

TABLE 9.4   Cost Visibility of Prevention per Case

Cost Visibility				
Total Cost = $41.75/case		Cost Elements		
Item No.	Name	Material	Labor	Burden	Total Cost ($)
1	Client Contact		1.17		1.17
2	Organization Contact		2.90		2.90
3	Secretarial Center Office		3.45		3.45
4	Secretarial Field Office		8.56		8.56
5	Case Management		0.28		0.28
6	General Administration		2.07		2.07
7	Grant Administration		0.28		0.28
8	Others		2.55		2.55
9	Travel Time		2.90		2.90
10	Administration Meetings		1.86		1.86
11	Supervisory Meetings		4.90		4.90
12	Training Meetings		0.41		0.41
13	Statistical Meetings		0.14		0.14
14	Evaluation Meetings		0.21		0.21
15	Advisory Council Meetings		2.76		2.76
16	Citizen Committee Meetings		1.24		1.24
17	Citizen Subcommittee Meetings		6.07		6.07

identification sheet, the present cost and target for improvement, and the time available for the study, we can define the scope of the project.

In a value-engineering project, the analysis of function should first be performed upon the system level or the whole process level. If the objectives of the value-engineering study are not achieved at that level, the second level should be studied, and so on down to the lowest possible level of detail. The lower the level of system hierarchy it reaches, the more detailed and complex the study might become.

9.2.2   Function Determination

Once you have defined the initial scope of the project, it is possible to start defining the functions to be performed or that are being performed by the product or service.

TABLE 9.5   Cost-Visibility Worksheet for a Car Hood Latch

		Cost Visibility for Hood Latch			
			Cost Elements		
Total Cost = $2.616		Material (Total $1.545)	Labor (Total $0.713)	Burden (Total $0.358)	Total Cost ($) (Total: 2.616)
Item No.	Part Name				
1	Primary Spring	0.219	0	0.035	0.254
2	Detent Spring	0.09	0	0.015	0.1046
3	Hook Spring	0.09	0	0.015	0.1046
4	Pivot Rivets	0.09	0.005	0.015	0.1104
5	Hook Pivot	0.08	0.005	0.015	0.0988
6	Fork Bolt	0.04	0.096	0.014	0.158
7	Mounting Bracket	0.426	0.198	0.101	0.7253
8	Back Plate	0.08	0.149	0.037	0.2661
9	Secondary Hook	0.26	0.151	0.067	0.4776
10	Detent Level	0.11	0.099	0.034	0.2429
11	Grease	0.02	0.005	0.004	0.0291
12	Sleeve	0.04	0.005	0.007	0.0523

9.2.2.1 What Is a Function?   As has been mentioned, a function is the property that makes something work or sell. The function is the end result desired by the consumer; it is what consumers pay for. Function is a requirement, a goal, or an objective.

A function is not an action; it is the objective of an action. For example, "file papers" is an action, but what is the purpose of the "file papers" action? We "file papers" not because we enjoy putting papers in folders or cabinets, but because we want to keep a record that we can use later. Therefore, the objective of "file papers" is actually to "store information," so in this case, the function should be "store information." The function is the desired result accomplished by an action—the action is just one method of accomplishing the objective.

Functions should be defined in two words, a verb and a noun. The resultant definition should not define a method for achieving the result. Abstract definitions offer more opportunity for creative questions that may produce a number of alternative solutions. For example, using "file papers" as the function might limit our thoughts to using papers, folders, cabinets. By using a more generic function definition, such as "store information," we are free to think about other solutions involving computers, CDs, and so on to read, retrieve, and catalog information.

It is also important that the function be measurable in some way—weight, cost, volume, time, space, etc. In some cases, satisfaction, desire, or some other abstract measure may require more subjective analysis, but it can still be measured by comparative techniques.

The following are some examples of function definitions and their measures:

Verb	Noun	Unit of Measure
Create	Design	Time
Confirm	Design	Time
Authorize	Program	Cost
Measure	Performance	Manhours

In his book *Techniques of Value Analysis,* Miles recognized the difficulty of applying this technically simple concept: "While the naming of functions may appear simple, the exact opposite is the rule. In fact, naming them articulately is so difficult and requires such precision in thinking that real care must be taken to prevent the abandonment of the task before it is accomplished." He also said, "Intense concentration, even what appears to be over concentration of mental work on these functions, forms the basis for unexpected steps of advancement of value in the product or service."

Defining functions in two words is the most difficult naming method. There is some feeling among value-engineering practitioners that it is unnecessary to struggle for two-word definitions; three-word definitions or short statement definitions should do the job. However, value-engineering practitioners also found that if the goal of a value-engineering study is to generate creative design solutions, two-word functions are imperative. If the function cannot be defined in two words, the function needs to be better understood. It is a struggle to define good functions, but the result is worth the effort.

Defining functions in two words is a forcing technique that requires consensus among team members, eliminates confusion, creates in-depth understanding of the requirement, clarifies overall knowledge of the project, and ultimately breaks down barriers to visualization, which is necessary to help define the creative questions that will lead to new, outstanding solutions to the project.

9.2.2.2 Types of Functions   There are two types of functions: work functions and sell functions.

- Work functions are those that do the job the customer wants. Work functions provide use value. They are always expressed in action verbs and measurable nouns, which result in quantitative definitions. This provides us with a quantitative means of measuring the work functions.

- Sell functions are functions that add appeal for the customer and make them want to buy. Sell functions usually provide prestige value to customers. They are always expressed in passive verbs and nonmeasurable nouns, which create qualitative definitions. Their measurement is extremely difficult.

For example, these are some work functions:

Verb	Noun	Unit
Collect	Payment	Dollar
Remove	Kidney	Time and/or cost
Transfer	Fund	Dollar/Time
Sell	Assets	Dollar/Cost

These are sell functions:

Verb	Noun
Increase	Beauty
Improve	Style
Increase	Prestige

Separating work and sell functions will help us to define the functions more precisely. It will also help us to identify the proportion of cost allocated to use value and prestige value.

All functions can also be divided into two levels of importance: basic functions and secondary functions. The basic functions are those that fulfill the primary purpose of a product or service. Secondary functions do not directly fulfill the primary purpose of the product or service, but support the primary purposes.

The result of the function determination step should be a completed function list as illustrated in Table 9.6.

TABLE 9.6   Function List

Project Name	Scope includes:					
	Scope does not include:					
List All Functions	Function Types					
Verb	Noun	Basic	Second	Work	Sell	Remarks
1.						
2.						
3.						
4.						

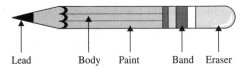

Figure 9.1   The parts of a pencil

Lead   Body   Paint   Band   Eraser

Example 9.3 Pencil Function Determination   Figure 9.1 shows a pencil. A pencil has five parts: lead, body, paint, band, and eraser. Even a very simple product such as a pencil will have quite a few functions, and it is easier to figure out a pencil's functions by looking at its five parts. Table 9.7 summarizes the functions performed by the parts of a pencil. Table 9.8 presents the function list for the pencil.

Example 9.4 Function List of Oakland County Youth Assistance Program   This example is a continuation of Example 9.1, which listed all the organizational activities that support youth assistance programs and their cost calculation. Reeve (1975) identified 41 functions in this program, some of which are listed in Table 9.9.

Reeve also provided a "glossary of functions," which provides detailed definitions for each function. In a value-engineering project, it is highly recommended that such a glossary be developed and that consensus be secured from group members on the definitions of each function. That way, in later discussions, every team member will be "on the same page" when each function is discussed. Here are some of the function definitions for this youth assistance program project:

- **Identify need**   Time spent in written and oral communication, that is, conferences, letters, interviews, etc., with school personnel and/or other referral sources regarding potential referrals in order to determine the need for the referral process.

TABLE 9.7   **Functions Performed by Pencil Parts**

	Function	
Part	Verb	Noun
Lead	Make	Marks
Eraser	Remove	Marks
Band	Secure	Eraser
	Improve	Appearance
Body	Support	Lead
	Transmit	Force
	Accommodate	Grip
Paint	Display	Information
	Protect	Wood
	Improve	Appearance

TABLE 9.8 Function List of a Pencil

Project Name: Pencil		Scope includes:				
		Scope does not include:				
List All Functions		Function Types				
Verb	Noun	Basic	Second	Work	Sell	Remarks
1. Make	Marks	√		√		
2. Remove	Marks		√	√		
3. Secure	Eraser		√	√		
4. Improve	Appearance		√		√	
5. Support	Lead		√	√		
6. Transmit	Force		√	√		
7. Accommodate	Grip		√	√		
8. Display	Information		√		√	
9. Protect	Wood		√	√		
10. Improve	Appearance		√		√	

TABLE 9.9 A Partial List of Functions for a Youth Assistance Program

List All Functions				
Verb	Noun	Basic	Second	Remarks
1. Identify	Need	√		
2. Define	Problem		√	
3. Plan	Treatment		√	
4. Diagnose	Problems		√	
5. Obtain	Information		√	
6. Involve	Client		√	
7. Identify	Client		√	
8. Utilize	Resource		√	
9. Assist	Client	√		
10. Improve	Process		√	
11. Indicate	Trend		√	
12. Maintain	Record		√	
13. Establish	Standard		√	
14. Analyze	Data		√	
15. Terminate	Contact		√	
16. Evaluate	Process		√	
17. Eliminate	Deviancy	√		

TABLE 9.9 A Partial List of Functions for a Youth Assistance Program (*continued*)

List All Functions				
Verb	Noun	Basic	Second	Remarks
18. Plan	Activities		√	
19. Determine	Needs		√	
20. Set	Goals		√	
21. Secure	Action		√	
22. Provide	Alternatives		√	
23. Develop	Programs		√	
24. Establish	Trust		√	
25. Exhibit	Concern		√	
26. Improve	Programs		√	
27. Evaluate	Programs		√	

- **Assist client**   This action includes counseling, offering alternatives, providing a referral service, indicating community programs, helping kids get to camp, talking to teachers or police or other authorities on client's behalf, aiding parents and children.

- **Eliminate deviancy**   (Measured by time spent) To return to a homeostatic position and develop modification techniques to reach normative behavior patterns.

- **Define problems**   All communication with client, parents, and referral sources for the purpose of describing the client's problem behavior.

9.2.2.3 Creativity and Function Definitions   The ultimate objective of value engineering is to create a better product or service design. Creativity is very important in coming up with new and better designs. But what makes people more creative? What is the ingredient that allows some people to break the barriers to visualization, to be able to look at something and immediately think of new exciting possibilities for new products, services, methods, or other useful or satisfying subjects?

The consensus seems to be that to be creative, one must be able to see beyond the obvious, the existing, but there is no clear-cut formula for producing creative people. It is known that creative people are somewhat different and that they exhibit certain characteristics. However, given two people with the same characteristics, one may prove to be creative, and another may not. Many people feel that the seeds for creativity exist in everyone.

In all probability, at least one of the ingredients of creativity is the ability to visualize, to detach oneself from reality and see beyond the stated problem, the object, or the material facts. A creative person must be able to create concepts, broaden and develop them, analyze and examine them, and out of all this select a new idea, new approach, or new solution to a requirement or problem.

According to L. D. Miles, creative thinking is constrained by the physical shape or concept of existing products and services. Concentrating on function helps to break down the barriers to visualization and offers outstanding opportunities for creativity.

The conventional approach to improving a product or process is to try to make the existing product work better, cost less, or meet some other objective. In this situation, creativity is stifled because the existing form constrains thinking. The function approach is truly different. It breaks the project into requirements called functions. The process of defining functions becomes a method for removing the barriers to visualization, making entirely new solutions possible.

The concept is disarmingly simple. It is easy to understand without learning complex systems or studying complex technology. However, the ability to use the system comes only from a thorough understanding of the principles and the determination and discipline to use them.

Function analysis is basic to value engineering and it starts with an understanding of the term "function" and of how to define functions that will offer creative opportunities. Defining and analyzing functions can help a person or group visualize alternative solutions. In fact, the struggle necessary to define functions properly makes it possible to see new and different things in subjects you may have seen many times before. It can help you imagine solutions that go beyond the stated problem, as outstanding people have done throughout the ages.

This means that not only are functions the basic ingredient of value engineering, but they provide the opportunity for people to break down barriers and create new things, to eliminate prejudices and allow insights never seen before.

9.2.2.4 Defining Functions for Creativity   In defining a function, it is important that these key questions be kept in mind at all times:

- What are we really trying to do when we perform this action? Why is it necessary to do this?
- Why is this part or action necessary?

Specific answers to these questions will aid in zeroing in on a useful function definition.

In function definition, it is helpful to think as if you are the product or service, and play their roles. The idea is to "let the job be the boss," as Kettering said. Be the crankshaft. What do you do? Act the part of the customer. How do you feel? What do you see? What does it do for you? If you were the plant manager, what would you want? How would you get it?

This system helps to eliminate bias in that functions can be defined from all viewpoints and sorted out in the FAST diagram through its cause-and-effect relationship for maximum understanding and subject evaluation.

9.2.2.5 Process for Defining Functions   In defining functions, start with the total project, whether it is an assembly, a complete process, a program, or an organization. Define the functions. Don't haggle over whether the function has been properly defined at this stage; it can be redefined later. Write every thought down so it will not be forgotten.

After all functions of the assembly have been defined, take each part or segment of the system and define the function of each. There will be some duplication, but this will be screened out later.

After all functions have been defined, screen the list to eliminate duplicate functions and redefine functions for clarity. Screen the list again to define the basic functions. The basic functions are the functions upon which all other system functions depend. If the basic function were not needed, none of the other functions would be needed either.

In many cases, a number of functions beyond the system scope will be defined. These are called higher-order functions, and they are the functions that cause the basic functions to be performed. A detailed discussion of system scope and high- and low-order functions is beyond the intent of this orientation; the scope will become clear during the construction of a FAST diagram, and it may cause a team to reconsider and redefine the scope because of the new understanding of the overall project.

The end result of applying these function definition principles will be clearly understandable, measurable functions for use in cost-function analysis and function evaluation, and will lead to outstanding opportunities in the creative phase.

9.2.3   Function Analysis and Evaluation

After the functions have been determined, identify the basic function or functions, as well as supporting functions. It is now time to create a Functional Analysis System Technique (FAST) diagram. The Functional Analysis System Technique was developed by Charles Bethway in 1964, and was first presented and published as a paper at the Society of American Value Engineers Conference in 1965. FAST contributed significantly to the most important activity in value-engineering projects, the function analysis and evaluation.

9.2.3.1 FAST Diagrams A FAST diagram is a logic chart that organizes the functions of a project and arranges them in a cause-and-effect relationship. By constructing a FAST diagram you can ensure that the functions have been properly defined and that nothing has been overlooked. FAST diagrams are simple in concept, but creating a FAST diagram is often difficult and frustrating, and it forces people to think about their project in a detailed and precise manner. Constructing a FAST diagram will create a focal point for the entire project, because eventually, all important information on the project will be precisely defined and displayed in the FAST diagram.

FAST diagrams are especially useful for projects where there are widely different opinions, fuzzy understandings, or cloggy definitions among team members, which are very common in the analysis of organizations, operations, and services. The construction of a FAST diagram tends to pull together the thinking of a group to create a dynamic, enthusiastic team.

Determining the basic function is the first step in constructing a FAST diagram. The basic function is the function that cannot be eliminated unless the product can be eliminated. There may be more than one, but an effort should be made to determine the one most basic function.

We will use Example 9.5 as a starting point to discuss FAST diagrams.

Example 9.5 Portion of FAST Diagram for Youth Assistance Program As you saw in Example 9.4, "identify need," "assist client," and "eliminate deviancy" are considered to be basic functions for the youth assistance program, because if any of these functions are not performed, the whole youth assistance program will not perform as intended. We can easily see that the functions are not working in isolation; they are related to each other and together perform the overall mission of the system. We can ask "why do we need these three basic functions ("identify need," "assist client," and "eliminate deviancy")"? If we think really hard, we may come to the conclusion that "because this is a youth assistance program, we want to change the life of these troubled youth so they can become better kids." Then we may come to another function, "modify behavior"; that is, the three basic functions are needed because we want to "modify behavior." Figure 9.2 illustrates this relationship.

At the top of Figure 9.2, the Why-How arrow indicates the relationships among these four functions:

- How do you "modify behavior"? You have to "identify needs," "assist clients," and "eliminate deviancy."

- Why we need to perform these three functions? Because we want to "modify behavior."

Why ◄────────► How

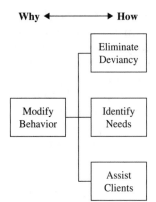

Figure 9.2 Relationships between basic functions and higher-order function

Similarly, we can expand Figure 9.2 by adding more functions, as illustrated in Figure 9.3.

Again, the function on the left gives the reason why the function on the right should be performed, and the function on the right tells how the function on the left can be accomplished. For example, the function on the right of "eliminate deviancy" is "plan activities," and the function on the right side of "plan activities" is "determine needs." We plan activities in order to eliminate deviancy, and we determine needs in order to plan activities. How do we eliminate deviancy? By planning activities! How do we plan activities? By determining needs!

9.2.3.2 Understanding the FAST Diagram Now let's look at how to establish a FAST diagram for a value-engineering project. The general format of a FAST diagram is illustrated in Figure 9.4.

The following terms are used in FAST diagrams:

- **Scope of the project** Scope of the project is defined by two vertically dotted lines. The portion between the two lines is the scope of the project.

Figure 9.3 Part of FAST diagram for youth assistance program

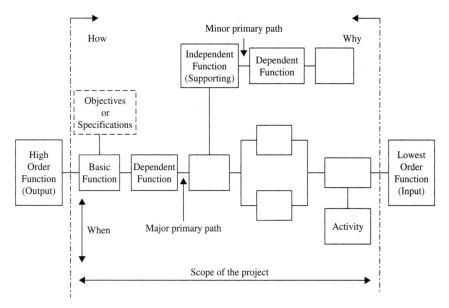

Figure 9.4   FAST diagram format

- **Highest-order function(s)**   This is the highest level of function. It is the overall objective of the product or service under study. The highest-order function is the output of the basic function(s). The highest-order function should be positioned just outside of the left scope line and to the left of the basic function(s). In general, any function to the left of another function on the primary path is a "higher-order" function.

- **Lowest-order function(s)**   The functions on the rightmost position and outside of the right scope line are the lowest order functions. They represent the initial inputs to all other functions. In general, any function to the right of another function on the critical path is a "lower" order function. The higher- or lower-order functions do not carry any meanings of relative importance.

- **Basic functions**   Basic functions are the functions located to the immediate right of the left scope line. The basic functions represent the purpose of the system under study. In value engineering, the basic functions cannot change. Secondary functions can be changed, combined, or eliminated.

- **Concept**   All the functions that are located to the right of the basic function(s) are used to describe the means to accomplish basic function(s). The concepts could be either existing conditions or proposed approaches to accomplish basic functions(s).

- **Objectives or specifications** Objectives or specifications are particular parameters or requirements that must be achieved in order to satisfy the highest-order functions. Objectives or specifications are not functions. In value engineering, objectives and specifications are studied because they will influence the method selection to achieve the basic functions and satisfy the user's requirements.

- **Primary path functions** Any function on the How or Why logic is a primary path function. If the function is along the Why direction that enters the basic function(s), then it is a major primary path. Otherwise, it will be an independent (supporting) function, and it is on a minor critical path. Supporting functions are usually secondary functions. Supporting functions are needed to help the system to achieve the objectives or specifications of the basic functions; they could also be needed because of a particular technical approach.

- **Dependent functions** From the first function to the right of the basic function(s), each successive function is dependent on the function that is on its immediate left, or higher-order function.

- **Independent functions** Independent functions do not depend on another function or method. Independent functions are located above the critical path functions.

- **Activity** The activity is the method selected to perform a function, or a group of functions.

- **Primary path functions** Any function on the How or Why logic is a primary path function. If the function along the Why direction enters the basic function(s), it is a major primary path; otherwise, it will be identified as an independent (supporting) function and be a minor critical path. Supporting functions are usually secondary. They exist to achieve the performance levels specified in the objectives or specifications of the basic functions, or because a particular approach was chosen to implement the basic function(s).

Independent functions (above the critical path) and activities (below the critical path) are the result of satisfying the When question.

- **Dependent Functions** Starting with the first function to the right of the basic function, each successive function is "dependent" on the one to its immediate left, or higher-order function, for its existence. That dependency becomes more evident when the How question and direction is followed.

- **Independent (or Supporting) Function(s)** Independent (or supporting) functions do not depend on another function or method selected to perform that function. Independent functions are located above the critical path function(s) and are considered secondary with respect to the scope, nature, and level of the problem, and its critical path.

- **Activity** The method selected to perform a function (or a group of functions) is an activity.

9.2.3.3 Symbols and Graphs Used in FAST Diagram Construction

Figure 9.5 shows the directions in a FAST diagram. The How and Why directions are always along the primary path, whether it is a major or minor primary path. The When direction indicates an independent or supporting function (up) or an activity (down).

We have already discussed the How and Why directions in Example 9.5—the lower-order function in the How direction (immediate right) always explains how a particular function can be accomplished; the higher-order function in the Why direction (immediate left) always tells the reason why a particular function should be performed. All the functions or activities along the When direction with a particular function will happen at the same time. We can detect these functions or activities by asking the question "When the function occurs, what else happens?" The independent functions and supporting functions are listed above the particular function; the activities will be listed under the particular function.

AND and OR Symbols Along the Primary Path   In the primary path of a FAST diagram, it is possible that several functions have to be performed simultaneously as the preconditions for lower-order function(s). Sometimes, these functions are related by a logical AND, and sometimes they are related by a logical OR. Figures 9.6–9.9 illustrate such cases.

In both Figure 9.6 and Figure 9.7, the fork is read as "AND." In Figure 9.6, how do you "build swim club"? By "construct pool" AND "construct club house." "Construct pool" and "construct club house" are

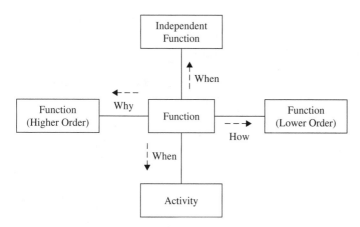

Figure 9.5 Directions in FAST diagram

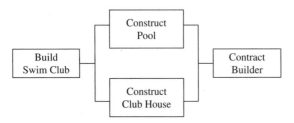

Figure 9.6 Two equally important functions in an AND relation

equally important. In Figure 9.7, how do you "determine compliance deviations"? By "analyze design" AND " review proposals." However, "analyze design" is more important than "review proposals."

In both Figure 9.8 and Figure 9.9, the multiple exit lines represent an "OR." In Figure 9.8, how do you "convert bookings (to delivery)"? By "extend bookings" OR "forecast orders," not both. "Extend bookings" and "forecast orders" are equally important. In Figure 9.9, how do you "identify discrepancies"? By "monitor performance" OR "evaluate design." However, "evaluate design" is less important than "monitor performance."

AND Symbols Along the When Direction   In a FAST diagram, the When direction is vertical. When several functions are located along the same vertical line, it means that these functions will be performed at the same time. In addition, when these functions are connected by lines, it means that there is an "AND" relationship among them. Figure 9.10 illustrates such an example.

In Figure 9.10, when you "influence the customer," you "inform customer" AND "apply skills." If it is necessary to rank the AND functions, those closest to the primary path should be the most important.

9.2.3.4 Completing a FAST Diagram   Now that you know the symbols and notation used in FAST diagrams, we can look at the step-by-step procedure for completing a FAST diagram.

Step 1   List all functions by using the function list as illustrated in Table 9.6. Be sure to identify each function with a verb and noun. Identify basic functions and secondary functions.

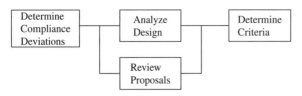

Figure 9.7 Two unequally important functions in an AND relation

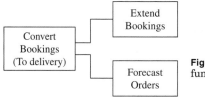

Figure 9.8 Two equally important functions in an OR relation

Step 2   Prepare a 1" × 2" card for each function. Take a close look at all functions and try to identify the relationships among them.

We can use the following logical questions for this purpose:

- **How?**   How is this function accomplished?
- **Why?**   Why is this function performed?
- **When?**   When is this function performed?

Select the function that you think is the basic function and apply the logic questions to the right and left of the basic function. Ask "How is this function performed?" to determine the function to the right. Ask "Why is this function performed?" to determine the function to the left. Repeat this process until the lowest-order functions are included. The path of functions thus created is called a primary path. You may end up with multiple primary paths.

Step 3   When a primary path has been selected and positioned on the chart, position all secondary functions that did not fit into the primary path by applying the When question, and add them above or below the primary path depending on whether they are supporting functions, independent functions, or actions. If the secondary functions are actually objectives or specifications, put them into the upper-left corner of the FAST diagram.

Example 9.5 illustrates this step-by-step process.

Example 9.5 Cigarette Lighter FAST Diagram   In this example we will create a FAST diagram for a typical cigarette lighter, as shown in Figure 9.11. Assume that we have compiled the function list in Table 9.10.

To create the FAST diagram, we first pick up the basic function "produce flame," and ask why and how, as illustrated by Figure 9.12.

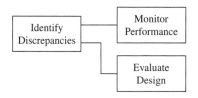

Figure 9.9 Two unequally important functions in an OR relation

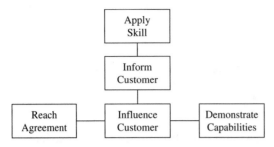

Figure 9.10   AND relationship on the When direction

The basic function of the cigarette lighter is to "produce flame." By answering the question "why produce flame?", we get the higher-order function "ignite cigarette." By answering the question "how to produce flame?" we get the lower-order function "ignite fuel."

TABLE 9.10   Function List for Cigarette Lighter

Verb	Noun	Basic	Second	Remarks
1. Produce	Flame	√		
2. Protect	Flame		√	
3. Manage	Flame		√	
4. Ignite	Fuel		√	
5. Release	Fuel		√	
6. Produce	Spark		√	
7. Control	Flow		√	
8. Restrict	Exit		√	
9. Energize	Particles		√	
10. Strike	Flint		√	
11. Generate	Heat		√	
12. Contain	Fuel		√	
13. Open	Valve		√	
14. Depress	Lever		√	
15. Enclose	Fuel		√	
16. Rub	Material		√	
17. Rotate	Wheel		√	
18. Apply	Force		√	
19. Activate	Thumb		√	
20. Accommodate	Hand		√	
21. Stimulate	Muscle		√	

Figure 9.11   A cigarette lighter

We then can ask the further question, how to ignite fuel? By answering this question, we will find that we need two lower-order functions to be performed, "produce spark" AND "release fuel." These two functions are of equal importance, so we add these two functions in the FAST diagram as illustrated in Figure 9.13.

Now we can continue to ask "why" and "how" questions to find lower-order functions for "release fuel" and "produce spark," and continue this process. We will end up with the diagram in Figure 9.14. The functions shown in this figure form the primary path of the FAST Diagram of the cigarette lighter.

There are still many functions in the function list that cannot be fit into the primary path. By asking when, we can fit the rest of the functions into the FAST diagram, as illustrated in Figure 9.15.

9.2.3.5   Cost-Function Relationship   When the FAST diagram is finished, it is possible to complete the cost-function worksheet. The cost-function worksheet lists all functions versus all parts of a product or actions of a system, procedure, or administrative activity. The objective is to convert product cost to function cost.

The cost of each piece of hardware or service activity is redistributed to the function performed. This proportional redistribution of cost to function requires information, experience, and judgment, and all team members must contribute their expertise.

After the cost of each part or action has been redistributed to the functions performed, the cost columns are totaled to obtain the function cost. This cost is then placed on the FAST diagram. The FAST diagram then becomes a very valuable tool. It tells what is happening, why, how, when, and what it costs to perform the function. It is now possible to evaluate the functions to determine if they are worth what is being paid for them. In other words, a value must be set on each function.

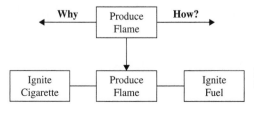

Figure 9.12   Start of FAST diagram construction for cigarette lighter

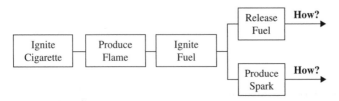

Figure 9.13 Partial FAST diagram for cigarette lighter

Determining the value of each function is a subjective process. However, it is a key element in the value process. Comparing the function cost to function value provides an immediate indication of the benefit being obtained for expended funds. The ratio of value cost to function cost is the performance index. The sum of all values is the value of the system or the lowest cost to reliably provide the basic function. It should be compared to the preliminary goal set earlier.

It may be that the new goal is considerably higher than the original. If this is the case, an evaluation of the diagram will indicate what must be done to achieve the original goal. It may indicate that an entirely new concept is required or it may be that it will be acceptable to settle for less. It is often the case that the original goal and the new value are close. An analysis of the function costs will again indicate necessary action.

This analysis clearly defines the task for product improvement. It breaks the problem down to functions that must be improved, revised, or eliminated to achieve the goal. The FAST diagram clearly identifies functions and their relationship to each other. Cost-visibility analysis can identify high cost areas.

We now are ready to identify the relationship between cost and function. Specifically, we are ready to identify the cost for each function. Also, after clearly defining each function, we are able to identify the best cost for each function. The difference between the current cost and the best cost is the profit improvement target, which will provide us with an estimate of profit improvement potential.

Example 9.6 Cost-Function Worksheet for a Pencil   Table 9.11 provides an example of a cost-function worksheet for the pencil discussed previously in Example 9.3.

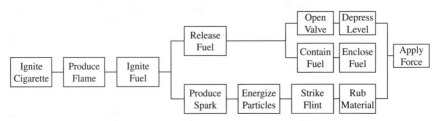

Figure 9.14 Primary path of FAST diagram for cigarette lighter

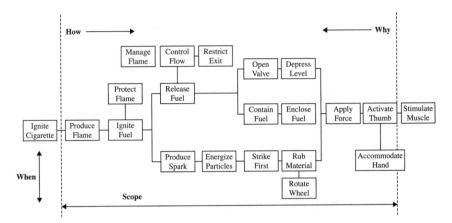

Figure 9.15 FAST diagram for cigarette lighter

We need to determine the cost of each function by distributing the cost of each part to its related function. For example, the cost for the pencil body is 0.94 cent, and 50 percent of pencil body cost is used to perform the function "transmit force," 40 percent of its cost is used to perform the function "support lead," and 10 percent of its cost is used to perform the function "display information." This breakdown of cost is subjective; hopefully, it is not too biased because it is based on the consensus of the team. By adding all the cost portions from all relevant parts for a function, we can get the cost for performing that function. For example, in Table 9.11, the cost of the function "transmit force" consists of 25 percent of metal band cost, which is 0.06 cent, and 50 percent of pencil body cost, which is 0.47 cent, therefore the cost of "transmit force" is 0.53 cent.

We also need to determine the best cost for each function. By definition, the best cost is "the lowest cost to adequately and reliably provide the function." The best way to determine the best cost of a function is by comparison to another function that we know is a best deal. For example, if a function is "tell time," we need to know the time precision requirement; for example, the required precision might be +/- 30 seconds after a month of use. Next, we can find a watch that can provide time with this precision reliably, but that doesn't provide any other functions, such as decoration, brand name recognition, and so on. In this way, a cheap, no-brand, plain, 99-cent electronic watch might be adequate. In this case, the best cost for "tell time" is 99 cents.

To make sure we determine the best value, we can ask the following questions:

1. Can we do without it? (If yes, the best cost is zero.)

2. Does it need all its features? (If no, get rid of all unnecessary features and then figure out the best cost.)

TABLE 9.11 Cost-Function Worksheet for Pencil

Functions

Pencil Components	Cost (Cents)	Remove %	Marks Cost	Secure %	Eraser Cost	Improve %	Appearance Cost	Make %	Marks Cost	Transmit %	Force Cost	Accommodate %	Grip Cost	Display %	Information Cost	Support %	Lead Cost	Protect %	Wood Cost
Eraser	.43	100	0.43																
Metal Band	.25			50	0.13	25	.06			25	.06								
Lead	1.2							100	1.2										
Body	.94									50	.47			10	.09	40	.38		
Paint	0.10					50	.05											50	.05
Total Cost	2.92	16	.43	5	.13	4	.11	40	1.2	17	.53			3	.09	13	.38	2	.05
Best Cost			.34		.10		.10		0.8		.30				.09		.28		.04
Profit Improvement Potential			.09		.03		.01		0.4		0.03				0		0.1		0.01

326

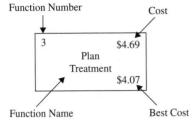

Figure 9.16 A fully-marked function block in a FAST diagram

3. Is anyone buying it for less?
4. Is there something better that can do the job?
5. Can it be made by a less costly method?
6. Can a standard item be used?
7. Can another dependable supplier provide it for less?
8. Would you pay the price if you were spending your own money?

Best cost is not always lower than the current cost. The best cost is the lowest cost to adequately and reliably provide this function, and it is possible that this function is not adequately and reliably provided in the current system. In this case, we might have to increase the cost for this function. This is also why Question 4 "Is there something better that can do the job?" is asked.

The cost and the best cost for functions are also often marked in the FAST diagram. A fully-marked function block in a FAST diagram has the format shown in Figure 9.16. Figure 9.17 shows a portion of a FAST diagram for the youth assistance program with fully-marked function blocks.

After the FAST diagram is fully developed and the cost-function worksheet fully filled, it is time for the creative phase of the value-engineering job plan.

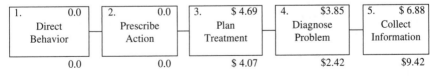

Figure 9.17 A portion of a FAST diagram with fully-marked function blocks for youth assistance program

9.3 Creative Phase

The difference between the cost and the best cost of each function is the profit-improvement potential, and functions that have high profit-improvement potentials are the perfect candidates for cost saving. The creative phase of the value-engineering project uses team members' creativity to develop alternative solutions to perform the functions that have high profit-improvement potential.

The creative phase is where free development of ideas is fostered. These ideas will form the basis for concepts that will lead to recommendations for improvement.

9.3.1 Brainstorming

Brainstorming is extremely helpful in the creative phase of value-engineering projects. It helps to loosen the mental barriers to creativity in order to create a great volume of ideas. In the beginning of brainstorming, the quantity of ideas is important—many obvious wrong ideas will be generated, but the more ideas that are generated, the better chance that some really brilliant but not obvious ideas will be among these ideas. The ideas will be screened and evaluated in the next phase.

In the brainstorming process, an atmosphere is generated that permits each person to lower his or her mental barriers. This may feel uncomfortable, but it is a necessary stretch that pushes you to think more freely and more creatively.

The following ground rules for brainstorming must be followed to create the proper environment for developing ideas:

1. No criticism is allowed during the session.

2. A peer group is desired. Never have high-level management and their assistants attend.

3. Quantity of ideas is desired. The more ideas, the more likelihood of at least one outstanding item.

4. A group size of six to ten participants is best.

5. No publicity on the session after its completion.

6. Combine ideas.

7. Wild ideas wanted. Usually the first 90 percent of all ideas will be those that have come up before.

8. Record all ideas; recording on paper is best.

The first step in brainstorming is to select questions for discussion. These questions are often selected based on the functions that have high profit-improvement potentials. The question is often in the form:

TABLE 9.12   Idea Generation Form

	What else can perform the function?
	Function: Enhance Appearance
1	Use laminate instead of paint.
2	Use stainless steel parts.
3	Use plastic material.
4	Use curves instead of sharp edges.
5	Delete complicated features.
6	Paint parts individually before assembly.
7	Use chrome plating.
8	Use multicolor paint.
9	Use gold material.

"What else can perform this function?" The questions are then presented to the group and the group will toss out ideas regarding the question. Any idea is acceptable and no discussion is allowed. For each question, it is desirable that at least one new idea should be generated that has never been thought of before.

Table 9.12 shows an example of ideas generated in a brainstorming session. The function under discussion is "enhance appearance" for a decoration.

9.4   Evaluation Phase

Evaluating the ideas developed during the creative phase is a critical step in the value-engineering job plan. The ideas generated will include practical suggestions as well as wild ideas. Each and every idea must be evaluated without prejudice to determine if it can be used or what characteristics of the idea may be useful.

Proper evaluation of the ideas is a critical step. Remember, if an idea is discarded without thorough evaluation, the key to a successful solution may be lost. The time to create ideas is in the creative phase. If an idea is discarded, there may not be another opportunity to develop it again.

During the screening process, it must be kept in mind that the objective is not to discard ideas but to look for the good in them. All too frequently, a new idea will create a negative reaction. For example, "That's a great idea but let me tell you what is wrong with it." We should say, "That's a great idea. What can we do to make it work?" There never seems to be any problem thinking of reasons why something will not work. However, developing ways to make an idea work takes ingenuity.

The state of mind during the screening process should be "How can we make it work or what is there about this idea we can use?" Evaluation processes can range from the simple to the complex. The method selected depends to some degree on the quantity and quality of ideas generated. The number of ideas can run from less than a hundred to over a thousand depending on the scope of the project. The first screening of the list should be to eliminate the ones that obviously are of no use to the project. However, each idea must be reviewed with a positive attitude. Look for the good rather than the bad and don't be too critical.

9.4.1 Relatively Simple Evaluations

The following process is suggested for initial screening.

Step 1   Each item on the idea generation form will be read out loud to the team. Each team member will vote whether to keep the idea for future evaluation or drop it. This is an impulse decision. Each person will decide by his initial reaction as to whether to keep the idea or not. However, if one person on the team wants to keep the idea, it must be kept on the list without question. During this initial screening, there should be no discussion of the idea; only a yes or no vote is acceptable. The result will be elimination of the obviously impractical ideas for this project.

Step 2   Each of the remaining items on the list will be read out and the group will discuss each idea. Table 9.13 shows a template that can be used for this step. The intent is to determine what there is about each idea that may be useful and decide whether to keep it on the list or to drop it. At this stage, many ideas will combine with other ideas to form basic groups or categories, such as materials, methods, organization, etc. The discussion may also result in new ideas that can be added to the list.

TABLE 9.13   Idea Screening Worksheet 1

Ideas	Implementation Cost	Development Cost	Total Cost
1.			
2.			
3.			
4.			
5.			
6.			
7.			
8.			
9.			

Step 3   After the initial screening process has been completed, it will be necessary to resort to systems designed to aid in identifying the best choice and an alternative, or to rank and weigh alternatives. It is always important to have a second choice to fall back on just in case the first choice cannot be implemented for reasons that may not become apparent until detailed development is underway.

When the initial list of ideas has been reduced to a choice of only a few alternatives, the simple system illustrated in Table 9.14 may be used. This sheet identifies the advantages and disadvantages of each alternative concept. In most cases, an idea listing more advantages than disadvantages will be the first choice. However, there may be an overpowering disadvantage that creates a serious roadblock. Can it be eliminated? If it can, the choice may be clear. If it cannot, the second alternative may be the best choice.

9.4.2   More Complex Evaluations

There may also be situations where the choice of alternatives will require more complex systems to aid in the evaluation process. Two systems that are favored because of their convenience, simplicity, and effectiveness are Pareto voting and paired comparisons. They may be used separately or in sequence depending on the situation. Each of these systems is described in detail in the following sections. They have been found applicable in a large number of cases and are extremely useful.

TABLE 9.14   Idea Screening Worksheet 2

Idea 1		Idea 2	
Advantage	Disadvantage	Advantage	Disadvantage

Idea 3		Idea 4	
Advantage	Disadvantage	Advantage	Disadvantage

There are also cases involving high risk or a substantial amount of money where even more detailed analysis is required. These may be situations where risk is critical and alternatives and trade-offs are necessary. In these cases, a matrix analysis may be necessary.

Experience has shown that this evaluation process is a difficult task. The impulse to quickly screen the list to zero in on the best ideas must be controlled. The mass of data must be handled systematically to obtain maximum benefit from the creative phase. Careful screening is essential to isolating the best concept to carry over into the planning phase where the idea will be developed into a practical recommendation for action.

9.4.3   Selection and Screening Techniques

A difficult problem that frequently confronts decision makers is the need to organize a large amount of data so that one or several of the most important items may be identified. It may be necessary to determine which of several alternatives appears to be the best, or it may be necessary to select a number of items so that they can be ranked and weighted by order of importance or some other criteria. Most people are not able to handle this task quickly and effectively.

However, experience has shown that a combination of two simple methods, Pareto voting and paired comparisons, will satisfy a majority of requirements. (A literature search identified 13 other methods for evaluating data to aid in decision making; these are listed in Tables 10.3 and 10.4 in Chapter 10).

9.4.3.1 Pareto Voting   Pareto voting is based on Pareto's Law of Maldistribution. Alfredo Pareto (1846–1923), a political economist, observed a common tendency of wealth and power to be unequally distributed, and this observation has been refined to the degree that it can be said that there is an 80/20 percent relationship between similar elements.

For example, 20 percent of the parts in an assembly contain 80 percent of the cost. This is most useful information in cost estimating; however, the relationship holds for many diverse examples, such as the following:

- Twenty percent of the states use 80 percent of the fuel oil.

- Twenty percent of the activities create 80 percent of the budgeted expense.

- Twenty percent of the items sold generate 80 percent of the profit.

In value engineering it is frequently necessary to select the best ideas, the highest-value functions, the highest-potential projects, or any of a number of other requirements. It has been found that the application of

Pareto voting can help to simplify the list and will in most cases ensure that the most important items have been selected. It also produces results quickly and can be incorporated into the value-engineering process to allow continuous operations without undue disruptions.

Pareto voting is conducted by requesting each team member to select what they believe are the items or elements that have the greatest effect on the system. This list of items is limited to 20 percent of the total number of items. For example, each team member would be allowed to select 6 items out of a list of 30. The vote is on an individual basis to obtain as much objectivity as possible.

The resultant lists are then compared and arranged into a new consolidated list in descending order based on the number of votes each item received. Usually, several items will have been selected by two or more team members. The top 10 to 15 items are then ranked and weighted in a second step by using paired comparisons.

Example 9.7 Pareto Voting   This example is based on the idea generation form that was used in Table 9.12. A team of six people conducted a Pareto vote on the nine ideas. Each member can only vote for two ideas, so a total of 12 votes will be received. The number of votes for each idea will be tallied, and the results are summarized in Table 9.15.

9.4.3.2 Paired Comparisons   Paired comparison, or numerical evaluation, as it is sometimes called, compares a list of items to rank and weights them in order of importance or some other criteria. Ranking is the assignment of a preferred order of importance to a list of items. Weighting is the determination of the relative degree of difference between items.

TABLE 9.15   Pareto Voting

Rank	What else will perform the function? Function: Enhance Appearance	Vote Received
1	Use laminate instead of paint.	5
2	Paint parts individually before assembly.	4
3	Use curves instead of sharp edges.	2
4	Use plastic material.	1
5	Use stainless steel parts.	0
5	Delete complicated features.	0
5	Use chrome plating.	0
5	Use multicolor paint.	0
5	Use gold material.	0

TABLE 9.16   Pencil Improvement ideas

Key Letter	Idea
A	Eliminate paint.
B	Reduce the length of lead.
C	Remove eraser.
D	Stain wood in lieu of paint.
E	Make body out of paper.

In paired comparisons each item is compared to every other item on the list in turn, using a simple matrix. It is most convenient for up to 15 items.

A comparative decision is made between any two items on a two- or three-level basis. In a two-level comparison, 2 = major difference; 1=minor difference. In a three-level comparison, 3 = significant difference, 2 = moderate difference, 1 = minimal difference.

Example 9.8 illustrates how paired comparisons work.

Example 9.8 Paired Comparison for Pencil Improvement   This example is based on the pencil improvement project presented in Examples 9.3 and 9.6. After some discussion about how to improve the pencil, several ideas about cost reduction for the pencil are proposed as shown in Table 9.16.

The next step will be evaluating idea A with respect to B, A versus C, and so on, for all possible pairs. Is A or B the better idea based on cost, benefit, customer satisfaction, etc? When comparing A and B, a B-2 result indicates that the team thinks that B is moderately better than A. Similarly, when comparing A and C, an A-1 result indicates "A is minimally better than C." Table 9.17 summarizes the comparisons for all possible pairs.

After comparing all pairs, all the boxes in Table 9.17 will be filled in. By adding the values for each idea (for example, A = 1, B = 2+3+2+1= 8), we get the results shown in Table 9.18. Clearly, B and E are the top choices.

TABLE 9.17   Paired Comparison of Pencil Ideas

	B	C	D	E
A	B-2	A-1	D-2	E-2
B		B-3	B-2	B-1
C			D-1	E-3
D				E-2

3   Significant
2   Moderate
1   Minimal

TABLE 9.18   Final Evaluation Results for Paired Comparison

Key Letter	Idea	Value
A	Eliminate paint.	1
B	Reduce the length of lead.	8
C	Remove eraser.	0
D	Stain wood in lieu of paint.	3
E	Make body out of paper.	7

The whole evaluation phase may go through several screening steps. Table 9.19 is a convenient template for recording the whole evaluation phase.

9.5   Planning Phase

After the evaluation phase, we have a final list of ideas that can be recommended to management for implementation. Now is the time to develop the best ideas in detail so that recommendations can be made convincingly. At this stage, we need to determine costs more accurately and discuss proposed solutions with relevant people. We need to get the latest material, labor, process and cost data. We must develop a cooperative atmosphere with everyone able to contribute to a successful solution, refine the cost of each solution, and determine the best and alternate recommendation for the performance of basic functions.

For a successful project completion, we also need to determine potential roadblocks, where they may come up, and how they may be eliminated. Table 9.20 can be used as a template.

TABLE 9.19   Idea Screening Result

What else will perform the function ? Function: Enhance Appearance	First Screening	Second Screening	Final Screening
1 Use laminate instead of paint.	√		
2 Use stainless steel parts.	√		
3 Use plastic material.	√		
4 Use curves instead of sharp edges.	√	√	
5 Delete complicated features.	√		
6 Paint parts individually before assembly.	√	√	√
7 Use chrome plating.			
8 Use multicolor paint.			
9 Use gold material.			

TABLE 9.20   Identify Roadblocks

Best Idea: Reduce the Length of Lead		
Roadblock	Where/Why	Action Required
Differ from traditional design practice	Design/out of specification Marketing/bad customer image	1. Show that people seldom use the full pencil length 2. Show that good style and low price is more important to customers
Alternative Idea: Use Body Out of Paper		
Roadblock	Where/Why	Action Required
Effect on strength and durability unknown	Design/no previous experience with this design	Show strength/durability test results
Perceived as a risk idea	Marketing/no idea if customer will buy in	Show this new design can make pencil body self-peeling to expose lead, no need for pencil sharpeners

In the planning phase, it is also very important to discuss how this project can be sold and implemented. Table 9.21 is a planning form that can be used to list the names of everyone who will be involved in accepting and implementing the proposal. We need to figure out possible problem areas and decide how they can be eliminated.

9.6   Reporting Phase

The object of the value-engineering study is to develop a successful recommendation for improvement in products, systems, organizations, etc., and therefore, in turn, for profits. It must be presented to management so that it will be accepted and implemented if it is to be worth anything to you and your company.

Your best recommendation must be prepared for presentation. Before and after costs and potential savings must be shown and clearly defined.

TABLE 9.21   Action Plan for Selling Value-Engineering Ideas

Department	Supervisors	Action Required	Problem	How to Solve Problems
1.				
2.				
3.				
4.				
5.				
6.				
7.				

Sketches should show the basic charges in whatever detail is necessary to prove results. You may need to provide simple models in some cases. You should list all advantages and disadvantages and show how the disadvantages were considered in your decision. If the procedure has been followed, all necessary data should be available in your notes and records.

The importance of the reporting phase should not be overlooked. If the recommendations are not presented properly and effectively, a good idea or an excellent recommendation may not be taken seriously. It is necessary to present the recommendation in a manner that will clearly demonstrate its advantages from the standpoint of the organization required to implement it.

The worksheets used during this chapter have been developed to provide all of the information necessary to prepare an effective recommendation. It is complete and concise. The next step is to arrange the material so that it will sell your idea. One of the most important considerations here is to provide complete information.

Failure to provide complete information has been proven to be a major cause for rejection of a proposal. Persons who are required to review or approve proposals of one type or another will verify that it is rare for complete information to be provided.

Keep these points in mind when preparing the final recommendation and report:

1. Cover all of the facts. Do not skip an important consideration thinking that it can be considered later. Do not plan surprises.

2. Justify the recommendation on both technical and economic grounds. Show the risk and the rewards involved, and show the cost to verify the idea as well as total lifetime program costs, such as design and developmental expense, capital investments necessary for buildings, tools, etc.

3. Indicate the effect on corporate profit, competitive position. or other important factors.

4. Discuss the proposal with people who will be affected by the idea.

9.7   Implementation Phase

The implementation phase of a value-engineering project deals with changing the product or service designs, based on the findings of the value-engineering analysis, and implementing the design changes.

The objective of a value-engineering study is the successful incorporation of recommendations into the product or operations. However, the success of a project often depends on the beginning of the study.

Each project must be thoroughly analyzed at the outset to determine its potential for benefit and the probability of implementation. This is as important as the knowledge and skill required to apply the system to attain successful results.

An excellent idea is worthless unless it can be properly implemented, and it must be implemented in the manner intended. There are many cases on record where the idea could not be implemented because of the high cost of making the change, and other cases where the recommendations were not properly understood and the implementation resulted in increased cost. This often results in disillusionment or the feeling that value engineering doesn't work for those sorts of problems. However, in most cases this is the result of inefficient preliminary analysis and preparation.

Selection of projects is a part of the entire value-engineering implementation process. Often management will assume that any project will prove profitable. This is not always the case. The project must be practical in relationship to its effect on the organization. It is not reasonable to expend effort and funds on a value study without first having done the necessary work to ensure that the project is practical, that it can be implemented, and that the necessary funds and manpower will be available.

To begin with, we will look at the overall organization and the implementation of value-engineering operations. Then we will look at some of the details that make for success.

9.7.1 Setting a Goal

What do we want to get from the implementation of value engineering work? What will be the objective? This is the first question to answer.

The goal should be very specific. Whatever the goal is, it should be defined in specific terms, such as: increase productivity by a specific percent; reduce product cost by a specific number of dollars per unit, etc.

The goal should be known to everyone. It can be product-oriented or directed towards manufacturing or administrative operations. It need not be company-wide. However, the scope can be broadened at any time. Once the goal has been determined, the means to achieve the objective can be developed.

9.7.2 Develop An Implementation Plan

The first task in the implementation is to create an implementation leader or coordinator. The coordinator should work with a trained value-engineering expert. The coordinator and the value-engineering expert should develop a training program for the implementation team members.

After the training, a carefully designed implementation plan should be developed jointly by the coordinator, the value-engineering expert, and team members.

From what we have noted here, it is obvious that the problem is complex from the standpoint of options. However, successful operations do not have to be extensive. Starting small and developing successfully is preferred to a lot of noise and a big crash because of poor planning.

One of the most important factors in value engineering is attitude on the part of management and the people on task teams. A positive, cooperative, supportive attitude is required.

In most cases whenever a new idea is presented to a group the initial reaction could be negative. The first remarks are, "It's interesting but let me tell you what's wrong with it." The best approach to this reaction is to listen carefully. They may have some ideas you overlooked. After all negative reaction has run out, be prepared to ask some specific positive questions of the group that will develop positive responses. For example, "I understand your difficulty in producing this part in the plant. What do you think we would have to do to make this practical? Do you see any changes we might make to satisfy our methods?" This will usually bring about a positive result.

Never argue. In many cases it is beneficial to solicit negative ideas, but be prepared to develop positive questions. Our attitude is that we must begin to ask, "What's good about this idea"? "How will it help us to do a better job?"

Changing people's attitudes is difficult and may never happen, but understanding the reasons behind the negative reaction should make it possible to persuade most people that they can benefit from success. Remember, there is a risk of failure in new ideas. New ideas require change and they may not work. People want proof that a new idea will work before they will support it. However, you may be able to show that the benefits are greater than the risk.

9.8 Automobile Dealership Construction (Park 1999)

A large realty company built and maintained many automobile dealership facilities all over the country. One major problem faced by this company was the long time between dealership project authorization and dealer occupancy. History showed that this long construction cycle caused tremendous lost sales, so the company wanted to use the value-engineering technique to shorten this cycle.

At the start of this project, it was found that the average duration from site selection to leasing was 502 days, or about 1.5 years. A review of the project process flow chart identified the activities that filled the

502 days: selecting and obtaining the options on the land, topographical surveys, soil borings, facility layouts, bid estimates and analysis, budget reviews, design, construction, and many, many others.

In this value-engineering project, a FAST diagram was developed, as illustrated in Figure 9.17. Out of the 30 functions, three (resolve restrictions, obtain data, and construct facility) took 85 percent of the time. This evaluation was obtained by using time, instead of cost, as a measure in the FAST diagram.

As a result of project recommendations, the project process procedure was revised to make it possible to conduct several long-term activities in parallel with other activities. For example, approval for early site work was obtained from property owners before ownership was transferred so that topographic surveys and soil boring could be done as soon as possible. Standard designs were developed for several parts of the facility to reduce overall design and development time, and a single-source contracting procedure was developed to reduce contractor project interface.

The result of these recommendations was a potential average saving of 262 days, or a 47 percent average saving in time per project. Based on the average annual construction program, the yearly benefit in increased rent would be over $1,250,000 per year. The additional increased vehicle sales were not included in the benefit.

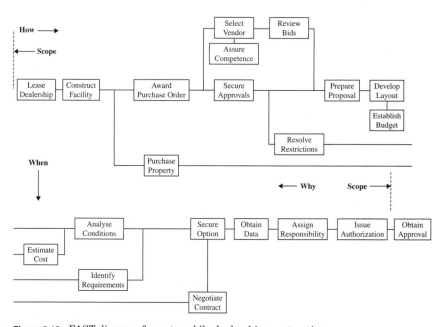

Figure 9.18 FAST diagram for automobile dealership construction

9.9   Engineering Department Organization Analysis (Park 1999)

A leading automobile company spent about $200 million dollars per year and employed 4,000 people, including engineers, designers, technicians, technical specialists, financial analysts, and so on. An economic downturn forced the company to cut costs. Painful lessons had previously been learned from across-the-board budget cuts, where an equal share of budget was cut across all departments. The result of that kind of budget cut was that some vital operations were seriously damaged, and other areas were okay. This time, the situation was critical; the budget had already been cut several times, and no one knew where to look next. A value-engineering project was initiated in order to identify hidden, unnecessary cost.

In this project, with 72 hours of total effort by a team of six people, a FAST diagram was developed that had 72 functions. The chart was then thoroughly discussed to ensure that it covered all aspects of the operation, and a glossary of the functions was made to ensure future understanding.

Table 9.22 gives a partial list of the functions for this engineering operation.

Here is a sample from the glossary of functions:

Create Design: To generate a new system, assembly, or component, measured by time, which includes time to come up with design ideas, design and layout time, engineers' working time, programming time, and so on.

The FAST diagram provided some interesting information. Most important, it showed that many functions were performed to satisfy functions outside the scope of engineering responsibilities. Many of these functions contributed to higher-order functions to support other company operations, such as the purchasing and legal department.

The next step was to determine how much each function costs and how funds were distributed among all the functions. To do that, departmental managers were asked to distribute their departmental cost by function. Cost-function worksheets were filled out. One portion of a cost-function worksheet is illustrated in Table 9.23.

TABLE 9.22   Partial List of Functions in Engineering Operation

Verb	Noun	Verb	Noun
Create	Design	Prepare	Plan
Transmit	Information	Negotiate	Alternatives
Evaluate	Information	Evaluate	Capabilities
Confirm	Design	Allocate	Resources
Model	Concept	Appropriate	Funds

TABLE 9.23 Cost/Function Worksheet

	Cost (Hours)		Functions						
Item No	Activity	Hours	Trans Info	Create Design	Auth Prog	Conf Design	Evalu Info	Collect Data	Make Model
1	Manager	1300	100	60	40	150	40		
2	Secretary	1736	1000			60	60		
3	Design Supervisor	1438	40	40		60	40	40	
4	Engineering Supervisor	1344	20	100	40	80	80	40	
5	Development Supervisor	2270				160	280		
6	Technical Specialist	1078		200	40				
7	Sr Design Engineers	2790	40	200	40	160	200	120	
8	Sr Develop Engineers	2790				320	1290	560	120
9	Design Engr	11109	280	3360	140	420	560	560	
10	Develop Engr	22560				1200	4200	3000	750
11	Design leader	4909	190	480		180	180		
12	Technician	22204		8320		1040	780		
13	Modeler	17580		4800					
14	Clerk	3392	880			800	400		
15	Mechanics	26528						4800	8600
16	Material	$198,000							
Total	Material								
Total	Hours								

A partial FAST diagram is illustrated in Figure 9.19.

As you can see in the FAST diagram, 43 percent of the available funds went to "confirm design," and only 14 percent went to "create design." This was considered to be a poor distribution of funds. Team members thought that this lopsided distribution of funds was a major source of problems, and across-the-board budget cuts would likely create big problems in new product design.

It was recognized that "confirm design" was a required function. However, changing the way that this function was performed would offer opportunities for major improvement in productivity, could improve the overall engineering operation, and could make more efficient use of engineering funds. The immediate recommendation was to review all areas involved in "confirm design." By following the value-engineering job plan, major changes were made in several areas that substantially increased output, cost savings, and avoidance of major capital investment. Substantial cost reduction was achieved without affecting the vital design functions.

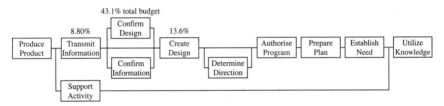

Figure 9.19   FAST diagram for engineering operation

Customer Value Creation Through Creative Design (TRIZ)

In developing any product or service, creative design combined with the right customer value position will usually bring huge success in the marketplace. Creative design can make your product the "first of its kind" in the marketplace, and it can make your product difficult for competitors to copy. A company with a good reputation in creativity can often build a strong brand image, and the brand's power further enhances the customer value of the company's products.

The Theory of Inventive Problem Solving (TRIZ) is an effective methodology that can help companies and product development people improve their creativity. The nature of TRIZ is to shortcut the creative process, and to effectively reuse the knowledge base developed in similar inventions in order to avoid reinvention. TRIZ is an indispensable tool for any customer-value-centric company. In this chapter we will cover the philosophical aspects of TRIZ in order to lay a foundation. Then we will look at the four-step TRIZ problem definition and solving process, together with the tools used in TRIZ.

10.1 Theory of Inventive Problem Solving (TRIZ)

TRIZ (Teoriya Resheniya Izobreatatelskikh Zadatch) was developed in the Soviet Union starting in the late 1940s and in English is known as the Theory of Inventive Problem Solving (TIPS). TRIZ has been developed based on more than 1500 person-years of research and study over many of the world's most successful solutions of problems from science and engineering, and systematic analysis of successful patents from around the world, as well as the study of the psychological aspects of human creativity.

Dr. Genrich S. Altshuller, the creator of TRIZ, started his investigation of invention and creativity in 1946. After initially reviewing 200,000 patent abstracts, Altshuller selected 40,000 as representatives of inventive solutions. He separated the patents into five levels according to their degree of inventiveness, with Level 1 being the lowest and Level 5 being the highest. He found that almost all inventions contain at least one contradiction, which means that an attempt to improve one feature of a system detracts from another feature. He found that the level of inventiveness often depends on how well the contradiction is resolved.

- **Level 1: Apparent or Conventional Solution** Inventions at Level 1 represent 32 percent of patent inventions and employ obvious solutions drawn from only a few clear options. These are not real inventions but narrow extensions or improvements of existing systems, which are not substantially changed due to the application of the invention. Usually a particular feature is enhanced or strengthened. Examples of Level 1 invention include increasing the thickness of walls to allow for greater insulation in homes, or increasing the distance between the front skis on a snowmobile for greater stability. These solutions may represent good engineering, but contradictions are not identified and resolved.

- **Level 2: Small Invention Inside Paradigm** Inventions at Level 2 offer small improvements to an existing system by reducing a contradiction inherent in the system but still requiring obvious compromises. These solutions represent 45 percent of the inventions. A Level 2 solution is usually found through a few hundred trial-and-error attempts, and it requires knowledge of only a single field of technology. The existing system is slightly changed, and includes new features that lead to definite improvements. The new suspension system between the track drive and the frame of a snowmobile is a Level 2 invention. The use of an adjustable steering column to increase the range of body types that can comfortably drive an automobile is another example at this level.

- **Level 3: Substantial Invention Inside Technology** Inventions at Level 3, which significantly improve the existing system, represent 18 percent of the patents. At this level, a contradiction is resolved within an existing system, usually through introducing some new elements. The Level 3 invention may involve hundreds of trials and errors. Examples of Level 3 invention include automatic transmission of cars. These Level 3 inventions usually borrowed some ideas from other industries, which are not known in the industry where the invention is applied.

- **Level 4: Invention Outside Technology** Inventions at Level 4 are found in science, not in technology. Such breakthroughs represent about 4 percent of inventions. Tens of thousands of random trials are usually required for these solutions. Level 4 inventions usually lie outside the technology's normal paradigm and involve using a completely different principle for the primary function. In Level 4 solutions, the contradiction is eliminated because its existence is impossible within the new system. That is, Level 4 breakthroughs use physical effects and phenomena that had previously been little known within the area. A simple example involves using materials with thermal memory (shape-memory metals) for a key ring. Instead of taking a key on or off a steel ring by forcing the ring open, the ring is placed in hot water. The metal memory causes it to open for easy replacement of the key. At room temperature, the ring closes.

- **Level 5: Discovery**   Inventions at Level 5 exist outside the confines of contemporary scientific knowledge. Such pioneering works represent less than 1 percent of inventions. These discoveries require lifetimes of dedication for they involve the investigation of tens of thousands of ideas. This type of solution occurs when a new phenomenon is discovered and applied to the invention problem. Level 5 inventions, such as lasers and transistors, create new systems and industries. Once a Level 5 discovery becomes known, subsequent applications or inventions occur at one of the four lower levels. For example, the laser, technological wonder of the 1960s, is now used routinely as a lecturer's pointer and a land surveyor's measuring instrument.

Other major findings of TRIZ are also based on extensive studies of inventions:

- A very small number of inventive principles and strategies summarize most innovations.

- Outstanding innovations often feature complete resolution of contradictions, not merely compromising on contradictions.

- Outstanding innovations often transform wasteful or harmful elements in the system to useful resources.

- Technological innovation trends are highly predictable.

10.1.1   What Is TRIZ?

TRIZ is a combination of methods, tools, and a way of thinking (Mann 2002). The ultimate goal of TRIZ is to achieve absolute excellence in design and innovation. In order to achieve absolute excellence, TRIZ has five key philosophical elements:

- **Ideality**   This is the ultimate criterion for system excellence; ideality is the maximization of the benefits provided by the system, and minimization of the harmful effects and costs associated with the system.

- **Functionality**   This is the fundamental building block of system analysis; it builds models about how a system works and how it creates benefit, harm, and costs.

- **Resource**   Maximum utilization of resources is one of the keys to achieving maximum ideality.

- **Contradictions**   Contradictions are a common inhibitor for increasing functionality, and removing contradictions usually greatly increases the functionality and raises the system to a totally new performance level.

- **Evolution**   The evolutionary trend of technological development is highly predictable, and it can be used to guide further development.

Based on these five key philosophical elements, TRIZ researchers developed a process of problem-solving. This is a four-step process, consisting of problem definition, problem classification and tool selection, solution generation, and evaluation.

10.1.1.1   Problem Definition   This is a very important step in TRIZ. If you define the right problem and do it accurately, this represents 90 percent of the solution. The problem-definition step includes the following tasks:

- **Function analysis**   This includes function modeling and analysis of the system. This is the most important task in the "definition" step. TRIZ has very well-developed tools for function modeling and analysis.

- **Technological evolution analysis**   This task looks into the relative technological maturity of all subsystems and parts. If a subsystem or part is technically "too" mature, it may reach its limit in performance and thus become a bottleneck for the whole system.

- **Ideal final result**   The ideal final result is the virtual limit of the system in TRIZ. It may never be achieved but it provides us with an "ultimate dream" and will help us to think "out of box."

10.1.1.2   Problem Classification and Tool Selection   TRIZ has a large array of tools for inventive problem solving; however, we must select the right tool for the right problem. In TRIZ, we must first classify the problem type and then select the tools accordingly.

10.1.1.3   Solution Generation   In this step, we apply TRIZ tools to generate solutions for the problem. Because TRIZ has a rich array of tools, it is possible to generate many solutions.

10.1.1.4   Evaluation   In any engineering project, we need to evaluate the soundness of the new solution. TRIZ has its own evaluation approach.

However, other non-TRIZ methods might also be used at this stage, such as axiomatic design and design vulnerability analysis.

10.2   TRIZ Fundamentals

Functionality, use of resources, ideality, contradictions, and evolution are the pillars of TRIZ. These elements make TRIZ distinctly different from other innovation and problem-solving strategies. In this section, we will look at all five elements.

10.2.1   Function Modeling and Analysis

Function modeling and functional analysis originated in value engineering (Miles 1961). A function is defined as the natural or characteristic action performed by a product or service.

However, products or services often provide many functions. For example, an automobile provides customers with the ability to get from point A to point B, but it also provides a comfortable riding environment, air conditioning, music, and so on. There are several types of functions:

- **Main basic function**   Among all the functions, the most important function is called the main basic function—it is the primary purpose or the most important action performed by the product or service. The main basic function must always exist, although methods or designs for achieving the function may vary. For example, "the ability to get from A to B" is the main basic function of an automobile.

Besides the main basic function, there are other useful functions as well; these are the secondary useful functions. There are several kinds of secondary useful functions:

- **Secondary basic functions**   These are not main basic functions, but customers definitely need them. For example, "providing a comfortable riding environment" is a must-have for automobiles.

- **Nonbasic but beneficial functions**   These are functions that provide customers with esteem value, comfort, and so on. For example, the paint finish on an automobile provides both basic and nonbasic functions; it protects the automobile from corrosion and rust, as well as creating a sleek look for the car.

Besides secondary useful functions, there are two other types of functions:

- **Supporting functions**   These are functions that support the main basic function or other useful functions. Supporting functions result from a design approach chosen to achieve the main basic function or other useful functions. If the design approach used to achieve the main basic function and other useful functions is changed, supporting functions may also change. There are at least two kinds of supporting functions:

■ **Assisting functions**   These are functions that assist other useful functions. For example, the engine suspension system provides the function of "locking the position of the engine in the automobile" so the engine can provide power without falling off the car.

■ **Correcting functions**   These are functions that correct negative effects of another useful function. For example, the main basic function of the engine is "to provide power for the automobile," but internal combustion engines also create heat, which is a negative effect—the water pump's function is to circulate water through the engine to correct this negative effect. If we change the design and use electricity as the power source of the automobile, the function of a water pump will no longer be needed.

■ **Harmful functions**   These are unwanted, negative functions caused by the method used to achieve useful functions. For example, an internal combustion engine not only provides power, but it also generates noise, heat, and pollution, which are harmful functions.

In summary, the main basic function and secondary useful functions provide benefits for the customer. Supporting functions are useful, or at least they are not harmful, but they do not provide benefits directly to the customer and incur costs. Harmful functions are not useful and provide no benefits at all.

10.2.1.1 Functional Statement   A function can be usually fully described by a statement consisting of three parts: a subject, a verb, and an object. For example, an automobile's main basic function can be described as follows:

Car	moves	people
(Subject)	(Verb)	(Object)

A toothbrush's main basic function can be described as

Toothbrush	brushes	teeth
(Subject)	(Verb)	(Object)

10.2.1.2 Functional Analysis Diagrams   The functional analysis diagram is a graphical tool used for describing and analyzing functions. The following graph is a typical template for a functional analysis diagram.

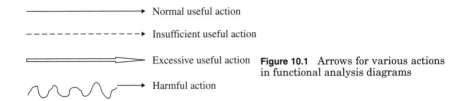

Normal useful action

Insufficient useful action

Excessive useful action

Harmful action

Figure 10.1 Arrows for various actions in functional analysis diagrams

The *subject* is the source of the action, and the *object* is the action receiver. *Action* is the verb in a functional statement, and it is represented by an arrow. In technical systems, the action is often accomplished by applying some kind of *field*, such as a mechanical, electrical, or chemical field. For example, the function "brush teeth" can be described by the following functional analysis diagram:

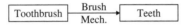

In this diagram, "Mech" stands for "mechanical field." Clearly, brushing teeth is an application of one kind of mechanical field, force.

In functional analysis diagrams, there are four types of actions, and they are represented by four types of arrows as illustrated in Figure 10.1.

Example 10.1 Brushing Teeth   If we use a toothbrush correctly, and clean our teeth properly, we call this brush action a "normal useful action." We can illustrate that with the following functional analysis diagram:

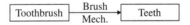

However, if we use the toothbrush too gently and do not brush long enough, or we use a worn-out toothbrush, our teeth will not get enough cleaning. This is shown in the following functional analysis diagram:

Clearly, this is a case of "insufficient useful action."

If we use a very strong toothbrush, and brush our teeth with great force and big strokes, our gums and teeth will be hurt. We can use the following functional analysis diagram to describe this situation:

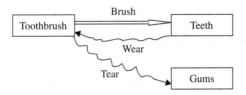

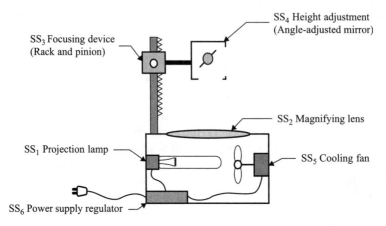

Figure 10.2 Overhead projector

That is, the toothbrush delivers excessive brushing action to the teeth, which will deliver a harmful action, "tear the gums," and make them bleed; and the teeth also may deliver a harmful action, "wear the toothbrush."

Example 10.2 Functional Modeling and Analysis   Figure 10.2 shows a schematic view of an overhead projector.

The following graph is the functional modeling and analysis diagram for the whole system:

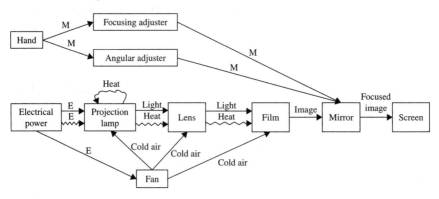

In this functional analysis graph, E stands for "electrical field," and M stands for "mechanical field."

There are many chains of action in this graph; that is, an object can be another object's subject. This produces a sequence of subject-action-object-action chains. Each chain describes a complete function. We can identify the following functions:

■ From "Electrical power" to "screen" is the function "to project image in the film to screen." This is the main basic function.

- From "Hand" to "focusing adjustor" to "mirror" is the function "to focus the image." That is a secondary basic function.

- From "Hand" to "angular adjuster" to "mirror" is the function "to project image to right position in screen." That is also a secondary basic function.

- From "Electrical power" to "film" is a harmful function chain, which, without correction, will damage the film and device.

- From "Electricity" to "fan" to projection lamp, lens, and film is a correcting function to compensate for the negative effect of the harmful function.

10.2.2   Use of Resources

Using resources effectively is very important in TRIZ. We also need to make use of resources in creative ways.

The primary mission for any product or process is to deliver functions. Because substances and fields are basic building blocks of functions, they are important resources from the TRIZ point of view. However, substances and fields are not sufficient to build and deliver functions; space and time are also important and necessary resources. From the TRIZ point of view, information and knowledge about how to use available resources are also important resources.

We can divide resources into the following categories:

- Substance resources
  - Raw materials and products
  - Waste
  - By-product(s)
  - System elements
  - Substance from surrounding environments
  - Inexpensive substance
  - Harmful substance from the system
  - Altered substance from system
- Field resources
  - Energy in the system
  - Energy from the environment
  - Energy/field that can be built upon existing energy platforms
  - Energy/field that can be derived from system waste
- Space resources
  - Empty space

- Space at interfaces of different systems
- Space created by vertical arrangement
- Space created by nesting arrangement
- Space created by rearrangement of existing system elements
- Time resources
  - Pre-work period
  - Time slot created by efficient scheduling
  - Time slot created by parallel operation
  - Post-work period
- Information/knowledge resources
  - Knowledge about all available substances (material properties, transformations, etc.)
  - Knowledge about all available fields (field properties, utilizations, etc.)
  - Past knowledge
  - Other people's knowledge
  - Knowledge on operation
- Functional resources
  - Unutilized or underutilized existing system main functions
  - Unutilized or underutilized existing system secondary functions
  - Unutilized or underutilized existing system harmful functions

In TRIZ, it is more important to look into cheap, ready-to-use, abundant resources than expensive, hard-to-use, and scarce resources, as demonstrated in Example 10.3.

Example 10.3 Cultivating Fish in Farmland   The southeast part of China is densely populated, so land is a scarce resource. Much land is used to plant rice. Agriculture experts suggest that farmland can be used to cultivate fish while it is used to grow rice, because water is a free and ready resource in rice paddies, and the waste produced by fish can be used as fertilizer for the rice.

10.2.3 Ideality

Ideality is a measure of excellence. In TRIZ, ideality is defined by the following equation

$$\text{Ideality} = \frac{\sum Benefits}{\sum Costs + \sum Harm} \tag{10-1}$$

Where

- Σ *Benefits* is the sum of the values of the system's useful functions. Here the supporting functions are not considered useful functions, because they will not bring benefits to customers directly. Supporting functions are considered part of the cost to make the system work.
- Σ *Costs* is the sum of the expenses for the system.
- Σ *Harm* is the sum of harms created by harmful functions.

In equation (10-1), a higher ratio indicates a higher ideality. When a new system is able to achieve a higher ratio than the old system, it is considered a real improvement.

In TRIZ, there is a law of increasing ideality, which states that the evolution of all technical systems proceeds in the direction of increasing ideality. The ideality of the system will increase in the following cases:

- Benefits are increased.
- Costs are reduced.
- Harms are reduced.
- Benefits are increased faster than costs and harms.

From the TRIZ point of view, technical systems or products are not goals in themselves. The real value of the product/system is in its useful functions. Therefore, the better system is the one that consumes fewer resources in both initial construction and maintenance.

The ideal final result (IFR) is when the ratio becomes infinite. The IFR system requires no material, consumes no energy and space, needs no maintenance, and will not break.

10.2.4 Contradictions

From the TRIZ standpoint, a challenging problem can be expressed as either a technical contradiction or physical contradiction.

10.2.4.1 Technical Contradictions A technical contradiction is a situation where efforts to improve technical attributes of a system lead to deterioration of other technical attributes. For example, as a container becomes stronger, it becomes heavier. Faster automobile acceleration reduces fuel efficiency.

A technical contradiction can be resolved either by finding a trade-off between the contradictory demands, or by overcoming the contradiction. Trade-off or compromise solutions do not eliminate the technical contradictions, but rather soften them, thus retaining the harmful (undesired) action or shortcoming in the system.

Altshuller analyzed thousands of inventions and formulated typical technical contradictions, such as productivity versus accuracy, reliability versus complexity, shape versus speed, etc. It was discovered that despite the immense diversity of technological systems and even greater diversity of inventive problems, there are only about 1250 typical system contradictions. These contradictions can be expressed as a table of contradiction of 39 design parameters (see Table 10.1).

From the TRIZ standpoint, overcoming a technical contradiction is very important both because attributes in the contradiction can be improved drastically, and the performance of the system will be raised to a whole new level. TRIZ has many tools for eliminating technical contradictions.

TABLE 10.1   Thirty-Nine Contradictory Design Parameters

1	Weight of moving object	21	Power
2	Weight of nonmoving object	22	Waste of energy
3	Length of moving object	23	Waste of substance
4	Length of nonmoving object	24	Loss of information
5	Area of moving object	25	Waste of time
6	Area of nonmoving object	26	Amount of substance
7	Volume of moving object	27	Reliability
8	Volume of nonmoving object	28	Accuracy of measurement
9	Speed	29	Accuracy of manufacturing
10	Force	30	Harmful factors acting on object
11	Tension, pressure	31	Harmful side effects
12	Shape	32	Manufacturability
13	Stability of object	33	Convenience of use
14	Strength	34	Repairability
15	Durability of moving object	35	Adaptability
16	Durability of nonmoving object	36	Complexity of device
17	Temperature	37	Complexity of control
18	Brightness	38	Level of automation
19	Energy spent by moving object	39	Productivity
20	Energy spent by nonmoving object		

10.2.4.2 Physical Contradictions   A physical contradiction means that a subject or an object has to be in mutually exclusive physical states.

A physical contradiction typically has this pattern: "To perform function F_1, the element must have property P, but to perform function F_2, it must have property $-P$, or the opposite of P." For example, an automobile has to be light in weight (P) to have high fuel economy (F_1), but it also has to be heavy in weight ($-P$) in order to be stable for driving (F_2).

Example 10.4   **Problem:** Some buildings are supported by piles. The pile should have a sharp tip to facilitate the driving process. However, sharp piles have reduced support capability. For better support capacity, the piles should have blunt ends. However, it is more difficult to drive a blunt-tipped pile.

Contradiction: A pile should be sharp to facilitate the driving process and it should be blunt to provide better support of the foundation.

TRIZ Solution: The situation clearly calls for a solution providing separation of contradictory properties in time. The pile is sharp during the driving process, and then its base is expanded, which could be realized by a small explosive charge.

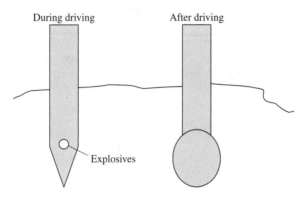

Conventional design philosophy is based on compromises (trade-offs). Contrary to this approach, TRIZ offers several methods of overcoming physical contradictions completely; these methods will be discussed thoroughly later in this chapter.

10.2.5 Evolution

TRIZ researchers have found that the "trends of evolution of many technical systems are similar and predictable." They found that many technical systems will evolve through five stages: pregnancy, infancy, growth, maturity, and decline. If we plot a time line on the horizontal axis (X-axis), and plot performance index, level of inventiveness, number of inventions (relating to the system), and profitability of inventions on the vertical axis (Y-axis), we will get the four curves shown in Figure 10.3. Because the shape of the first curve (performance versus evolutionary stages) has a S shape, it is also called an S-curve.

10.2.5.1 Pregnancy For a technical system, the pregnancy stage is the time between an idea's inception and its birth. A new technological system emerges only after the following two conditions are satisfied:

- There is a need for the function of the system.
- There are means (technology) to deliver this function.

The development of an airplane can be used as an example. The need for the function of airplane, that is, "to fly," existed long ago in many people's dreams and desires. However, the technical knowledge of aerodynamics and mechanics was not sufficient for the development of human flight until the 1800s.

The technologies for the airplane were available since the development of glider flight in 1848 and the gasoline engine in 1859. It was the Wright brothers who successfully integrated both technologies in their aircraft in 1903—and a new technology got off the ground.

10.2.5.2 Infancy The birth of a new technical system is the starting point of the infancy stage. This is the first stage of an S-curve. The new system appears as a result of a high-level invention. Typically,

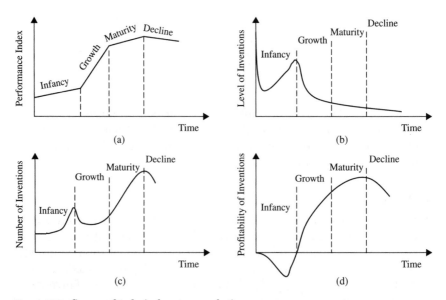

Figure 10.3 Curves of technical system evolution

the system is primitive, inefficient, unreliable, and has many unsolved problems. It does, however, provide some new functions, or the means to provide the function.

In the infancy stage, the performance level is low and its improvement is slow (Figure 10.3a), due to the lack of human and financial resources. Most people may not be convinced of the usefulness of the system, but a small number of enthusiasts who believe in the system's future continue to work towards its success. The level of inventiveness is usually high, because the initial concept is often very inventive and patentable. It is usually Level 3, 4, or even 5 (Figure 10.3b). But the number of inventions in this system is usually low (Figure 10.3c), because the system is fairly new. The profit is usually negative (Figure 10.3d), because at this stage of the technology the customers are usually few and the expense is high.

10.2.5.3 Growth (Rapid Development) The growth stage begins when society recognizes the value of the new system. By this time, many problems have been overcome; efficiency and performance have improved in the system, and people and organizations invest money in developing the new product or process. This accelerates the system's development, improving the results and in turn, attracting greater investment. Thus, a positive feedback loop is established, which serves to further accelerate the system's evolution.

In this stage, the improvement of performance is fast (Figure 10.3a) because of the rapid increase in investment and the removal of many technical bottlenecks. The level of inventiveness is lower, because most inventions in this stage are incremental improvements. They are mostly Level 1 or Level 2 (Figure 10.3b). But the number of inventions is usually high (Figure 103c). The profit is usually growing fast (Figure 10.3d).

10.2.5.4 Maturity In the maturity stage, system development slows as the initial concept upon which the system was based nears exhaustion of its potential. Large amounts of money and labor may be expended, but the results are usually very marginal. At this stage, standards are established. Improvements occur through system optimization and trade-offs. The performance of the system still grows but at a slower pace (Figure 10.3a). The level of invention is usually low (Figure 10.3b), but the number of inventions in the form of industrial standards is quite high (Figure 10.3c). The profitability is usually dropping because of the saturation of the market and increased competition (Figure 10.3d).

10.2.5.5 Decline At this stage, the limits of the technology have been reached and no fundamental improvement is available. The system

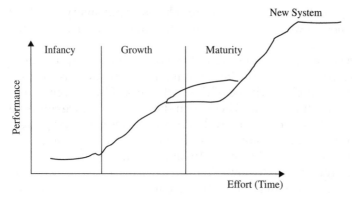

Figure 10.4 S-curve for two generations of a system

may no longer be needed, because the function provided may no longer be needed.

It is really important to start the next generation of technical systems long before the decline stage in order to keep the company from failing. Figure 10.4 illustrates the S-curves of two successive generations of a technical system.

10.3   The TRIZ Problem-Solving Process

TRIZ has a four-step problem-solving process. The four steps are problem definition, problem classification and problem tool selection, problem solution generation, and problem concept evaluation. We will look at each step in detail.

10.3.1   Problem Definition

Problem definition is a very important step. The quality of the solution is highly dependent on problem definition.

Defining the problem starts with several questions:

1. What is the problem?
2. What is the scope of the project?
3. What subsystem, system, and components are involved?
4. Do we have a current solution for the problem? Why is the current solution not good?

These are common questions that are asked in any engineering project. By answering these questions, we can define the scope of the project and focus on the right problem area.

Besides answering these common questions, several TRIZ methods are also very helpful in the problem definition stage:

- Functional modeling and functional analysis
- Ideality calculations
- S-curve analysis
- Contradiction analysis

10.3.1.1 Functional Modeling and Functional Analysis   After identifying the project scope, it is very helpful to establish a functional model of the subsystem involved in the project. Functional modeling and analysis enables us to see the problem more clearly and precisely.

We will take another look at the toothbrush example to illustrate how functional analysis can help the problem definition.

Example 10.5 Toothbrush Problem Revisited   Assume that we are a toothbrush manufacturer and that the current regular toothbrush does not perform satisfactorily, that is, teeth cannot be adequately cleaned. We can first draw the following functional diagram:

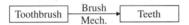

By analyzing the functional diagram, we may come up with the following possibilities:

- The current lack of performance may be caused by "inadequate action," that is, the actual functional diagram is the following:

If that is the case, it belongs to the problem of "inadequate functional performance"; we can use the TRIZ standard solution technique to resolve this problem.

- We may find that the current functional model is too limiting, because the function statement "toothbrush brush teeth" limits our solution to using only a brush and to use "mechanical action only." We can develop the following alternative functional model:

The subject "toothbrush" is replaced by a more general "tooth cleaning device." The object "teeth" is changed to "dirt in teeth," which is more precise.

The action "brush" is changed to a more general term "remove." Under this alternative functional modeling, many possible subjects and actions are open for selection. For example, we can use hydraulic action to clean teeth, or chemical action to clean teeth. We can even consider "pretreatment of teeth" to keep them dirt free, and so on. Clearly this alternative functional modeling opens a gate for problem solving and innovation.

10.3.1.2 Ideality and Ideal Final Result   After functional modeling and functional analysis, we can evaluate the ideality of the current system:

$$\text{Ideality} = \frac{\sum Benefits}{\sum Costs + \sum Harm}$$

The ideal final result (the ultimate optimal solution) for a system is one where

$$\sum Benefits \to \infty, \text{ and } \sum Costs + \sum Harm \to 0$$

By comparing the ideality of the current system with the ideal final result, we can identify where the system improvements should go and what aspects of the system should be improved. This will definitely help with the problem definition and identify what problem should be solved.

10.3.1.3 S-Curve Analysis   It is very beneficial to evaluate the evolutionary stage of the current technical system involved in any TRIZ project. For example, if the current subsystem is at the growth stage, we should focus our attention on gradual improvement. If our subsystem is near maturity, we will know that it is time to develop the next generation of this subsystem.

10.3.1.4 Contradiction Analysis   By using the method described in Figure 10.4, we can identify whether there are any physical or technical contradictions in our current system. TRIZ has many methods to resolve contradictions.

10.3.2 Problem Classification and Tool Selection

After the problem has been defined, we should be able to classify the problem into one of the following categories. For each category, there are many TRIZ methods available to resolve the problem:

- **Physical contradiction**
  Method: Separation principles

- **Technical contradiction**
  Method: Inventive principles

- **Imperfect functional structures**

  This problem occurs when
  - There are inadequate useful functions or a lack of needed useful functions.

  - There are excessive harmful functions.
  Method: Functional improvement methods

- **Excessive complexity**

  This problem occurs when the system is too complex and costly and some of its functions can be eliminated or combined.
  Method: Trimming

- **System improvement**

  This problem occurs when the current system is doing its job but enhancement is needed to beat the competition.
  Method: Evolution of technological systems

- **Develop useful functions**

  This problem occurs when we can identify what useful functions are needed to improve the system but we do not know how to create these functions.
  Method: Physical, chemical, and geometric effects database

10.3.3 Problem-Solution Generation

After the problem is classified, there are usually many TRIZ methods available for solving the problem, so many alternative solutions can be found. These solutions will be evaluated in the next step.

10.3.4 Problem Concept Evaluation

There are many concept evaluation methods that can be used to evaluate and select the best solution. These methods are often not TRIZ-related. Frequently used concept evaluation methods include Pugh concept selection, and value engineering.

10.4 Technical Contradiction Elimination and Inventive Principles

Genrich Altshuller analyzed more than 40,000 patents and identified about 1250 typical technical contradictions. These contradictions are further expressed in a matrix of 39×39 engineering parameters (contradictions). To resolve these contradictions, Altshuller compiled 40 principles. Each of 40 principles contains subprinciples, totaling 86 subprinciples.

It should be noted that the 40 principles are formulated in a general way. If, for example, the contradiction table recommends principle 30, "flexible shell and thin films," it means that the solution of the problem relates somehow to changing the degree of flexibility or adaptability of a technical system being modified. The contradiction table and the 40 principles do not offer direct solutions to problems; they only suggest the most promising directions for searching for a solution. The problem solver has to interpret these suggestions and find their own way to a particular situation.

Usually people solve problems by analogical thinking. We try to relate the problem confronting us to some familiar standard class of problems (analogs) for which a solution exists. If we draw upon the right analog, we arrive at a useful solution. Our knowledge of analogous problems is the result of educational, professional, and life experiences.

What if we encounter a problem for which we have no analog? This obvious question reveals the shortcomings of the usual approach to invention problems. The contradiction table and 40 principles offer us clues to solving the problems with which we are not familiar.

When using the contradiction table and 40 principles, the following simple procedure will be helpful.

1. Decide on the attribute to be improved, and use one of the 39 parameters in the contradiction table to standardize or model this attribute.

2. Answer the following questions:
   a) How can this attribute be improved using conventional means?
   b) Which attribute would suffer if conventional means were used?

3. Select an attribute in the contradiction table corresponding to step 2b.

4. Use the contradiction table, identify the principles in the intersection of the row (attributes improved) and column (attribute deteriorated) for overcoming the technical contradiction.

Here are the 40 principles for reference.

Principle 1: Segmentation

- Divide an object into independent parts.
- Make an object easy to disassemble.
- Increase the degree of fragmentation (or segmentation) of an object.

Principle 2: Taking out

- Separate an "interfering" part (or property) from an object, or single out the only necessary part (or property) of an object.

Principle 3: Local quality
- Change an object's structure from uniform to nonuniform; change an external environment (or external influence) from uniform to nonuniform.
- Make each part of an object function in conditions most suitable for its operation.
- Make each part of an object fulfill a different and useful function.

Principle 4: Asymmetry
- Change the shape of an object from symmetrical to asymmetrical.
- If an object is asymmetrical, increase its degree of asymmetry.

Principle 5: Merging
- Bring closer together (or merge) identical or similar objects; assemble identical or similar parts to perform parallel operations.
- Make operations contiguous or parallel, and bring them together in time.

Principle 6: Universality
- Make a part or object perform multiple functions, to eliminate the need for other parts.

Principle 7: "Nested doll"
- Place one object inside another, place each object, in turn, inside the other.
- Make one part pass through a cavity in the other.

Principle 8: Anti-weight
- To compensate for the weight of an object, merge it with other objects that provide lift.
- To compensate for the weight of an object, make it interact with the environment (for example, use aerodynamic, hydrodynamic, buoyancy, and other forces).

Principle 9: Preliminary anti-action
- If it will be necessary to do an action with both harmful and useful effects, this action should be replaced later with anti-actions to control harmful effects.
- Create beforehand stresses in an object that will oppose known undesirable working stresses later on.

Principle 10: Preliminary action
- Perform, before it is needed, the required change of an object (either fully or partially).

- Prearrange objects so that they can come into action from the most convenient place without losing time for their delivery.

Principle 11: Beforehand cushioning

- Prepare emergency means beforehand to compensate for the relatively low reliability of an object.

Principle 12: Equipotentiality

- In a potential field, limit position changes (for example, change operating conditions to eliminate the need to raise or lower objects in a gravity field).

Principle 13: "The other way around"

- Invert the action(s) used to solve the problem (for example, instead of cooling an object, heat it).

- Make movable parts (or the external environment) fixed, and fixed parts movable.

- Turn the object (or process) "upside down."

Principle 14: Spheroidality

- Instead of using rectilinear parts, surfaces, or forms, use curvilinear ones, move from flat surfaces to spherical ones, from parts shaped as a cube (parallelepiped) to ball-shaped structures.

- Use rollers, balls, spirals, domes.

- Go from linear to rotary motion, use centrifugal forces.

Principle 15: Dynamics

- Allow (or design) the characteristics of an object, external environment, or process to change them to optimal, or to find an optimal operating condition.

- Divide an object into parts capable of movement relative to each other.

- If an object (or process) is rigid or inflexible, make it movable or adaptive.

Principle 16: Partial or excessive actions

- If 100 percent of an effect is hard to achieve using a given solution method, the problem may be considerably easier to solve by using slightly less or slightly more of the same method.

Principle 17: Another dimension

- Move an object in two- or three-dimensional space.

- Use a multistory arrangement of objects instead of a single-story arrangement.

- Tilt or reorient the object; lay it on its side.
- Use "another side" of a given area.

Principle 18: Mechanical vibration

- Cause an object to oscillate or vibrate.
- Increase its frequency (even up to the ultrasonic).
- Use an object's resonance frequency.
- Use piezoelectric vibrators instead of mechanical ones.
- Use combined ultrasonic and electromagnetic field oscillations.

Principle 19: Periodic action

- Instead of continuous action, use periodic or pulsating actions.
- If an action is already periodic, change the periodic magnitude or frequency.
- Use pauses between impulses to perform a different action.

Principle 20: Continuity of useful action

- Carry on work continuously; make all parts of an object work at full load, all the time.
- Eliminate all idle or intermittent actions or work.

Principle 21: Skipping

- Conduct a process, or certain stages (e.g. destructive, harmful, or hazardous operations) at high speed.

Principle 22: "Blessing in disguise"

- Use harmful factors (particularly, harmful effects of the environment or surroundings) to achieve a positive effect.
- Eliminate the primary harmful action by adding it to another harmful action to resolve the problem.
- Amplify a harmful factor to such a degree that it is no longer harmful.

Principle 23: Feedback

- Introduce feedback (referring back, cross-checking) to improve a process or action.
- If feedback is already used, change its magnitude or influence.

Principle 24: "Intermediary"

- Use an intermediate carrier article or intermediary process.
- Merge one object temporarily with another (which can be easily removed).

Principle 25: Self-service
- Make an object serve itself by performing auxiliary helpful functions.
- Use waste resources, energy, or substances.

Principle 26: Copying
- Instead of an unavailable, expensive, fragile object, use simpler and inexpensive copies.
- Replace an object or process with their optical copies.
- If visible optical copies are already used, move to infrared or ultraviolet copies.

Principle 27: Cheap short-living
- Replace an expensive object with a multitude of inexpensive objects, compromising certain qualities (such as service life, for instance).

Principle 28: Mechanics substitution
- Replace a mechanical means with a sensory (optical, acoustic, taste, or smell) means.
- Use electric, magnetic, and electromagnetic fields to interact with the object.
- Change from static to movable fields, from unstructured fields to those having structure.
- Use fields in conjunction with field-activated (for example, ferromagnetic) particles.

Principle 29: Pneumatics and hydraulics
- Use gas and liquid parts of an object instead of solid parts (e.g. inflatable, filled with liquids, air cushion, hydrostatic, hydro-reactive).

Principle 30: Flexible shells and thin films
- Use flexible shells and thin films instead of three-dimensional structures.
- Isolate the object from the external environment using flexible shells and thin films.

Principle 31: Porous materials
- Make an object porous or add porous elements (inserts, coatings, etc.).
- If an object is already porous, use the pores to introduce a useful substance or function.

Principle 32: Color changes
- Change the color of an object or its external environment.
- Change the transparency of an object or its external environment.

Principle 33: Homogeneity

- Make objects interact with a given object of the same material (or a material with identical properties).

Principle 34: Discarding and recovering

- Make portions of an object that have fulfilled their function go away (discard by dissolving, evaporating, etc.) or modify these directly during operation.
- Conversely, restore consumable parts of an object directly during operation.

Principle 35: Parameter changes

- Change an object's physical state (e.g. to a gas, liquid, or solid).
- Change the concentration or consistency.
- Change the degree of flexibility.
- Change the temperature.

Principle 36: Phase transitions

- Use phenomena occurring during phase transitions (e.g. volume changes, loss or absorption of heat, etc.).

Principle 37: Thermal expansion

- Use thermal expansion (or contraction) of materials.
- If thermal expansion is being used, use multiple materials with different coefficients of thermal expansion.

Principle 38: Strong oxidants

- Replace common air with oxygen-enriched air.
- Replace enriched air with pure oxygen.
- Expose air or oxygen to ionizing radiation.
- Use ozonized oxygen.
- Replace ozonized (or ionized) oxygen with ozone.

Principle 39: Inert atmosphere

- Replace a normal environment with an inert one.
- Add neutral parts, or inert additives to an object.

Principle 40: Composite materials

- Change from uniform to composite (multiple) materials.

Example 10.6 Using 40 Principles and Contradiction Matrix to Improve Wrench Design   When we use a conventional wrench to undo an overtightened or corroded nut (as shown in the following image), one of the problems is that the corners of the nut take a concentrated load and may wear out quickly. You can reduce the clearance between wrench and nut, but it will be difficult to fit the wrench onto the nut. Is there anything we can do to solve this problem?

It seems obvious that we want to reduce the clearance between the wrench and nut to improve reliability, but this leads to the deterioration of operations—the wrench is harder to use. From the TRIZ standpoint, a technical contradiction is present when a useful action simultaneously causes a harmful action.

A problem associated with a technical contradiction can be resolved either by finding a trade-off between the contradictory demands, or by overcoming the contradiction. Trade-off or compromise solutions do not eliminate the technical contradictions, but rather soften them, thus retaining harmful (undesired) actions or shortcomings in the system. An engineering problem becomes an inventive one when the technical contradiction cannot be overcome by conventional means and trade-off solutions are not acceptable. The 40 principles and the contradiction matrix are important tools for overcoming contradictions.

1. Build contradiction model.

Look into the problems and find a pair of contradictions. The contradictions should be described using two of the 39 parameters for technical contradictions. In this problem, the contradictions are:

- Things we want to improve: Reliability (parameter 27)
- Things are getting worse: Ease of operation (parameter 33)

2. Check contradiction matrix.

Locate the parameter to be improved in the row, and the parameter to be deteriorated in the column in the contradiction matrix for inventive principles. The matrix offers the following principles: 27, 17, and 40 (see the partial matrix that follows; the full matrix is shown in Tables 10.3a–c).

3. Interpret principles.

Read each principle and construct analogies between the concepts of principle and your situation, then create solutions to your problem.

TABLE 10.2   Partial Contradiction Matrix

What should be improved?	25. Waste of time	26. Quantity of substance	27. Reliability	28. Measurement accuracy	29. Manufacturing precision	30. Harmful action at object	31. Harmful effect caused by the object	32. Ease of manufacture	33. Ease of operation	34. Ease of repair	35. Adaptation
25. Waste of time		35 38 18 16	10 30 4	24 34 28 32	24 26 28 18	35 18 34	35 22 18 39	35 28 34 4	4 28 10 34	32 1 10	35 28
26. Quantity of substance	35 38 18 16		18 3 28 40	3 2 28	33 30	35 33 29 31	3 35 40 39	29 1 35 27	35 29 10 25	2 32 10 25	15 3 29
27. Reliability	10 30 4	21 28 40 3		32 3 11 23	11 32 1	27 35 2 40	35 2 40 26		27 17 40	1 11	13 35 8 24
28. Measurement accuracy	24 34 28 32	2 6 32	5 11 1 23			28 24 22 26	3 33 39 10	6 35 25 18	1 13 17 34	1 32 13 11	13 35 2
29. Manufacturing precision	32 26 28 18	32 30	11 32 1			26 28 10 36	4 17 34 26		1 32 35 23	25 10	

In this case, Principle 17 (another dimension) indicates that the wrench problem may be resolved if you "move an object in two or three-dimensional space" or "use a different side of the given area." From Principle 27 (cheaper short-living) and Principle 40 (composition material), we may "replace an expensive object with a multitude of inexpensive objects" and "change from uniform material to composite material."

TABLE 10.3a Contradiction Table of Inventive Principles 1–13 (continued)

What should be improved? / What is deteriorated?	1. Weight of movable object	2. Weight of fixed object	3. Length of movable object	4. Length of fixed object	5. Area of movable object	6. Area of fixed object	7. Volume of movable object	8. Volume of fixed object	9. Speed	10. Force	11. Stress, pressure	12. Shape	13. Object's composition stability
1. Weight of movable object			15 8 29 34		29 17 38 34		29 2 40 28		2 8 15 38	8 10 18 37	10 36 37 40	10 14 35 40	1 35 19 39
2. Weight of fixed object				10 1 29 35		35 30 13 2		5 35 14 2		8 10 19 35	13 29 10 18	13 10 29 14	26 39 1 40
3. Length of movable object	8 15 29 34				15 17 4		7 17 4 35		13 4 8	17 10 4	1 8 35	1 8 10 29	1 8 15 34
4. Length of fixed object		35 28 40 29				17 7 10 40		35 8 2 14		28 10	1 14 35	13 14 15 7	39 37 35
5. Area of movable object	2 17 29 4		14 15 18 4				7 14 17 4		29 30 4 34	19 30 35 2	10 15 36 28	5 34 29 4	11 2 13 39
6. Area of fixed object		30 2 14 18		26 7 9 39						1 18 35 36	10 15 36 37		2 38
7. Volume of movable object	2 26 29 40		1 7 35 4		1 7 4 17				29 4 38 34	15 35 36 37	6 35 36 37	1 15 29 4	28 10 1 39

8. Volume of fixed object		35 10 / 19 14	19 14	35 8 / 2 14						2 18 / 37	24 35	7 2 / 35	34 28 / 35 40
9. Speed	2 28 / 13 38		13 14 / 8		29 30 / 34		7 29 / 34			13 28 / 15 19	6 18 / 38 40	35 15 / 18 34	28 33 / 1 18
10. Force	8 1 / 37 18	18 13 / 1 28	17 19 / 9 36	28 10	19 10 / 15	1 18 / 36 37	15 9 / 12 37	2 36 / 18 37	13 28 / 15 12		18 21 / 11	10 35 / 40 34	35 10 / 21
11. Stress, pressure	10 36 / 37 40	13 29 / 10 18	35 10 / 36	35 1 / 14 16	10 15 / 36 28	10 15 / 36 37	6 35 / 10	35 24	6 35 / 36	36 35 / 21		35 4 / 15 10	35 33 / 2 40
12. Shape	8 10 / 29 40	15 10 / 26 3	29 34 / 5 4	13 14 / 10 7	5 34 / 4 10		14 4 / 15 22	7 2 / 35	35 15 / 34 18	35 10 / 37 40	34 15 / 10 14		33 1 / 18 4
13. Object's composition stability	21 35 / 2 39	26 39 / 1 40	13 15 / 1 28	37	2 11 / 13	39	28 10 / 19 39	34 28 / 35 40	33 15 / 28 18	10 35 / 21 16	2 35 / 40	22 1 / 18 4	
14. Strength	1 8 / 40 15	40 26 / 27 1	1 15 / 8 35	15 14 / 28 26	3 34 / 40 29	9 40 / 28	10 15 / 14 7	9 14 / 17 15	8 13 / 26 14	10 18 / 3 14	10 3 / 18 40	10 30 / 35 40	13 17 / 35
15. Duration of moving object's operation	19 5 / 34 31		2 19		3 17 / 19		10 2 / 19 30		3 35 / 5	19 2 / 16	19 3 / 27	14 26 / 28 25	13 3 / 35
16. Duration of fixed object's operation		6 27 / 19 16		1 40 / 35		35 38		35 34 / 38					39 3 / 35 23
17. Temperature	36 22 / 6 38	22 35 / 32	15 19 / 9	15 19 / 9	3 35 / 39 18		34 39 / 40 18	35 / 6 4	2 28 / 36 30	35 10 / 3 21	35 39 / 19 2	14 22 / 19 32	1 35 / 32
18. Illumination	19 1 / 32	2 35 / 32	19 32 / 16		19 32 / 26		2 13 / 10		10 13 / 19	26 19 / 6		32 30	32 3 / 27
19. Energy expense of movable object	12 18 / 28 31		12 28		15 19 / 25		35 13 / 18		8 15 / 35	16 26 / 21 2	23 14 / 25	12 2 / 29	19 13 / 17 24
20. Energy expense of fixed object		19 9 / 6 27								36 37			27 4 / 29 18

TABLE 10.3b Contradiction Table of Inventive Principles 14–25 (continued)

What is deteriorated? → ↓ What should be improved?	14. Strength	15. Duration of moving object's operation	16. Duration of fixed object's operation	17. Temperature	18. Illumination	19. Energy expense of movable object	20. Energy expense of fixed object	21. Power	22. Waste of energy	23. Loss of substance	24. Loss of information	25. Waste of time	26. Quantity of substance
1. Weight of movable object	28 27 / 18 40	5 34 / 31 35		6 29 / 4 38	19 1 / 32	25 12 / 34 31	18 19 / 28 1	13 36 / 18 31	6 2 / 34 19	5 35 / 3 31	10 24 / 35	10 35 / 20 28	3 26 / 18 31
2. Weight of fixed object	28 2 / 10 27		2 27 / 19 6	28 19 / 32 22	35 19 / 35		18 19 / 28 1	15 19 / 18 22	18 19 / 28 15	5 8 / 13 30	10 15 / 35	10 20 / 35 26	19 6 / 18 26
3. Length of movable object	8 35 / 29 34	19		10 15 / 19	32	8 35 / 24		1 35	7 2 / 35 39	4 29 / 23 10	1 24	15 2 / 29	29 35
4. Length of fixed object	15 14 / 28 26		1 40 / 35	3 35 / 38 18	3 25			12 8	6 28	10 28 / 24 35	24 26	30 29 / 14	
5. Area of movable object	3 15 / 40 14	6 3		2 15 / 16	15 32 / 19 13	19 32		19 10 / 32 18	15 17 / 30 26	10 35 / 2 39	30 26	26 4	29 30 / 6 13
6. Area of fixed object	40		2 10 / 19 30	35 39 / 38				17 32	17 7 / 30	10 14 / 18 39	30 16	10 35 / 4 18	2 18 / 40 4
7. Volume of movable object	9 14 / 15 7	6 35 / 4		34 39 / 10 18	10 / 13 2	35		35 6 / 13 18	7 15 / 13 16	36 39 / 34 10	2 22	2 6 / 34 10	29 / 30 7
8. Volume of fixed object	9 14 / 17 15		35 34 / 38	35 / 6 4				30 6	6 28	10 39 / 35 34		35 16 / 32 18	35 3
9. Speed	8 3 / 26 14	3 19 / 35 5		28 30 / 36 2	10 13 / 19	8 15 / 35 38		19 35 / 38 2	14 20 / 19 35	10 13 / 28 38	13 26		10 19 / 29 38
10. Force	35 10 / 14 27	19 2		35 10 / 21		19 17 / 10	1 16 / 36 37	19 35 / 18 37	14 15	8 35 / 40 5		10 37 / 36	14 29 / 18 36

374

Parameter													
11. Stress, pressure	9 18 3 40	19 3 27		35 39 19 2		14 24 10 37		10 35 14	2 36 25	10 36 3 37		37 36 4	10 14 36
12. Shape	30 14 10 40	14 26 9 25		22 14 19 32	13 15 32	2 6 34 14		4 6 2	14	35 29 3 5		14 10 34 17	36 22
13. Object's composition stability	17 9 15	13 27 10 35	39 3 35 23	35 1 32	32 3 27 15	13 19	27 4 29 18	32 35 27 31	14 2 39 6	2 14 30 40		35 27	15 32 35
14. Strength		27 3 26		30 10 40	35 19	19 35 10	35	10 26 35 28	35	35 28 31 40		29 3 28 10	29 10 27
15. Duration of moving object's operation	27 3 10			19 35 39	2 19 4 35	28 6 35 18		19 10 35 38		28 27 3 18	10	20 10 28 18	3 35 10 40
16. Duration of fixed object's operation			19 18 36 40	19 18 36 40				16		27 16 18 38	10	28 20 10 16	3 35 31
17. Temperature	10 30 22 40	19 3 39	19 18 36 40		32 30 21 16	19 15 3 17		2 14 17 25	21 17 35 38	21 36 29 31		35 28 21 18	3 17 30 39
18. Illumination	35 19	2 19 6		32 35 19		32 1 19	32 35 1 15	32	19 16 1 6	13 1	1 6	19 1 26 17	1 19
19. Energy expense of movable object	5 19 9 35	28 35 6 18		19 24 3 14	2 15 19			6 19 37 18	12 22 15 24	35 24 18 5		35 38 19 18	34 23 16 18
20. Energy expense of fixed object	35				19 2 35 32					28 27 18 31			3 35 31

TABLE 10.3c Contradiction Table of Inventive Principles 27-39 (continued)

What should be improved? / What is deteriorated?	27. Reliability	28. Measurement accuracy	29. Manufacturing precision	30. Harmful action at object	31. Harmful effect caused by the object	32. Ease of manufacture	33. Ease of operation	34. Ease of repair	35. Adaptation	36. Device complexity	37. Measurement or test complexity	38. Degree of automation	39. Productivity
1. Weight of movable object	3 11 1 27	28 27 35 26	28 35 26 18	22 21 18 27	22 35 31 39	27 28 1 36	35 3 2 24	2 27 28 11	29 5 15 8	26 30 36 34	28 29 26 32	26 35 18 19	35 3 24 37
2. Weight of fixed object	10 28 8 3	18 26 28	10 1 35 17	2 19 22 37	35 22 1 39	28 1 9	6 13 1 32	2 27 28 11	19 15 29	1 10 26 39	25 28 17 15	2 26 35	1 28 15 35
3. Length of movable object	10 14 29 40	28 32 4	10 28 29 37	1 15 17 24	17 15	1 29 17	15 29 35 4	1 28 10	14 15 1 16	1 19 26 24	35 1 26 24	17 24 26 16	14 4 28 29
4. Length of fixed object	15 29 28	32 28 3	2 32 10	1 18		15 17 27	2 25	3	1 35	1 26	26		30 14 7 26
5. Area of movable object	29 9	26 28 32 3	2 32	22 33 28 1	17 2 18 39	13 1 26 24	15 17 13 16	15 13 10 1	15 30	14 1 13	2 36 26 18	14 30 28 23	10 26 34 2
6. Area of fixed object	32 35 40 4	26 28 32 3	2 29 18 36	27 2 39 35	22 1 40	40 16	16 4	16	15 16	1 18 36	2 35 30 18	23	10 15 17 7
7. Volume of movable object	14 1 40 11	25 26 28	25 28 2 16	22 21 27 35	17 2 40 1	29 1 40	15 13 30 12	10	15 29	26 1	29 26 4	35 34 16 24	10 6 2 34
8. Volume of fixed object	2 35 16		35 10 25	34 39 19 27	30 18 35 4	35		1		1 31	2 17 26		35 37 10 2
9. Speed	11 35 27 28	28 32 1 24	10 28 32 25	1 28 35 23	2 24 35 21	35 13 8 1	32 28 13 12	34 2 28 27	15 10 26	10 28 4 34	3 34 27 16	10 18	35 37 10 2
10. Force	3 35 13 21	35 10 23 24	28 29 37 36	1 35 40 18	13 3 36 24	15 37 18 1	1 28 3 25	15 1 11	15 17 18 20	26 35 10 18	36 37 10 19	2 35	3 28 35 37

11. Stress, pressure	10 13 / 19 35	6 28 / 25	3 35	22 2 / 37	2 33 / 27 18	1 35 / 16	32 15 / 26	2 13 / 1	1 15 / 29	16 29 / 1 28	15 13 / 39	15 1 / 32	10 14 / 35 37
12. Shape	10 40 / 16	28 / 32 1	32 30 / 40	22 1 / 2 35	35 1	1 32 / 17 28	32 35 / 30	2 15 / 10 16	35 30 / 34 2	2 35 / 22 26	35 22 / 39 23	1 8 / 35	17 26 / 34 10
13. Object's composition stability		13	18	35 24 / 18 30	35 40 / 27 39	35 19	32 40 / 28 2	27 / 11 3	15 3 / 32	2 13 / 28	27 3 / 15 40	15	23 35 / 40 3
14. Strength	11 3	3 27 / 16	3 27	18 35 / 37 1	15 35 / 22 2	11 3 / 10 32	12 27	29 10 / 27	1 35 / 13	10 4 / 29 35	19 29 / 39 35	6 10	29 35 / 10 14
15. Duration of moving object's operation	11 2 / 13	3	3 27 / 16 40	22 15 / 33 28	21 39 / 16 22	27 / 1 4	1	1	2		25 14 / 6 35	1	35 17 / 14 19
16. Duration of fixed object's operation	34 27 / 6 40	10 26 / 24		17 1 / 40 33	22	35 10	1	1	2				20 10 / 16 38
17. Temperature	19 35 / 3 10	32 19 / 24	24	22 33 / 35 2	22 35 / 2 24	26 27	26 27	4 10 / 16	2 18 / 27	2 17 / 16	3 27 / 35 31	26 2 / 19 16	15 28 / 35
18. Illumination		11 15 / 32	3 32	15 19	35 19 / 32 39	19 35 / 28 26	28 26 / 19	15 17 / 13 16	15 1 / 19	6 32 / 13	32 15	2 26 / 10	2 25 / 16
19. Energy expense of movable object	19 21 / 11 27	3 1 / 32		1 35 / 6 27	2 35 / 6	28 26 / 30	19 35	1 15 / 17 28	15 17 / 13 16	2 29 / 27 28	35 38	32 2	12 28 / 35
20. Energy expense of fixed object	10 36 / 23			10 2 / 22 37	19 22 / 18	1 4					19 35 / 16 25		1 6

4. Resolve the problem.

The working surface of a wrench can be redesigned in nonuniform shape by applying Principle 17 (see the illustration below). Principles 27 and 40 can be used together: Attach soft metal or plastic pads on the wrench's working surfaces when tightening or undoing expensive nuts.

Table 10.3(a–c) is the complete contradiction matrix, which can be used to select the most appropriate inventive principles.

Statistical Basics and Six Sigma Metrics

Six Sigma is a data-driven management system with near-perfect-performance objectives (Pande et al. 1998). In this context, *data-driven* means that in Six Sigma, the real data collected in the process under study is the only source for measuring the current performance, analyzing the root causes of the problem, and deriving improvement strategies. Near-perfect-performance objectives means that in Six Sigma, the process will be continuously improved until it achieves very low levels of defects and very high levels of performance. Clearly, it also needs real data from the process to verify that the desired performance requirements are met.

The theoretical basis for modern data analysis is statistics, and there are different methods in statistics that can be used to analyze the data. Some of them are very simple, such as descriptive statistics, which can provide intuitive display and analysis of the data. Some of them are more sophisticated, such as probability distribution models and statistical inferences; these analyses are more powerful and they can provide more insights and provide credible inferences and predictions about processes. All popular Six Sigma performance metrics are based on the theory of statistics, so familiarity with basic statistics is essential in understanding Six Sigma metrics.

In this chapter, we will review several descriptive statistical methods and several commonly-used probability distribution models, as well as Six Sigma metrics.

11.1 Six Sigma and Data Analysis

Data analysis is a very important part of Six Sigma. In real business and engineering processes, much of the data that we collect consists of random variables, that is, their value will vary with some degree of uncertainty. Consider the case in Example 11.1.

Example 11.1 A Data Set from a Semiconductor Manufacturing Process   In a semiconductor manufacturing process, we have a step where an oxide film is grown on a silicon wafer in a furnace. In this step, a cassette of wafers is placed in a quartz "boat" and the boats are placed in the furnace. A gas flow is created in the furnace and it is brought up to temperature and held there for a specified period of time. In this process, the most desirable oxide film thickness is 560 Angstroms, and the specification of the oxide thickness is 560 ± 100 Angstroms; that is, an oxidized wafer is out of specification if its thickness is either lower than 460 angstroms or higher than 660 Angstroms.

We collected the following film thickness data in the process:

547	563	578	571	572	575	584	549	546	584	593	567	548	606	607	539	554	533	535
522	521	547	550	610	592	587	587	572	612	566	563	569	609	558	555	577	579	552
558	595	583	599	602	598	616	580	575										

How well does this process satisfy our quality requirement?

In Example 11.1, the film thickness clearly varies from wafer to wafer, so it is a random variable. A random variable can be either discrete or continuous. If the set of all possible values is finite or countably infinite, the random variable is discrete; if the set of all possible values of the random variable is an interval, then the random variable is continuous. Clearly, the film thickness data is a continuous variable.

11.2   Descriptive Statistics

Descriptive statistics are a set of simple graphical and numerical methods that can quickly display some intuitive properties found in data. The commonly-used graphical descriptive statistical methods include dot plots, histograms, and box plots.

11.2.1   Dot Plot

The *dot plot*, as illustrated in Figure 11.1, is a simple yet effective diagram in which each dot represents a data point from Example 11.1. Dot plots can display distribution patterns and the spread of data points.

11.2.2   Histogram

A *histogram* is a diagram displaying a frequency distribution. In a histogram, the horizontal axis is partitioned into many small segments,

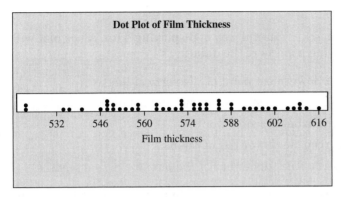

Figure 11.1   Dot Plot of film thickness data

and the number of data points (or the percentage of points) that fall in each segment is called the *frequency*. This determines the height of the bar for that segment.

For example, the histogram for the film thickness data in Example 11.1 is displayed in Figure 11.2. The leftmost segment is the bracket (515 to 525), and in the data set, there are two data points (521 and 522) in this range, so the height of the bar is 2. We can see that a large portion of the data falls in the 545 to 585 segment.

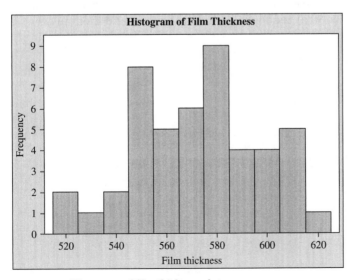

Figure 11.2   Histogram of film thickness data

11.2.3   Box Plot

A *box plot* is another very useful way of displaying data. A box plot will display the following aspects of the data:

- Minimum value (lowermost point of the vertical line)
- Maximum value (uppermost point of the vertical line)
- Median value (horizontal center line)
- Twenty-fifth percentile (lower bar of the box)
- Seventy-fifth percentile (upper bar of the box)

Figure 11.3 shows the box plot of the data in Example 11.1. The centerline of the box corresponds to 572, which is the median of the data. The lower bar of the box corresponds to 552, which is the 25th percentile of the data, and the upper bar corresponds to 592, which is the 75th percentile of the data. The uppermost point corresponds to 616, which is the maximum of the data, and the lowermost point corresponds to 521, which is the minimum of the data.

11.2.4   Numerical Descriptive Statistics

Numerical descriptive statistics are numbers calculated from a data set in order to help us create a mental image of the distribution pattern of the data. There are three types of numerical descriptive statistics:

1. The numerical measure that describes the central tendency of the data, that is, where the center of the data set is. Frequently used measures of central tendency include the mean and median.

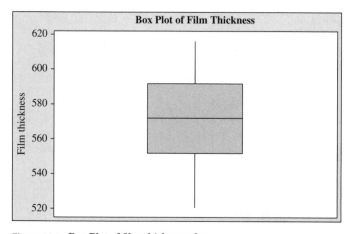

Figure 11.3   Box Plot of film thickness data

2. The numerical measures that describe the spread of data, also called the measures of variation. The frequently used measures of variation include variance, standard deviation, range, maximum, and minimum.

3. The numerical measures that describe the relative position of the data, often called the measures of relative standing. The frequently used measure here includes the percentile points.

11.2.4.1 Measures of Central Tendency The mean and median are the most frequently used measures for central tendencies. I will define both measures and give numerical examples in the following sections.

Mean The *mean*, $\bar{y}$, is also called the *arithmetic mean*. For a set of n measurements, $y_1, y_2, ..., y_n$, $\bar{y}$ is the average of the measurements, specifically:

$$\bar{y} = \frac{1}{n}\sum_{i=1}^{n} y_i \qquad (11.1)$$

Example 11.2 The Calculation of Mean The mean for the data in Example 11.1 is

$$\bar{y} = \frac{1}{47}(547 + 563 + ... + 575) = 572.02.$$

Clearly, the mean is simply a numerical average, which gives a good sense of where the center is for a data set. It is the most commonly used measure of central tendency. However, in some cases, it is not a preferred measure. For example, assume that in a subdivision, there are 20 families. Most of the families have an annual income around $40,000, but there is one family with an annual income of $1,000,000. If we use the arithmetic mean as the measure of central tendency, $\bar{y}$ will be approximately $90,000, which is by no means a typical income in this circumstance. The median will be a better measure in this case.

Median The *median* of a set of measurements, $y_1, y_2, ..., y_n$, is the middle number when the measurements are arranged in ascending (or descending) order. Specifically, let $y_{(i)}$ denote the ith value of the data set when $y_1, y_2, ..., y_n$ are arranged in ascending order; then the median m is the following:

$$y_{[(n+1)/2]} \quad \text{if } n \text{ is odd}$$

$$m = \frac{y_{(n/2)} + y_{(n/2+1)}}{2} \qquad \text{if } n \text{ is even.} \qquad (11.2)$$

Example 11.3 The Calculation of Median   The median of the data set in Example 11.1 can be calculated by first arranging the data in ascending order:

$$y_{(1)}, y_{(2)}, ..., y_{(n-1)}, y_{(n)} = 521, 522, 533, ..., 610, 612, 616$$

In this data set, $n = 47$, which is an odd number, so $n + 1/2 = 48/2 = 24$, and $y_{24} = 572$
Therefore, $m = 572$.

11.2.4.2 Measures of Variation   The most commonly used measures of variation are the range, the variance, and the standard deviation.

Range   The *range* is equal to the difference between the largest (maximum) and the smallest (minimum) measurements in a data set:

$$\text{Range} = \text{Maximum} - \text{Minimum} \qquad (11.3)$$

Example 11.4 The Calculation of Range   For the data set in Example 11.1, the maximum $= y_{(n)} = y_{(47)} = 616$, and the minimum $= y_{(1)} = 521$. Therefore, Range $= 616-521 = 95$.
The range is very easy to compute, but it only gives the distance between the two most extreme observations. It is not a good measure of variation for the whole data set. Variance and standard deviation are better measures in this respect.

Variance   The *variance* of a sample of n measurements, $y_1, y_2, ..., y_n$, is defined as:

$$s^2 = \frac{1}{n-1} \sum_{i=1}^{n} (y_i - \bar{y})^2. \qquad (11.4)$$

Example 11.5 The Calculation of Variance   For the data set of Example 11.1, the variance can be computed as:

$$s^2 = \frac{1}{47-1}[(547 - 572.02)^2 + (563 - 572.02)^2 + \cdots + (575 - 572.02)^2]$$

$$= 601.72$$

Sample variance, s^2, is obviously an average of the sum of the squared deviation from the mean of all observations. Squared deviation makes sense because whether an observation is smaller or larger than the mean, the squared deviation will always be positive. The average of squared deviation is a measure of variation for the whole data set. The drawback to variance, however, is that the measurement unit of variance is the square of the measurement unit of the original data. So if the original data is length in inches, the variance will be in square inches, which cannot be compared with the original data.

Standard Deviation The *standard deviation* is the square root of the variance. Specifically, the standard deviation of a sample of n measurements, $y_1, y_2, ..., y_n$, is defined as:

$$s = \sqrt{s^2} = \sqrt{\frac{1}{n-1} \sum_{i=1}^{n} (y_i - \bar{y})^2}. \tag{11.5}$$

One advantage of standard deviation over variance is that the standard deviation has the same measurement scale as that of the mean. Also, the spread of data is usually measured by the mean and standard deviation.

Example 11.6 The Calculation of Standard Deviation For the data in Example 11.1, the standard deviation can be computed as:

$$s = \sqrt{s^2} = \sqrt{601.72} = 24.53.$$

11.2.4.3 Measure of Relative Standing The *measure of relative standing* provides a numerical value or score that describes a predefined location relative to other observations in a data set. A very commonly used measure of relative standing is the $100p$th percentile, often simply called percentile points.

The $100p$th percentile of a data set is a value y located so that $100p$ percent of the data is smaller than y, and $100(1-p)$ percent of data is larger than y, where $0 \le p \le 1$.

Example 11.7 The Calculation of Percentile The median is the 50th percentile, because 50 percent of the data points are smaller than the median, and 50 percent of the data points are larger. The 25th percentile is often called lower quartile, Q_L, or Q_1. Twenty-five percent of the data points will be smaller than Q_L, and 75 percent of the data points will be larger in a given data set. The 75th percentile is often called the upper quartile, denoted by Q_U or Q_3. Seventy-five percent of the data points will be smaller than Q_U, and 25 percent of the data will be larger than Q_U.

MINITAB is a popular statistical software that can compute all types of descriptive statistics conveniently. The following MINITAB output is the descriptive statistics for the data set in Example 11.1.

Descriptive Statistics: Film Thickness

```
Variable          N  N*    Mean  SE Mean  StDev  Minimum      Q1  Median      Q3
Film Thickness   47   0  572.02     3.58  24.53   521.00  552.00  572.00  592.00
or
Variable          Maximum
Film Thickness     616.00
```

11.3   Random Variables and Probability Distributions

The data set collected in a process, such as the data set described in Example 11.1, is called a *sample* of data, because it only reflects a snapshot of the process. For example, the data set in Example 11.1 is only a small portion of production data. If we were able to collect all the film thickness data for all wafers in the whole life cycle of the oxidation furnace, we would have collected a whole *population* of data.

In real-world business decision making, the population is of more interest for the decision makers. The decision makers are definitely more interested in the overall quality level for the population. Random variables and probability distributions are the mathematical tools used to describe the behavior of populations.

11.3.1   Discrete and Continuous Random Variables

A *random variable* can be defined as a variable that takes different values according to some specific probability distribution. A random variable can be a discrete random variable if it can take only a countable number of values. A simple example of a discrete random variable is the values from rolling a fair six-faced die: it can only be 1, 2, 3, 4, 5, 6, with equal probabilities. A random variable can also be a continuous random variable, if it can take all real values in a given interval. For example, the height of a person you meet randomly on a street is a continuous random variable.

The probability structure of a random variable, y, is described by its probability distribution. If y is a discrete random variable, its probability distribution is described by the probability function, often denoted by $p(y)$. If y is a continuous random variable, its probability distribution is described by its probability density function, often denoted by $f(y)$.

The properties of the probability function $p(y)$ and probability density function $f(y)$ are summarized as follows:

For a discrete random variable y

$$0 \le p(y_i) \le 1 \qquad y_i \text{ where represents the possible values that } y$$

can take

$$P(y = y_i) = p(y_i)$$

$$\sum_{All-y_i} p(y_i) = 1.$$

For a continuous random variable y

$$f(y) \ge 0$$

$$P(a \le y \le b) = \int_a^b f(y)dy$$

$$\int_{-\infty}^{+\infty} f(y)dy = 1.$$

11.3.2 Expected Values, Variance, and Standard Deviation

Random variable and probability distributions deal with the population, and the population's mean, μ, is the most frequently used measure of central tendency for populations. μ is also called the expected value of the random variable y.

The expected value $E(y) = \mu$ is defined as follows:

$$\mu = E(y) = \sum_{All-y} yp(y) \qquad \text{if } y \text{ is discrete}$$

$$\mu = E(y) = \int_{-\infty}^{+\infty} yf(y)dy \qquad \text{if } y \text{ is continuous.}$$

Population variance is often denoted by σ^2, and it is often simply called the variance. The variance of a random variable y is defined as follows:

$$\sigma^2 = \sum_{All-y} (y - \mu)^2 p(y) \qquad \text{if } y \text{ is discrete}$$

$$\sigma^2 = \int_{-\infty}^{+\infty} (y - \mu)^2 f(y)dy \qquad \text{if } y \text{ is continuous.}$$

Variance is often denoted by $Var(y)$. From the definition of expected value, it is clear that:

$$\sigma^2 = Var(y) = E[(y - \mu)^2].$$

The population standard deviation is often called the standard deviation, σ, which is simply the square root of variance σ^2:

$$\sigma = \sqrt{\sigma^2}$$

11.3.3   Probability Distribution Models

In Six Sigma-related applications, there are several probability distribution models that are frequently used as the basis for data-driven decision making. They are the normal distribution, exponential distribution, binomial distribution, and Poisson distribution. We will look at each of these in turn.

11.3.3.1   Normal Distribution   The normal distribution (or Gaussian distribution) was first proposed by Gauss (1777–1855). *Normal distribution* is often used to model the probability distribution of continuous variables that have the following properties:

- There are many random factors that can affect the value of the random variable.
- Each of these random factors has a relatively small influence on the random variable; there is no dominant factor.

Normal distribution is the most popular distribution in quality engineering and Six Sigma. It is often used to model the following random variables:

- Quality characteristic of parts from suppliers
- Students' test scores or employee performance scores

The probability density function of the normal distribution is as follows:

$$f(y) = \frac{1}{\sqrt{2\pi}\sigma} e^{-\frac{1}{2}\left(\frac{y-\mu}{\sigma}\right)^2} \qquad -\infty < y < +\infty$$

For a normal distribution:

$$E(y) = \mu$$

$$Var(y) = \sigma^2$$

A normal random variable y with $E(y) = \mu$ and $Var(y) = \sigma^2$ is denoted by $N(\mu, \sigma^2)$.

The probability density function $f(y)$ displays a bell-shaped curve as illustrated in Figure 11.4. The distribution is centered at μ, and a smaller σ will result in a tighter curve, and vice versa.

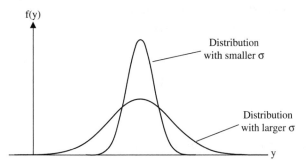

Figure 11.4 Normal probability density curve

An important special case of the normal distribution is the standard normal distribution. In the standard normal distribution, $\mu = 0$ and $\sigma^2 = 1$. The standard normal random variable is often denoted by $z \sim N(0,1)$. The standard normal distribution table is mainly used to calculate probability for all kinds of normal distribution. The shape and percentage distribution properties of the standard normal distribution are illustrated in Figure 11.5.

Figure 11.5 shows that if $y \sim N(\mu,\sigma^2)$, then $P(\mu - \sigma \le y \le \mu + \sigma) = P(-1 \le z \le 1) = 0.6826 = 68.27$ percent, that is, 68.27 percent of observations from a normal population will locate within one standard deviation distance from the mean. Similarly, $P(\mu - 2\sigma \le y \le \mu + 2\sigma) = P(-2 \le z \le 2) = 0.9545 = 95.45$ percent, that is, 95.45 percent of observations from a normal population will locate within two standard deviations' distance from the mean. $P(\mu - 3\sigma \le y \le \mu + 3\sigma) = P(-3 \le z \le 3) = 0.9973 = 99.73$ percent, that is, 99.73 percent of observations from a normal population will locate within three standard deviations' distance from the mean.

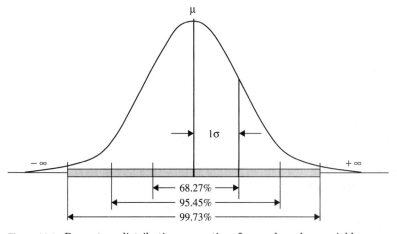

Figure 11.5 Percentage distribution properties of normal random variable

11.3.3.2 Exponential Distribution   An *exponential distribution* is featured by the following probability density function:

$$f(y) = \frac{e^{-\frac{y}{\beta}}}{\beta} \qquad \text{for } 0 \le y < \infty$$

with mean and variance

$$E(y) = \mu = \beta \qquad Var(y) = \beta^2.$$

Exponential distributions are often used to model:

- Lifetime of some electronic components
- Time between customers entering a service facility
- Time between consecutive machine failures or earthquakes

11.3.3.3 Binomial Distribution   A *binomial distribution* is a discrete probability distribution that characterizes a binomial random variable. Binomial random variables can be used for situations where all of the following apply:

- There are n successive trials, and each trial will only have two distinct outcomes, S (success) or F (failure).
- The probability of success $P(S) = p$, $P(F) = 1-P(S) = 1-p$.
- The result of each trial will not affect the results of any other trials.

If all these conditions are true, the number of successes, y, out of n trials, will be a binomial random variable and its probability function $p(y)$ will be:

$$p(y) = \frac{n!}{y!(n-y)!} p^y (1-p)^{n-y} \qquad y = 0,1,\ldots, n.$$

The mean and variance of the binomial random variable are:

$E(y) = \mu = np$ and $Var(y) = \sigma^2 = np(1-p)$.

The binomial distribution is often denoted by $y \sim B(n, p)$.
These are examples of binomial random variables:

- The number of defective parts, y, in a lot of n parts in a sequential quality inspection
- The number of positive customer responses, y, in a survey involving n customers

11.3.3.4 Poisson Distribution The Poisson probability distribution provides a model for the probability of occurrence of rare events that happen in a unit of time, area, volume, and so on. Actually, the *Poisson distribution* is an extreme case of binomial distribution, where n is very large, and p is very small. That is, the probability of a rare event occurrence, $p = P(S)$, is very small, but the number of trials, n, is very large.

In a Poisson distribution, the parameter λ ($\lambda = np$) is used, and the probability function of Poisson distribution, $p(y)$, is:

$$p(y) = \frac{\lambda^y e^{-\lambda}}{y!} \qquad y = 0, 1, 2, \ldots$$

The mean and variance of the Poisson distribution are:

$$E(y) = \lambda \quad \text{and} \quad Var(y) = \sigma^2 = \lambda.$$

11.3.4 Statistical Parameter Estimation

All probability distribution models depend on population parameters, such as μ, σ^2 in normal distributions, and p in binomial distributions. Without these parameters, no probability distribution model can be used. In real-world applications, these population parameters are usually not available, but statistical estimates of these population parameters can be computed based on a sample of data from the population.

The commonly-used statistical estimate for μ in a normal distribution is the sample mean $\bar{y}$, where

$$\bar{y} = \frac{1}{n} \sum_{i=1}^{n} y_i.$$

For a sample of n observations, $y_1, y_2, \ldots, y_n$, from $y \sim N(\mu, \sigma^2)$, the commonly-used statistical estimate for σ^2 in normal distribution is the sample variance s^2, where

$$s^2 = \frac{1}{n-1} \sum_{i=1}^{n} (y_i - \bar{y})^2.$$

For a sample of n observations $y_1, y_2, \ldots, y_n$ from $y \sim N(\mu, \sigma^2)$, and $\bar{y} = \frac{1}{n} \sum_{i=1}^{n} y_i$.

The commonly-used statistical estimate for p in a binomial distribution $B(n, p)$ is the sample ratio, $\hat{p}$, where

$$\hat{p} = \frac{y}{n}$$

where y is the actual number of successes (S) in n trials.

However, statistical estimates are only approximations of the true population parameters. When sample size is small, there will be substantial discrepancies between population parameters and statistical estimates. As the sample size gets larger, the discrepancies will get smaller.

11.4 Quality Measures and Six Sigma Metrics

For any product or business process, there are always performance metrics to measure and improve. For example, in a loan approval process, the cycle time (the time from loan application to loan decision) is a performance metric. In Example 11.1, the oxide film thickness is a performance metric and the ideal thickness is 560 Angstroms. Most process performance metrics are random variables, like the loan cycle time and the oxide's film thickness.

There are many *quality measures* that are developed to measure the performance of a process when the process involves randomness. Quality measures compare the degree of randomness in the process's performance and compare the degree of randomness with the process's performance specification. The most commonly used process performance quality measure is the process capability index. In Six Sigma practice, many other process performance-related metrics have also been developed, such as Sigma quality level, DPMO (defect per million opportunities), and so on.

In this section, we will look first at the process capability index, and then at other Six Sigma metrics.

11.4.1 Process Performance

Process performance is a measure of how well a process performs. It is measured by comparing the actual process performance level to the ideal process performance level. For the oxide film build process, its performance may be measured by its thickness, and its ideal performance level would be 560 Angstroms.

For most processes, performance level is not constant. We call this variation the *process variability*. If the process performance can be measured by a real number, then the process variability can usually be modeled by normal distribution, and the degree of variation can be measured by the standard deviation of that normal distribution.

If the process performance level is not a constant but a random variable, we can use process mean and process standard deviation as key performance measures. Mean performance can be calculated by averaging a large number of performance measurements.

If processes follow the normal probability distribution, a high percentage of the process performance measurements will fall between $\pm3\sigma$ of the process mean, where σ is the standard deviation. That is, approximately 0.27 percent of the measurements would naturally fall outside the $\pm3\sigma$ limits, and the balance of them (approximately 99.73 percent) would be within the $\pm3\sigma$ limits.

Since the process limits extend from -3σ to $+3\sigma$, the total spread amounts to about 6σ total variation. This total spread is often used to measure the range of process variability, also called the *process spread*.

For any process performance measure, there are usually some performance specification limits. For example, if the oxide film thickness in a wafer is too high or too low, it will not function well. For example, if deviation cannot be more than 100 Angstroms from the target value of 560 Angstroms, its specification limits would be 560 ± 100 Angstroms. We would say that its specification spread is (460, 660), where 460 Angstroms is the lower specification limit, or LSL, and 660 Angstroms is the upper specification limit, or USL.

11.4.2 Process Capability Indices

Capability indices are simplified measures that quickly describe the relationship between the variability of a process and the spread of the specification limits.

11.4.2.1 The Capability Index – C_p The equation for the simplest capability index, C_p, is the ratio of the specification spread to the process spread, the latter represented by six standard deviations or 6_σ. C_p assumes that the normal distribution is the correct model for the process.

$$C_p = \frac{USL - LCL}{6\sigma}$$

C_p can be translated directly to the percentage or proportion of nonconforming product outside specifications, if the mean of the process performance is at the center of the specification limits.

When $C_p = 1.00$, approximately 0.27 percent of the parts are outside the specification limits (assuming that the process is centered on the midpoint between the specification limits) because the specification limits closely match the process USL and LSL. We say this is about 2700 parts per million (ppm) nonconforming.

When $C_p = 1.33$, approximately .0064 percent of the parts are outside the specification limits (assuming the process is centered on the midpoint between the specification limits). We say this is about 64 parts per million (ppm) nonconforming. In this case, we would be looking at normal curve areas beyond $1.33 \times 3\sigma = \pm 4\sigma$ from the center.

When $C_p = 1.67$, approximately .000057 percent of the parts are outside the specification limits (assuming the process is centered on the midpoint between the specification limits). We say this is about .6 parts per million (ppm) nonconforming. In this case, we would be looking at normal curve areas beyond $1.67 \times 3_\sigma = \pm 5_\sigma$ from the center of the normal distribution.

11.4.2.1 The Capability Index $-C_{pk}$ The major weakness in C_p is that, for many processes, the mean performance of the process is not equal to the center of the specification limit. Also, many process means will drift from time to time. When that happens, the probability calculation for nonconformance will be wrong when we still use C_p. Therefore, one must consider where the process mean is located relative to the specification limits. The index C_{pk} is created to do exactly this.

$$C_{pk} = Min\left\{\frac{USL - \mu}{3\sigma}, \frac{\mu - LSL}{3\sigma}\right\} = Min\left\{C_{PU}, C_{PL}\right\}.$$

For example, suppose the process standard deviation is $\sigma = .8$ with USL = 24, LSL = 18, and the process mean $\mu = 22$.

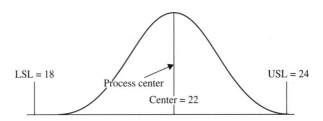

$$C_{pk} = Min\left\{\frac{24 - 22}{3 \times 8}, \frac{22 - 18}{3 \times 8}\right\} = Min(0.83, 1.67) = 0.83 .$$

It is also clear that

$$C_{PU} = 0.83, \quad C_{PL} = 1.67.$$

If the process mean were exactly centered between the specification limits:

$$C_p = C_{pk} = 1.25.$$

C_{pk} is a more accurate process performance metric when process mean is not at the nominal design point.

Example 11.8 Process Capability Calculation For the film thickness data given in Example 11.1, LSL = 460, USL = 660, and we do not know the exact value of μ and σ. However, we can calculate that $\bar{y} = 572.02$ and $s = 24.53$. Because the sample size of this data set is fairly large $(n = 47)$, we can substitute μ and σ by using $\bar{y}$ and s. This gives the following result:

$$C_{pk} = Min \left\{ \frac{660 - 572.02}{3 \times 24.53}, \frac{572.02 - 460}{3 \times 24.53} \right\} = Min\ (1.19, 1.51) = 1.19$$

MINITAB can be used to conduct a comprehensive process capability analysis. The following MINITAB output is the process capability analysis for the film thickness data.

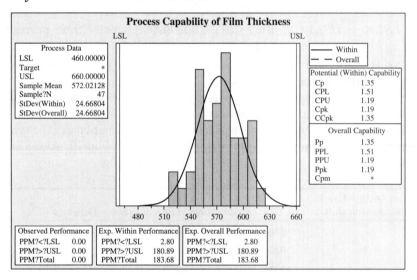

11.4.3 Sigma Quality Level
(Without Mean Shift)

In 1988, the Motorola Corporation was the winner of the Malcolm Baldridge National Quality Award. Motorola bases much of its quality effort on its Six Sigma Program. The goal of this program was to reduce the variation in every process to such an extent that a spread of 12σ (6σ on each side of the mean) fits within the process specification limits.

Figure 11.6 gives a graphical illustration of this Six Sigma quality. If the actual variation measured by standard deviation is σ, 6σ quality means that the total spread of the specification is six times the standard deviation on each side of the mean.

For a Six Sigma quality level, $C_p = \dfrac{USL - LSL}{6\sigma} = \dfrac{12\sigma}{6\sigma} = 2$.

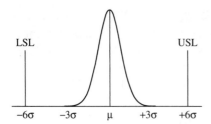

Figure 11.6 Six Sigma quality

By using normal probability distribution, it can be computed that

$$P(y) \text{ will be in specification if } P(y) =$$

$$P(LSL \le y \le USL) = P(-6 \le z \le 6) = 0.999999998 = 99.9999998 \text{ percent.}$$

Clearly, $P(y)$ will be out of specification if $P(y) = 1 - 0.9999999998 = 0.000000002$, or 0.002 defects per million parts (ppm).

Similarly, if the spread of the specification is five times σ on each side of the mean, it is called Five Sigma quality.

The following table summarizes the relationship between C_p, Sigma quality level (without mean shift), percentage in specification, and defective ppm.

Sigma Quality Level (without mean shift)	C_p	Percentage in Specification	Defective PPM
1	0.33	68.27	317300
2	0.67	95.45	45500
3	1.0	99.73	2700
4	1.33	99.9937	63
5	1.67	99.999943	0.57
6	2.00	99.9999998	0.002

11.4.4 Sigma Quality Level (With Mean Shift)

In most actual processes, the process mean, $\bar{y}$, is not a constant. The process mean will shift from time to time. For example, in manufacturing processes, with a change of raw material or operator, the process may suddenly change its mean level. In service processes, with a change of server or shift, the process mean may also change.

Motorola allocates 1.5σ on either side of the process mean to take into account this shifting of the mean. For a Six Sigma quality level, with the maximum possible mean shift of 1.5σ, the minimum distance from the process mean to one of the specification limits could be as small as 4.5σ. The following illustration shows the relationship between mean shift and Six Sigma quality level.

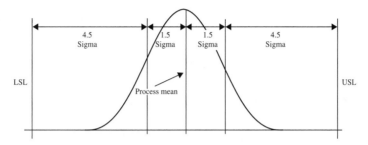

Thus, even if the process mean strays as much as 1.5σ from the process center, a full 4.5σ remains. This ensures a worst-case scenario of 3.4 parts per million (ppm) nonconforming on each side of the distribution. With the inclusion of a 1.5σ mean shift, for the same sigma quality level, the defective ppm will be much larger than when not considering a mean shift.

The following table summarizes the relationship between C_p, Sigma quality level (with mean shift), percentage in specification, and defective ppm.

Sigma Quality Level (with mean shift)	C_p	Percentage in Specification	Defective PPM
1	0.33	30.23	697700
2	0.67	69.13	308700
3	1.0	93.32	66810
4	1.33	99.3790	6210
5	1.67	99.97670	233
6	2.00	99.999660	3.4

References

Aaker, D. A. (1996) *Building Strong Brands*, New York: The Free Press.

Abrams, B. (2000) *Observational Research Handbook: Understanding How Consumers Live with Your Product, First Edition*, New York: McGraw-Hill.

Agar, M. (1996) *The Professional Stranger: An Informal Introduction to Ethnography, Second Edition*, Burlington, MA: Academic Press.

Allen, J., Robinson, C., and Stewart, D. eds., (2001) *Lean Manufacturing: A Plant Floor Guide*. Dearborn, MI: Society of Manufacturing Engineers.

Altshuler, G. S. (1988) *Creativity as Exact Science*, New York: Gordon & Breach.

Altshuler, G. S. (1990) "On the theory of solving inventive problems", *Design Methods and Theories* 24 (2): 1216–22.

Anderson, E., Rornell, C., and Lehmann, D. (1994) "Customer Satisfaction, Market Share and Profitability: Findings from Sweden", *Journal of Marketing*, July, pp. 53–66.

Arnold, D. (1992) *The Handbook of Brand Management*, Reading, MA: Addison-Wesley.

Ballard, B. (1987) "Academic Adjustment: The Other Side Of The Export Dollar", *Higher Education Research and Development*, Volume 6, Issue 2, pp. 109–119.

Batra, R., Lehmann, D., and Singh, D. (1993) "The Brand Personality Component of Brand Goodwill: Some Antecedents and Consequences", *Brand Equity and Advertising: Advertising's Role in Building Strong Brands*, Aaker D., and Biel, A. (eds.), Hillsdale, NJ: Lawrence Erbaum Associates Publishers, pp. 83–96.

Berry M. and Linoff G.S. (2000), *Master Data Mining*, New York: Wiley.

Biel A.L. (1993) "Converting Brand Image into Equity," *Brand Equity and Advertising: Advertising's Role in Building Strong Brands*, Aaker D., and Biel, A. (eds.), Hillsdale, NJ: Lawrence Erbaum Associates Publishers, pp. 67–82.

Clark, K. and Fujimori, T. (1991). *Product Development Performance: Strategy, Organization and Management in the World Auto Industry*, Boston, MA: Harvard Business School Press.

Clausing, D. P. (1994) *Total Quality Development: A Step by Step Guide to World-Class Concurrent Engineering*, New York: ASME Press.

Cohen, L. (1988) "Quality function deployment and application perspective from Digital Equipment Corporation", *National Productivity Review* 7 (3): 197–208.

Cohen, L. (1995) *Quality Function Deployment: How to Make QFD Work for You*. Reading, MA: Addison-Wesley.

Cooper, R.G. (1990) "Stage-gate systems: A new tool for managing new products", *Business Horizons*, Vol 33, issue 3, pp. 44–54.

Davenport, T.H., De Long, D.W., and Beers, M.C. (1998) "Successful Knowledge Management Projects", *Management of Technology and Innovation*, Vol. 39, No. 2, pp. 43–57.

Davis, S. M. (2000) *Brand Asset Management*. San Francisco, Jossey-Bass CA, Inc.

de Brentani, U. (1993) "The new product process in financial services: Strategy for success", *International Journal of Bank Marketing* 11 (3): 15–22.

Deming, E. (1982) *Out of the Crisis*, Cambridge, MA: Massachusetts Institute of Technology, Center for Advanced Engineering Study.

Dixon, J.R. (1966) *Design Engineering: Inventiveness Analysis and Decision Making*, McGraw-Hill, New York.

Edelstein, H.A. (1999) *Introduction to Data Mining and Knowledge Discovery, Third Edition*, Potomac, MD: Two Crows Corporation.

Farquhar, P.H. (1989) "Managing Brand Equity", *Marketing Research*, Vol. 3, pp. 24–33.

Fox, J.E., Mockovak, W., Fisher, S.K, and Rho, C. (2003) "Usability Issues Associated with Converting Establishment Survey to Web-Based Data Collection", *Proceedings of the Federal Committee of Statistical Methodology*, Washington, D.C.

Fredriksson, B. (1994) "Holistic systems engineering in product development", *The Saab-Scania Griffin*, November.

Gale, B. (1994) *Managing Customer Value*, New York: The Free Press.

George, M. (2003) *Lean Six Sigma for Service*, New York: McGraw-Hill.

Goetsch, D. L., and Davis, S. B. (2000) *Quality Management: Introduction to Total Quality Management for Production, Processing, and Services, Third Edition*, Upper Saddle River, NJ: Prentice Hall.

Gold, R.L. (1958) "Roles in Sociological Field Observations", *Social Forces*, Vol. 36, No. 3 pp. 217–223.

Hauser, J. R., and Clausing, D. (1988) "The House of Quality", *Harvard Business Review*, 66 (3): pp. 63–73.

Hubka, V. (1980) *"Principles of Engineering Design"*, (translated by W.E. Eder), London: Butterworth Scientific.

Huthwaite, B. (2004) *The Lean Design Solution*, Mackinac Island, MI: Institute for Lean Design.

Jacobson, R. and Aaker, D. (1987) "The strategic role of product quality", *Journal of Marketing* 51: pp. 31–44.

Kaufman, J. J. (1989) *Value Engineering for the Practitioner, Second Edition*, Raleigh, NC: North Carolina State Univ. Press.

Kawakita, J. (1977) *A Scientific Exploration of Intellect ("chi" no tankengaku)*, Tokyo: Kodansha.

Kawakita, J. (1991) *The Original kj Method*, Tokyo: Kawakita Research Institute.

Keller, K. L. (1993) "Conceptualizing, measuring, and managing customer-based brand equity", *Journal of Marketing* 57: pp. 1–22.

Kennedy, M. (2003) *Product Development for the Lean Enterprise*, Richmond, VA: The Oaklea Press.

Kim, W. and Mauborgne, R. (2006) *Blue Ocean Strategy: How to Create Uncontested Market Space and Make Competition Irrelevant*. Boston, MA: Harvard Business School Press.

Kochan, N. (Ed.) (1997) *The World's Greatest Brands*, New York: University Press.

Levine, M. (2003) *A Branded World*, Hoboken, NJ: John Wiley and Sons, Inc.

Liker, J. K. (2004) *The Toyota Way,* New York: McGraw-Hill.

Loftus, E. and Wells, G. (1984) *Eyewitness Testimony: Psychological Perspectives*, New York: Cambridge Univ. Press.

MacElroy, B. (2000) "Variables influencing dropout rates in web-based surveys", *Quirk's Marketing Research Review*, July/August. Retrieved October 9, 2002, from http://www.modalis.com/english/news/ newsitem000808.html.

Mann, D. (2002) *Hands-on Systematic Innovation*, Ieper, Belgium: CREAX Press.

Mann, D. (2004) *Hands-on Systematic Innovation for Business and Management*, Clevedon, UK: IFR Press.

Mann, D. and Domb, E. (1999) "40 Inventive (business) principles with examples", *TRIZ Journal*, September.

McCord, K. R., Eppinger, S. D. (1993) "Managing the integration problem in concurrent engineering", *Working paper #3594-93-MSA*, Cambridge MA: Sloan School of Management, Massachusetts Institute of Technology.

Miles, L.D. (1961) *Techniques for Value Analysis and Value Engineering*, New York, McGraw-Hill.

Miles, B. L. (1989) "Design for assembly: A key element within design for manufacture", *Proc. IMechE, Part D, Journal of Automobile Engineering* 203: pp. 29–38.

Monden, Y. (1993) *The Toyota Management System: Linking the Seven Key Functional Areas*, Portland, OR: Productivity Press.

Morgan, J. and Liker, J. (2006) *The Toyota Product Development System*. New York: Productivity Press.

Murphy, J. (1990) *Brand Valuation, Establish a True and Fair Value*, London: Hutchinson Business Books.

Nielsen, J. (2000) *Designing Web Usability: The Practice of Simplicity*, Indianapolis, IN, New Riders Publishing.

Nonaka, I. and Takeuchi, H. (1995) *The Knowledge-Creating Company: How Japanese Companies Create the Dynamics of Innovation*, New York: Oxford Univ. Press.

Ohno, T. (1990) *Toyota Production System: Beyond Large-Scale Production*, Portland, OR: Productivity Press.

Park, R. J. (1992) *Value Engineering*, Birmingham, MI: R. J. Park & Associates, Inc.

Park, R. J. (1999) *Value Engineering, a Plan for Invention*, Boca Raton, FL: Saint Lucie Press.

Penney, R. K. (1970) "Principles of Engineering Design", *Postgraduate*, No. 46, pp. 344–349.

Phal, G. and Beitz (1988) *Engineering Design: A Systematic Approach*, New York: Springer-Verlag.

Phaal, R., et al. (2001) *Technology Roadmapping: Linking technology resources to business objectives*, Cambridge, UK: Univ. of Cambridge, Centre for Technology Management.

Pimmler, T. U. and Eppinger, S. D. (1994) "Integration Analysis of Product Decomposition", *Design Theory and Methodology*, Vol 68, pp. 343–351.

Polanyi, M. (1967) *The Tacit Dimension*, New York: Anchor Books.

Pugh, S. (1991) *Total Design: Integrated Methods for Successful Product Engineering*, Reading, MA: Addison-Wesley.

Pugh, S. (1996) *Creating Innovative Products Using Total Design*, Reading, MA: Addison-Wesley.

Ramaswamy, R. (1996) *Design and Management of Service Processes*, Reading, MA: Addison-Wesley.

Rantanen, K. (1988) "Altshuler's methodology in solving inventive problems", *Proceedings of ICED* 88, Budapest, pp. 23–25 Aug.

Rea, L. M. and Parker, R. A. (1992) *Designing and Conducting Survey Research: A Comprehensive Guide*, San Francisco, CA: Jossey-Bass Inc.

Reeve, D. (1975) "Value engineering analysis of the Oakland County Youth Assistance Program, Master's thesis", Oakland Univ.

Reinertsen, D. (1997) *Managing the Design Factory*, New York: The Freedom Press.

Reis, A. (1981) *Positioning: The Battle for Your Mind*, New York: McGraw-Hill.

Rother, M. and Shook, J. (2003) *Learning to See: Value-Stream Mapping to Create Value and Eliminate Muda*, Brookline, MA: The Lean Enterprise Institute, Inc.

Rust, L. (1993) "Observations: Parents and Children Shopping Together" *Journal of Advertising Research*, July/August, pp. 65–70.

Sherden, W. A. (1994) *Market Ownership*, New York: American Management Association.

Shigeru, M. (1988) *Management for Quality Improvement: The Seven New QC Tools*, Cambridge, MA: Productivity Press.

Shingo, S. (1989) *The Toyota Production System from an Industrial Engineering Viewpoint*, Portland, OR: Productivity Press.

Smith, P. G. and Reinertsen, D. G. (1998) *Developing Products in Half the Time: New Rules, New Tools*, New York: Van Nostrand Reinhold (International Thomson Publishing Company).

Spradley (1980) *Participant Observation*. New York: Harcourt, Brace, Jovanovich.

Squires, S. and Byrne, B. (eds.) (2002) *Creating Breakthrough Ideas: The Collaboration of Anthropologists and Designers in the Product Development Industry*, New York: Bergin & Garvey.

Suh, N. P. (1984) "Development of the science base for the manufacturing field through the axiomatic approach", *Robotics & Computer Integrated Manufacturing* 1 (3–4).

Suh, N. P. (1990) *The Principles of Design, First Edition*, New York: Oxford Univ. Press.

Suh, N. P. (2001) *Axiomatic Design: Advances and Applications, First Edition*, New York: Oxford Univ. Press.

Tauber, F. M. (1988) "Brand Leverage: Strategy for Growth in a Cost Controlled World", *Journal of Advertising Research*, Vol 28, pp. 26–30.

Tsourikov, V. M. (1993) "Inventive machine: Second generation", *Artificial Intelligence & Society* 7: pp. 62–77.

Ullman, D. G. (1992) *The Mechanical Design Process*, New York: McGraw-Hill.

Ulrich K. T. and Eppinger S. D. (2000) *Product Design and Development*, New York: McGraw-Hill.

Ulwick, A. (2005) *What Customers Want: Using Outcome-Driven Innovation to Create Breakthrough Products and Services*, New York: McGraw-Hill.

Watkins, T. (1986) *Economics of the Brand*, London: McGraw-Hill.

Wellner, A. S. (2002) "The Test Drive-ethnographic research and marketing research", *American Demographics*, Vol. Oct 1, 2002 http://findarticles.com/p/articles/mi_m4021/is_2002_Oct_1/ai_92087420.

Wicker, A. W. (1969) "Attitudes versus actions: The relationship of verbal and overt behavioral responses to attitude objects", *Journal of Social Issues* 25 (4): pp. 41–78.

Womack, J. P., Jones, D. T., and Roos, D. (1990) *The Machine that Changed the World*, New York: Rawson Associates.

Womack, J. P. and Jones, D. T. (2003) *Lean Thinking*, New York: The Free Press.

Yang, K. and El-Haik, B. (2003) *Design for Six Sigma: A Roadmap for Product Development*, New York: McGraw-Hill.

Index